Cryohydrology

Yongjian Ding · Shiqiang Zhang · Rensheng Chen

Cryohydrology

Science Press
Beijing

Springer

Yongjian Ding
Northwest Institute of Eco-Environment
and Resources
Chinese Academy of Sciences
Lanzhou, China

Shiqiang Zhang
Northwest University
Xi'an, China

Rensheng Chen
Northwest Institute of Eco-Environment
and Resources
Chinese Academy of Sciences
Lanzhou, China

ISBN 978-981-96-7359-9 ISBN 978-981-96-7360-5 (eBook)
https://doi.org/10.1007/978-981-96-7360-5

Jointly published with Science Press.
The print edition is not for sale in China mainland. Customers from China mainland please order the print book from: Science Press.
ISBN of the Co-Publisher's edition: 978-981-96-7359-9

This Springer imprint is published by the registered company Springer Nature Singapore Pte Ltd.
The registered company address is: 152 Beach Road, #21-01/04 Gateway East, Singapore 189721, Singapore

Foreword by Dahe Qin October 2024, Beijing, China

The cryosphere, alongside the atmosphere, hydrosphere, lithosphere, and biosphere, is one of the five major components of the climate system. As a unique sphere within this system, the cryosphere encompasses glaciers, snow cover, and permafrost, representing the frozen parts of the Earth's surface. The cryosphere stores 77% of the world's freshwater resources, making it the largest reservoir of freshwater on land, often referred to as the "Earth's solid water reservoir." Changes in the cryosphere significantly affect this vast freshwater resource. Through its influence on watershed hydrological processes and regional and global water cycles, the cryosphere alters water supply capacities, affects hydrological regulation functions, and disrupts ocean circulation dynamics. The scientific and practical importance of cryospheric hydrology research is immense, and the publication of the English version of *Cryohydrology* will provide a valuable textbook and reference tool for promoting and understanding the fundamentals of cryospheric hydrology on a global scale.

As a crucial branch of cryosphere science, Cryohydrology has been redefined and restructured within the broader framework of cryosphere science. Chinese scientists, working under the "cryosphere science" discipline, have already published the Chinese edition of the *Cryosphere Science Series* (referred to as the "Series"). Led by the *Introduction to Cryosphere Science*, the Series includes volumes on *Cryosphere Physics*, *Cryosphere Chemistry*, *Cryosphere Geography*, *Cryosphere Climatology*, *Cryosphere Hydrology*, *Cryosphere Biology*, *Cryosphere Microbiology*, *Cryosphere Environmental Science*, *Quaternary Cryosphere*, *Cryosphere Engineering*, *Cryosphere Disasters*, *Cryosphere Humanities*, *Cryosphere Remote Sensing*, and *Planetary Cryosphere*. The volume *Cryohydrology* is one of the key books in this Series. The *Introduction to Cryosphere Science* was published in English in 2021, and now, this volume is the first comprehensive English monograph on *Cryohydrology*, making it a landmark achievement. Its publication not only signifies the growing international influence of the cryosphere science discipline but also marks a solid step forward by Chinese scientists in advancing the theories and methodologies of cryosphere hydrology.

In 2017, Yongjian Ding's team published *Introduction to Cold Region Hydrology*, which summarized hydrological processes and discipline characteristics of glaciers, permafrost, snow, river and lake ice, and sea ice. Cold region hydrology focuses on longitudinal studies of specific elements of the cryosphere. By 2020, the team had further developed the field of Crtohydrology from a systematic cryosphere science perspective. Compared to traditional cold region hydrology, Cryohydrology synthesizes the traditional element-based approach and provides a more comprehensive framework by connecting the hydrological processes of all cryospheric elements and highlighting their unique hydrological characteristics. This approach represents both an attempt and a significant advancement in the construction of a discipline for Cryohydrology.

As the English edition of *Cryohydrology* is about to be published, I am truly delighted by this systematic achievement of Chinese scientists! I sincerely hope that the publication of this Series will serve as a catalyst for the continued deepening of cryosphere hydrology research, leading to further important discoveries and contributions towards protecting our shared home—Earth.

Dahe Qin
Academician, Chinese Academy of Sciences
Former Director, China Meteorological Administration
Co-chair, Working Group I, IPCC Fourth and Fifth Assessment Reports

Foreword by Jianyun Zhang October 2024, Nanjing, China

Cryospheric meltwater is not only a unique hydrological phenomenon but also a critical water resource. In vast regions of western, northeastern, and northern China, snow and permafrost meltwater influences water resource utilization, ecological restoration, and agricultural production to varying degrees. This is particularly true in the arid inland river basins of northwestern China, where water resources originate in the high mountain regions and are consumed downstream in the oasis areas. Changes in the cryosphere dictate the variation and seasonal distribution of water resources in these high-altitude areas, making Cryohydrology essential for the effective utilization and management regulation of basin water resources. Globally, the cryosphere functions as both a high-latitude "wetland" and a "water tower" for many high-altitude mountain ranges, with widespread impacts. With the intensifying effects of climate change, the last decades have seen accelerated glacier retreat, deepening of the active layer in permafrost, increasing permafrost temperatures, and intensified snow cover fluctuations. These changes have led to significant alterations in runoff and sediment dynamics in cryospheric regions.

Cryohydrology is an important subfield of hydrology and water resource science. Water issues related to the cryosphere primarily involve glacier hydrology, snow hydrology, and permafrost hydrology. Since these hydrological phenomena occur in cold regions, they are often collectively referred to as "cold region hydrology," or, by some, as "permafrost hydrology." Both "permafrost hydrology" and "cold region hydrology" encompass the hydrological processes of glaciers, snow, permafrost, and river/lake ice in cold regions. These studies typically focus on individual cryospheric elements, discussing the formation, changes, and impacts of water. In contrast, Cryohydrology integrates all cryospheric elements, considering both their shared and unique hydrological characteristics to build a comprehensive disciplinary framework. From this perspective, *Cryohydrology* represents a pioneering and innovative approach.

Currently, nearly half of the world's population faces water scarcity for at least part of the year. A quarter of the global population is under "extremely high" water stress, and the worldwide demand for freshwater is growing at a rate of about 1% per year. The occurrence of record-breaking extreme rainfall events is increasing globally, as are the frequency, duration, and intensity of meteorological droughts. Climate change is intensifying the global water cycle, further increasing the frequency and severity of droughts and floods. The cryosphere contains nearly four-fifths of the world's freshwater resources and is undergoing rapid change under the continued impact of global warming. The implications of cryospheric meltwater on freshwater resource utilization, regional and global water cycles, and large-scale hydrological processes are not only hot topics in international scientific research but also critical issues for humanity's sustainable development. The publication of *Cryohydrology* provides an essential scientific reference for understanding the hydrological processes, functions, and impacts of the cryosphere.

Under the leadership of researcher Yongjian Ding, the team published *Introduction to Cold Region Hydrology* in 2017, and the Chinese version of *Cryosphere Hydrology* in 2020. Additionally, they published a paper titled "Hydrological Basis and Discipline System of Cryohydrology: From a Perspective of Cryospheric Science" in *Frontiers in Earth Science*, which was the first systematic introduction to Cryohydrology on an international journal. However, it is difficult to fully cover the research topics, methods, and basin-to-global scale impacts of Cryohydrology within a single paper. Building on the Chinese edition of *Cryohydrology*, the team incorporated the latest research advancements in Cryohydrology to produce the English edition. The completion of the English version of *Cryohydrology* offers readers an accessible pathway to quickly gain knowledge in this field and fills the critical gap in the availability of structured educational materials on this subject.

The publication of the English edition of *Cryohydrology* marks the maturation and international recognition of the discipline of Cryohydrology. I am delighted to see this achievement and celebrate this milestone. It is also gratifying to witness the emergence of a new generation of young scientists, under the guidance of researcher Yongjian Ding, who are taking on significant roles in this field. Their growth represents the future of the discipline, and I wish them continued success in their endeavors!

Jianyun Zhang
Academician, Chinese Academy of Engineering
Chief Editor, Advances in Water Science

Preface

The role of mountainous "water tower" basins, typically located within or originating from the cryosphere, is of critical importance. Because the cryosphere is highly sensitive to climate change, hydrological changes within mountainous cryospheric regions significantly impact the sustainable economic and social development of entire basins. With the growing effects of climate change, the global cryosphere has undergone accelerated changes in recent decades. From the Arctic to the Tibetan Plateau, from the Andes to the Alps, and from inland river basins to Siberia, these changes affect high-latitude oceans, global sea levels, and hydrological cycles. As a result, the study of hydrological changes in the cryosphere and their global impacts has become a major focus of climate change research. Simultaneously, from a systems-based perspective, the cryosphere's interactions with the hydrosphere and other Earth systems have emerged as critical surface processes with significant influence on the climate and the broader Earth system.

Key areas of research in Cryohydrology include glacial meltwater, snowmelt runoff, permafrost hydrology, and freshwater balance in high-latitude regions. Major research topics encompass observation and modeling methods, hydrological processes, runoff variability, watershed effects, and impacts on regional and global water cycles. Additionally, the melting of ice and snow alters sediment and nutrient dynamics and contributes to greenhouse gas emissions, such as methane (CH_4) and nitrous oxide (N_2O), particularly from water bodies in degrading permafrost regions. On a practical level, hydrological changes in the cryosphere play a critical role in hydropower development, agricultural irrigation, water resource management, ecological conservation, disaster risk mitigation, and the sustainable development of society. The demand for scientific and technological solutions to address these challenges is both extensive and urgent. Cryohydrology has thus emerged as a new discipline, shaped by global change research and advances in cryosphere science.

Historically, hydrological research related to cryosphere elements, such as glacier and snow hydrology, has developed its own methodological frameworks. These subfields are either branches of glaciology or specialized areas within hydrology. In earlier studies, water issues related to multiple cryospheric elements were often categorized under "cold region hydrology" or "permafrost hydrology." Internationally, the term "permafrost hydrology," primarily developed by Canadian scholar, encompasses glacier, snow, river, and lake ice hydrology. Whether termed "permafrost hydrology" or "cold region hydrology," these approaches typically focused on individual cryospheric elements. For example, within cold region hydrology, disciplines such as glacier, permafrost, snow, river, and lake ice hydrology are analyzed in terms of water formation (production and melting), variations (ice accumulation and reduction), and their effects on runoff and water resources at the watershed, regional, and global scales.

In contrast, Cryohydrology integrates all cryospheric elements into a unified framework. It synthesizes the traditional discipline-specific approaches into a broader, holistic perspective, centered on the method-process-role-impact framework. In this model, methodology serves as the foundation, process as the core, role as a focus on the cryosphere's hydrological function at the basin scale, and impact as the assessment of cryosphere hydrology's consequences at regional and global levels. This conceptual framework underpins the guiding principles of this monograph.

Cryohydrology is the scientific discipline dedicated to studying the hydrological processes, mechanisms, effects, and impacts of cryospheric elements. It addresses two main aspects: first, the hydrological mechanisms and changes within the cryosphere, including research methodologies, production and melting processes, runoff variability, and sediment and hydrochemical processes; and second, the broader impacts of Cryohydrology on basins, regions, and even the global scale. Synthesizing the shared and distinct hydrological characteristics of cryospheric elements, Cryohydrology aims to establish itself as an independent discipline, with no prior precedent for such a comprehensive framework. Theoretically, energy and water balance serve as the physical foundation of Cryohydrology, providing the essential principles that support its development. The core foundational disciplines of Cryohydrology are cryosphere science and hydrology, while water resources science, geography, and atmospheric science also play key roles. In terms of energy balance, Cryohydrology closely intersects with cryosphere science and atmospheric science, while water balance is more directly related to hydrology and water resources science.

This book was organized by researcher Yongjian Ding, who established the general outline of the discipline and engaged in extensive discussions with a team of scholars specializing in Cryohydrology. After seven drafts and multiple revisions, the final version was completed. Throughout the writing process, five group meetings and three editorial board meetings were held to review and refine the manuscript. The contributors invested significant time and effort into shaping this work, and we are profoundly grateful to the numerous reviewers whose valuable feedback helped finalize the text.

The contributors to this book include Yongjian Ding (Chaps. 1, 7 and 8), Shiqiang Zhang (Chaps. 1 and 4), Rensheng Chen (Chap. 7), Tianding Han (Chaps. 3 and 5), Jinkui Wu (Chaps. 5 and 6), Xiangying Li (Chaps. 2, 5 and 6), Haidong Han (Chap. 2), Hongkai Gao (Chaps. 3 and 4), Qiudong Zhao (Chaps. 2 and 4), Donghui Shangguan (Chaps. 3 and 8), Xiaobo He (Chap. 2), Jia Qin (Chap. 6), Xiqiang Wang (Chap. 7), Junfeng Liu (Chap. 7), Yong Yang (Chap. 7), Yaping Chang (Chap. 1), Shengxia Wang (Chap. 8), and Shaoping Wang (Chap. 1). We extend our special thanks to Yaping Chang for serving as the academic secretary, and to Shaoping Wang for assisting with conference logistics and contributing to the successful completion of this monograph.

As this monograph is published, I feel a mix of relief and apprehension. I am pleased that, following the release of the Chinese version of *Cryohydrology* in 2020, we have completed this English version. However, this edition is not simply a translation. It includes substantial revisions, with content rearranged, added, and removed from the original Chinese version. The early stages of writing were disrupted by the global pandemic, which resulted in long periods of inactivity and a significant impact on our progress and morale. At times, the project seemed in jeopardy, and I even considered abandoning it altogether. Fortunately, the writing team's enthusiasm reignited in early 2023, and through our collective efforts, we finally completed the manuscript. I extend my deepest gratitude to the team members for their dedication, perseverance, and collaboration.

At the same time, I feel some unease as this book goes to press. As a summary of the discipline of Cryohydrology, it may contain limitations, particularly given the rapid advances in this field. New understandings, theories, and methods related to the fundamental principles, processes, and changes in Cryohydrology are constantly emerging. With this in mind, I hope this book serves as a starting point for further exploration, and I sincerely invite readers to offer their critiques and guidance to help advance the development of this discipline.

This book has been supported by the Strategic Priority Research Program of the Chinese Academy of Sciences (XDA2310040309), the National Natural Science Foundation of China under the project "Research on Coupling Mechanisms of Permafrost Changes and Hydrological Processes" (42330512), the Tibet Science and Technology Program (XZ202301ZY0001G, XZ202401JD0007), as well as by the National Key Laboratory of Cryospheric Science and Permafrost Engineering of the Northwest Institute of Ecology and Environmental Resources, Chinese Academy of Sciences. Additional support was provided by the National Key Laboratory of Ecological Security and Sustainable Development in Arid Zones of the Northwest Institute of Ecology and Environmental Resources, and by the Tanggula Mountain Cryosphere and Environment Tibet Autonomous Region Field Scientific Observatory. We are deeply grateful to all these institutions for their support.

Lanzhou, China
October 2024

Yongjian Ding

Contents

Chapter 1
Introduction

Yongjian Ding, Shiqiang Zhang, Yaping Chang, and Shaoping Wang

This chapter introduces the concept, evolution, and characteristics of Cryohydrology. Initially, it explores the connections between the cryosphere, water and society, encompassing the geographical distribution of cryospheric components worldwide, the cryosphere water resource, and the relationship between cryosphere and global water cycle. Subsequently, the development of the Cryohydrology discipline is discussed, including its origins, its relationship with Cryospheric Science, and its connection to hydrology in cold regions. The chapter concludes by outlining the disciplinary system of Cryohydrology, including its research content, discipline characteristics, and its interrelationships with other fields.

1.1 Cryosphere, Water and Society

The cryosphere is defined as a continuous sphere within the climate system where temperatures are at or below the freezing point of pure water. The global cryosphere can be divided into three categories: continental cryosphere, marine cryosphere, and

Y. Ding · Y. Chang · S. Wang
Northwest Institute of Eco Environment and Resources, Chinese Academy of Sciences, Lanzhou, China
e-mail: dyj@lzb.ac.cn

Y. Chang
e-mail: changyaping@lzb.ac.cn

S. Wang
e-mail: w1215lin@163.com

S. Zhang (✉)
Northwest University, Xi'an, China
e-mail: zhangsq@nwu.edu.cn

Y. Ding et al., *Cryohydrology*, https://doi.org/10.1007/978-981-96-7360-5_1

aerial cryosphere. The continental cryosphere, which develops and exists on land surfaces, includes glaciers, ice sheets, permafrost (excluding subsea permafrost), snow, river ice, and lake ice. The marine cryosphere encompasses sea ice, overlying snow cover, ice shelves, icebergs, and subsea permafrost. The aerial cryosphere consists of solid water present in the troposphere and stratosphere where temperatures are below freezing point (<0 °C). In this space, solid forms of water primarily appear as snow, hail, graupel, and ice pellets. Considering Cryohydrology primarily investigates hydrological processes on land, this discussion will focus on the distribution of the continental cryosphere.

1.1.1 Distribution of Cryosphere

The continental cryosphere occupies 52–55% of the global land area. Mountain glaciers and ice sheets cover approximately 10% of the global land surface (Antarctic ice sheet and Greenland ice sheet for 9.5%, mountain glaciers for 0.5%). According to the IPCC's Sixth Assessment Report (AR6), glaciers in 2000 covered a total area of 705,700 ± 33,200 km^2, with an ice volume of 324 ± 84 mm to sea level equivalent. These glaciers span an elevation range from sea level, for example in south-east Alaska, to >8,000 m a.s.l. in the Himalaya and Karakoram, and situated across various climatic regions (Fig. 1.1).

In the Northern Hemisphere, snow cover reached its maximum extent in January, averaging 45.2×10^6 km^2 from 1964 to 2004, and its minimum in August, averaging 1.9×10^6 km^2. From November to April, snow covered more than 33% of the Northern Hemisphere's land area on average, and even 49% in January. In the Southern Hemisphere, snow is primarily distributed in the southern regions of South America, the southern island of New Zealand, and the high mountains of eastern Australia.

The global permafrost region, excluding permafrost beneath glaciers and ice sheets, comprises 9–12% of the global land area. Including the Greenland and Antarctic ice sheets, the permafrost area expands to approximately 36.67×10^6 km^2, which represents 24% of the global land area. Seasonally frozen ground, excluding the active layer over permafrost, accounts for 33% of the land area in the Northern Hemisphere (Fig. 1.1). Including the active layer, seasonally frozen ground covers more than 56% of the land area in the Northern Hemisphere on average and can extend up to 80% during extremely cold years.

On average, 7% of the ocean's surface is covered by ice. During midwinter, the maximum Arctic sea ice extent reaches 15×10^6 km^2, whereas in summer, the minimum is 6×10^6 km^2. The Antarctic sea ice extent (SIE) exhibits significant seasonal variation, with the winter maximum exceeding 18×10^6 km^2 and the summer minimum is only 3×10^6 km^2. Rivers and lakes at high latitudes are also covered by ice during winter, though their ice volume and surface area are relatively smaller compared to other cryosphere components.

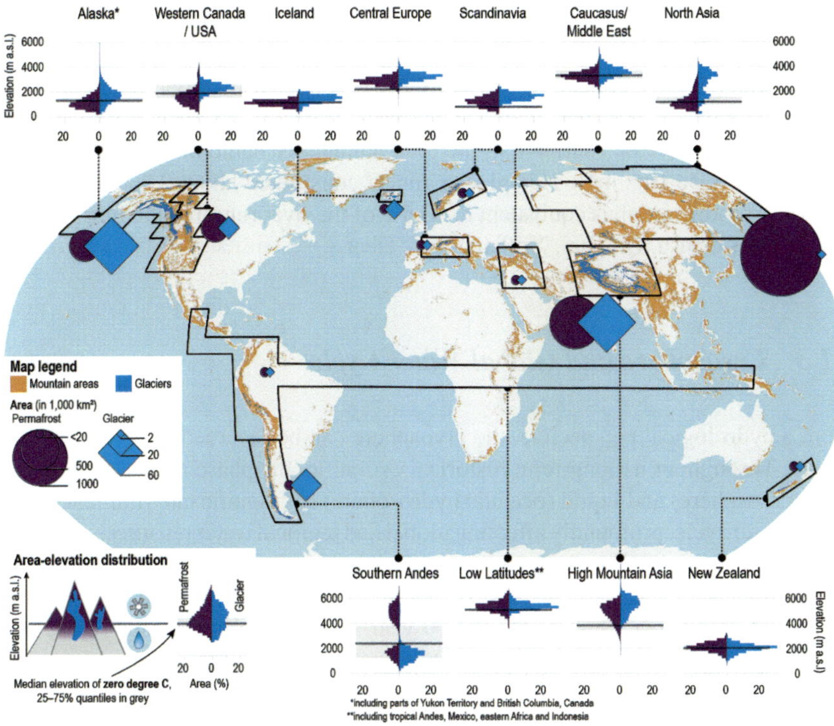

Fig. 1.1 Distribution of mountain areas (orange shading) and glaciers (blue) as well as regional summary statistics for glaciers and permafrost in mountains (IPCC 2021)

1.1.2 Cryosphere Water Resource

The cryosphere, as a solid body of water, serves as a significant water resource, characterized by its total reserves and annual replenishment. The continental cryosphere holds about 75% of the world's fresh water, with more than 70% of the global freshwater resources contained in glaciers and ice sheets. Should all the ice from the Greenland and Antarctic ice sheets melt, the resultant sea-level rise would be about 7.36 m and 58.30 m, respectively. Annually, the maximum water equivalent of snow in the Northern Hemisphere amounts to approximately 3×10^6 km^3. Additionally, about 5.3–7.3% of Earth's ocean surface is covered by sea ice and ice shelves, making the cryosphere the largest freshwater resource reservoir globally.

The cryosphere, prevalent at high altitudes and latitudes, serves as the origin for many of the world's major rivers. Notably, the cryosphere on the Qinghai-Tibet Plateau, which is the primary component of the "Asian Water Tower," sources significant rivers such as the Yangtze River, the Yellow River, the Indus River and the Ganges River. Unlike rainfall, the cryosphere supplies water by converting solid water into liquid, thereby releasing water from long-term glacial storage. This process ensures a continuous water supply even during droughts and periods of low rainfall. The depletion of this water source typically requires extensive and prolonged climatic fluctuations.

The annual replenishment of rivers by the cryosphere is an important component of surface runoff. In 2010, the volume of meltwater from China's glaciers was approximately 7.80×10^{10} m^3, surpassing the total annual discharge of the Yellow River into the sea (6×10^{10} m^3). This glacier runoff constituted about 2.2% of Chinese total river runoff, which is equivalent to 10.5% of the river runoff in the four western provinces of Gansu, Qinghai, Xinjiang, and Tibet.

1.1.3 Cryosphere and Global Water Cycle

From a hydrological perspective, the cryosphere can be regarded as a solid hydrosphere. Throughout its long-term historical evolution, the phase transition between solid (cryosphere) and liquid (oceanic hydrosphere) has significantly influenced the global water cycle, profoundly affecting global and regional water resources, ecology, and climate systems (Fig. 1.2). In terms of global water balance, an expansion of the cryosphere means the reduction of liquid water, thereby weakening the water cycle, and vice versa. During the 10,000-year-scale glacial-interglacial cycle and millennial-scale Dansgaard-Oeschger (DO) fluctuations, marked changes in global terrestrial ice volume and sea levels have occurred. These variations link the atmosphere, oceans, land, and ecosystems through the phase transitions of the water cycle, becoming pivotal in climate system changes. As human impact on the climate intensifies, the global cryosphere is undergoing significant transformations. The hydrological impact of the cryosphere on both global and regional water cycle processes is increasingly affecting regional sustainable development, besides altering the global hydrosphere.

The water stored in the Greenland and Antarctic ice sheets could contribute to sea level rises of approximately 7.36 m and 58.3 m, respectively. Changes in continental ice volumes have been a major factor in sea level variations over the past century. However, assessing the cryosphere's impact on sea level change remains highly uncertain. Excluding the influence of changes in land water storage, the contributions to sea level rise from thermal expansion due to ocean warming and the cryosphere have been nearly equivalent since the onset of industrialization. The projected contribution from the cryosphere is expected to surpass that of thermal expansion.

1.1.4 Cryosphere and Society

The mountain cryosphere is a critical source of freshwater, within mountainous regions and in downstream areas, sustaining ecosystems and supporting livelihoods both nearby and far beyond the mountain ranges. Immerzeel et al. (2019) introduced

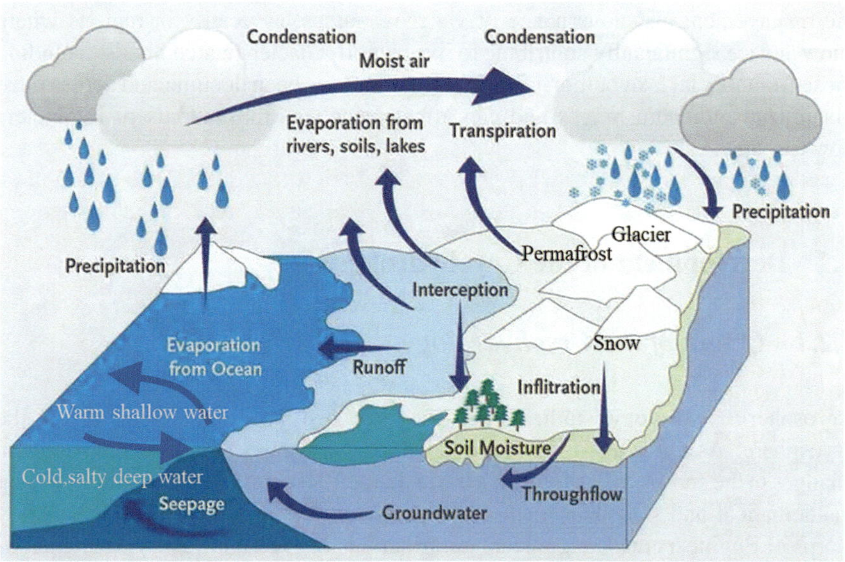

Fig. 1.2 Schematic map of global water cycle and cryosphere

the global Water Tower Index (WTI) to evaluate all water towers based on their role in water supply and the dependence of downstream ecosystems and societies on these resources. This study also assessed their vulnerability to water stress, governance challenges, hydro-political tensions, and future climatic and socio-economic changes. The findings highlighted that the Indus Water Tower Unit (WTU) plays a pivotal role, characterized by a balanced contribution of precipitation, glacier melt, snow, and surface water.

On the other hand, glacier shrinkage can degrade water quality by accelerating the release of stored anthropogenic legacy pollutants, thereby impacting downstream ecosystem services. Changes in the mountain cryosphere are likely to cause significant shifts in downstream nutrients such as dissolved organic carbon (DOC), nitrogen, and phosphorus, and increase the presence of heavy metals, particularly mercury, along with other legacy contaminants. This poses a potential threat to human health. The sediments and a large array of chemical species in the meltwater from glaciers, snow, and permafrost affect the water environment of streamflows and the primary productivity of aquatic ecosystems, ultimately impacting the elemental cycles of mountainous regions and downstream areas.

Moreover, due to the reliance on water resources as a critical input, hydropower operations are likely to be impacted by changes in runoff from glaciers and snow cover. Changes in glacier and moraine-dammed lakes, as well as changes in sediment supply, can influence hydropower generation. Many glacier lakes, having increased in volume, pose a risk to hydropower infrastructure when they outburst. Glacier and snowmelt also provide essential irrigation water to adjacent lowlands. The effects of climate change on the hydrological regime and water availability are crucial for the management and governance of water resources, especially in regions where snow and ice significantly contribute to river runoff. Glacier-related floods, including those from glacial lake outburst floods (GLOFs), have been documented across most glacierized mountain ranges and can affect areas tens to hundreds of kilometers downstream.

1.2 Development of the Cryohydrology

1.2.1 Origin of the Cryohydrology

Cryospheric hydrology studies have long focused on single elements of the cryosphere. As early as 1836, Jervis conducted forecasts of snowmelt based on changes in the snow-covered area of a basin. In the 1940s, researchers began applying mathematical and statistical methods to predict snowmelt runoff. The U.S. Army Corps of Engineers made significant contributions in 1956 and 1960 with the publication of the monographs "Snow Hydrology" and "Snowmelt Runoff," which detailed the theory, methods, and applications of snow hydrology and snowmelt runoff forecasting. During the International Hydrological Decade from 1965 to 1974, glacier hydrology attracted widespread interest. The first International Hydrological Conference in 1969, held in Cambridge, England, marked the emergence of modern glacier hydrology. Throughout this decade, over 60 glacial hydrological observation and experimental basins were established, transitioning glacier hydrology from qualitative descriptions to a phase combining observational experiments with quantitative analysis. In 1982, the "High Mountain Hydrology" session at the first International Hydrological Science Conference in Exeter, UK, recognized snow and ice hydrology as a new research field of hydrological research. At the 19th Geophysical Conference (IUGG) in Canada in 1987, the topic "The Impact of Climate Change on Hydrology and Water Resources" was set up. Ice and snow hydrology has received high attention due to its sensitivity to climate change. Subsequently, hydrological research on glaciers, snow, and frozen ground progressed, and in-depth simulation analysis was carried out on the basis of observational experiments. In 2012, Canadian scholar Woo Ming-ko synthesized existing findings and published "Permafrost Hydrology," a foundational monograph on cold region hydrology processes including glaciers, snow, frozen ground, and river and lake ice. This is a foundational monograph on hydrology in cold regions. At present, hydrological research in cold regions is penetrating into the socio-economic field, and more emphasis is placed on the impact of various hydrological elements of the cryosphere on sustainable development.

In these studies, water issues related to the cryosphere elements were often attributed to cold region hydrology or permafrost hydrology (Woo 2012). However, "cold region hydrology" is broadly defined and lacks a precise scientific delineation. It is sometimes specifically linked to permafrost regions but generally encompasses all hydrological phenomena in cold environments, including glaciers, frozen ground, snow cover, and river and lake ice. Research on multi-element hydrological phenomena within the cryosphere is also referred to as permafrost hydrology, especially by Canadian scholars (Woo 2012). Regardless of the terminology—whether permafrost or cold region hydrology—the studies focus on the hydrological issues related to specific elements of the cryosphere. The discipline system inherits the respective disciplinary characteristics of each single cryosphere element and focuses on the intersection of individual cryosphere elements. In terms of research content and disciplinary structure, these studies align with the subject on a single element of the cryosphere, such as glacier hydrology and snow hydrology. Yet, the collective impact of various cryospheric elements at basin, regional, and global levels has seldom been explored.

Recent advancements in Cryospheric Science, coupled with increasing socioeconomic demands for cryospheric water resources to support global sustainability, necessitate a shift in Cryohydrology research from a traditional single-element perspective to a multi-element, integrated approach. Consequently, it is essential to consolidate cryospheric water issues within a unified research framework and establish a disciplinary system from the perspective of Cryospheric Science. This system should encompass the hydro-thermal processes of glaciers, frozen ground, snow cover, and lake ice, as well as the hydrological functions of various cryospheric elements, all considered within the same framework.

Cryohydrology can be viewed as both a traditional and an emerging discipline. Traditionally, it has been grounded in the theories and methodologies of mass balance, ablation, hydrothermal processes, and the generation and convergence mechanisms of individual cryosphere elements. However, with the increasing impacts of global warming on the cryosphere, the hydrological processes and influences of cryosphere elements are garnering broader attention. Water issues related to the cryosphere can no longer be addressed solely within the traditional framework of single-element hydrology. It is necessary to transfer to a Cryospheric Science perspective, thereby promoting the formation of the Cryohydrology discipline.

1.2.2 Cryohydrology and Cryospheric Science

Cryospheric Science examines the formation, evolution, and interactions of the cryosphere with other environmental spheres, particularly the impacts of the cryosphere on the rest of the environment and adaptation mechanisms (Qin et al. 2018). This field includes: (i) the mechanisms of formation, development, and evolution of cryospheric elements and their interactions; (ii) the interactions among individual cryospheric elements, the entire cryosphere, and other Earth spheres such as the atmosphere, hydrosphere, lithosphere, and biosphere, with a particular focus on the intersection of the cryosphere and hydrosphere regarding water cycles and water availability (Fig. 1.3a); and (iii) the impacts and risks of cryospheric changes on other Earth spheres, including the links between the cryosphere and social sustainable development, especially adaptation strategies and countermeasures at global and

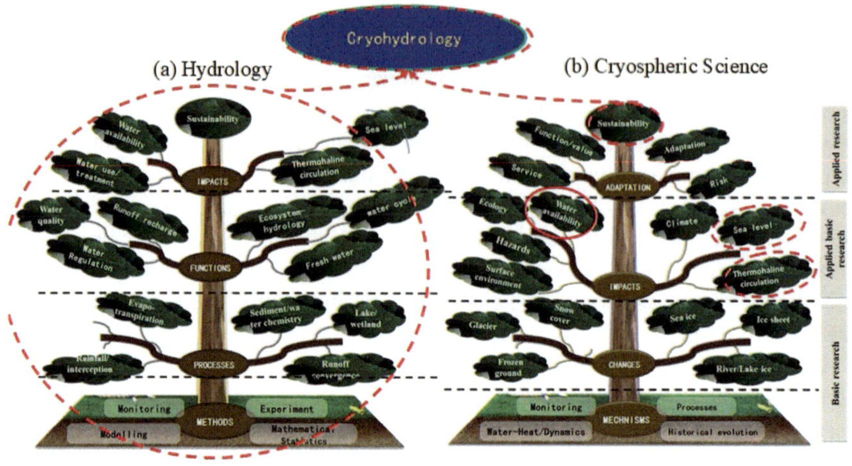

Fig. 1.3 Research content tree of Hydrology (**a**) and **b** Cryospheric Science (after Qin et al. 2018) as well as and the relationship between Hydrology, Cryohydrology and Cryospheric Science. Note that Cryohydrology focuses on the water cycle and water availability, and the intersection between Cryospheric Science and Hydrology (Ding et al. 2020)

regional scales (Qin et al. 2018). Cryohydrology is accompanied by the simultaneous development of Cryospheric Science, and several of its research fields are closely related to the elements of Cryospheric Science. From a hydrological perspective, the hydrological effects, water cycle, and water resources of the cryospheric elements are central to Cryohydrology (Ding et al. 2020).

The research framework of Cryohydrology closely mirrors that of Cryospheric Science, with the former's research contents forming a part of the latter. Within the Cryospheric Science discipline system, Cryohydrology is aligned with studies on water availability, sea level, and thermohaline circulation at the applied basic research level (Fig. 1.3b). The formation mechanism and changes in cryosphere elements at the basic research level are the scientific basis for Cryohydrology. The service, adaptation, and sustainability of water availability, and sea level change at the applied research level are the outputs of Cryohydrology (Fig. 1.3b).

1.2.3 Cryohydrology and Cold Region Hydrology

Among the cryosphere elements, most are stable and persist long-term, with the exception of some unstable snow cover (lasting fewer than 2 months) and short-term frozen ground (freezing periods of fewer than 15 days). There is no universally accepted definition of cold regions, which are typically identified as high-latitude or high-altitude areas. Consequently, hydrological research in these regions is commonly referred to as cold region hydrology.

The cold region is considered the core area influenced by the cryosphere, yet it does not encompass its entire extent, which is significantly broader. Theoretically, Cryohydrology research encompasses all hydrological phenomena and principles across the entire cryosphere, including both its core and impact areas. While cold region hydrology focuses primarily on water issues within the cryosphere's core areas. Both Cryohydrology and cold region hydrology concentrate on elements of the cryosphere with active hydrological functions and their impacts. Therefore, from this perspective, cold region hydrology and Cryohydrology are fundamentally similar.

The primary distinction between cold region hydrology and Cryohydrology is that cold region hydrology focuses on the special hydrological processes of different cryosphere elements, encompassing various research methods, processes/mechanisms, basinal functions, and global/regional impacts (Fig. 1.4). While Cryohydrology emphasizes the common characteristics across different cryosphere elements, including the research methods applicable to all elements such as ice shelves/icebergs, sea ice, snow cover, frozen ground, and glaciers/ice sheets. Moreover, Cryohydrology investigates the combined impact of all these elements at both basinal and global scales. For example, the study of the importance and vulnerability of the world's water towers (Immerzeel et al. 2019) concerns not only their water-supplying role in the cryosphere's core area, but also the downstream dependence of ecosystem and society in the cryosphere's impact areas. It is a typical Cryohydrology study, but not a typical study of cold region hydrology.

Cold region hydrology focuses on individual cryosphere elements from a vertical perspective of method, process, and impact, while Cryohydrology aims to broadly apply common hydrological principles across various associated cryosphere elements. For instance, in cold region hydrology, elements such as glaciers, permafrost, snow, river/lake ice, and sea ice are studied in terms of water formation (generation and convergence processes), changes (variations in ice volume), and impacts (influences on runoff and water resources in basins, regions, and globally) (Fig. 1.4). However, Cryohydrology integrates the study of hydrothermal processes and hydrological functions related to glaciers, frozen ground, snow, and river ice within a unified framework (Fig. 1.4). Additionally, the concept of a 'cold region' has geographic attributes, and its boundaries are often indistinct, making the scientific definition of cold region hydrology less precise than that of Cryohydrology.

1.3 Discipline System of Cryohydrology

1.3.1 Research Content

The discipline mainline of Cryohydrology along method-process-function-impact, and its core research issues can be sorted out along this mainline (Fig. 1.5).

The research methodologies employed include field observations and experiments, laboratory experiments and analyses, applications of remote sensing and

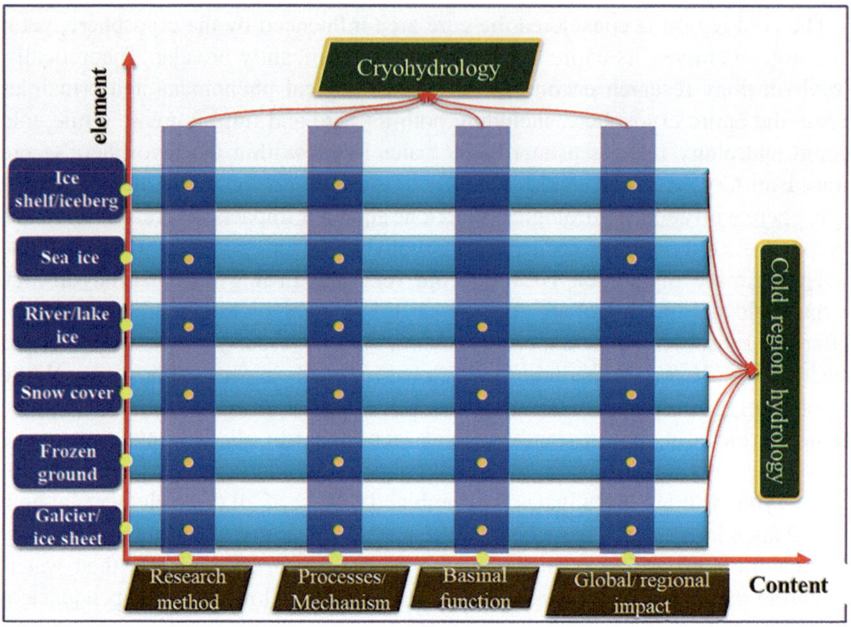

Fig. 1.4 Scope of Cryohydrology and cold region hydrology. Note that Cryohydrology focuses on the common hydrological law of different elements of cryosphere, while cold region hydrology focuses on the content of research method, processes/mechanism, basinal function, global/regional impact of single element of cryosphere (Ding et al. 2020)

geographic information systems, as well as mathematical statistics and model simulations.

The processes and mechanisms of Cryohydrology research encompasses three primary aspects: (1) the processes of melting, runoff generation, and convergence, which include snow and ice accumulation, meltwater convergence, and freeze-melt runoff dynamics in frozen ground, as well as river, lake, and sea ice formation and migration; (2) changes in runoff processes, detailing supply characteristics to cryospheric rivers, meltwater runoff dynamics, and projections of cryospheric meltwater runoff; (3) sediment and water chemistry, which covers river sediment dynamics, water chemistry processes, biogeochemical processes, and applications of water chemistry (Ding et al. 2020).

Cryohydrology is hydrologically crucial at the basin scale, including water conservation, river runoff recharge, regulation of water resources, and mitigation of extreme hydrological events.

Cryohydrology's main impact is in its output to other spheres. The above research issues in method, processes, and function are primarily related to the hydrological characteristics of the cryosphere elements. For example, Musselman et al. (2018) noted an increased frequency of rain-on-snow events at higher elevations where seasonal snow cover persists, attributing this to a transition from snowfall to rainfall

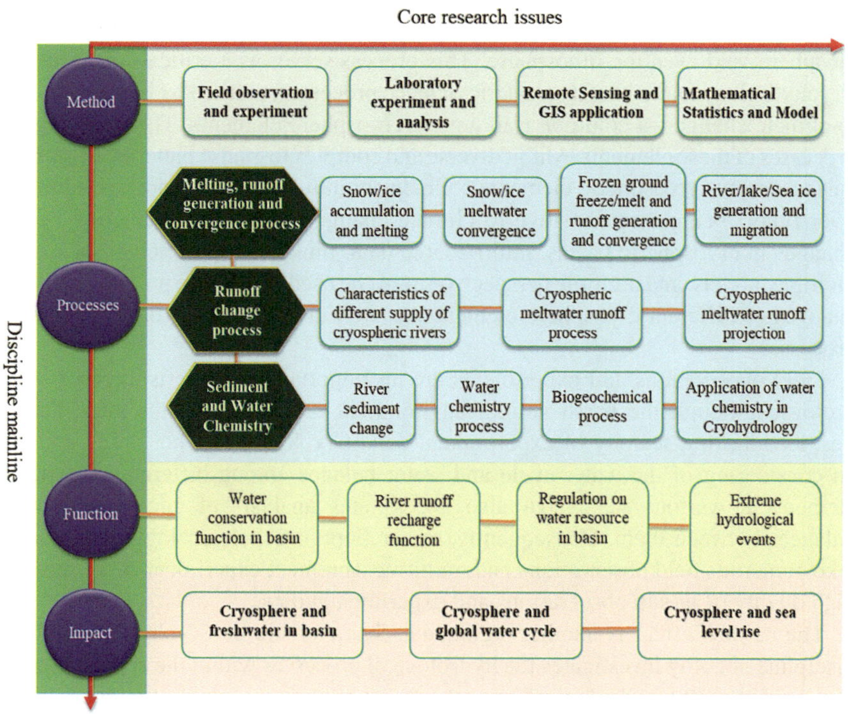

Fig. 1.5 The core research issues of Cryohydrology along the discipline mainline which primarily include method (in-situ observation, laboratory experiment, remote sensing application, and statistics/model), processes (melting, runoff generation and convergence, runoff changes, sediment and chemistry), function (water conservation, runoff recharge, regulation, extreme events), and impact (freshwater in basin, global water cycle, and sea level rise) (Ding et al. 2020)

and noting a corresponding increase in flood risk by 20–200% across North America. Additionally, when meltwater enters a catchment and mixes with rainfall runoff, the influence of cryospheric elements on river runoff varies across different parts of the basin, depending on the proportion of meltwater runoff and the extent of frozen ground coverage. The impact of Cryohydrology includes cryospheric and freshwater dynamics within river basins, the global water cycle, and contributions to sea level rise.

1.3.2 Discipline Characteristics

The core characteristic of Cryohydrology is the phase change of water, setting it apart from traditional hydrology primarily through the frequent phase transitions between the vapor, liquid, and solid states of water and the associated changes in water, energy, mass, and solutes. These phase transitions involve significant energy exchanges, impacting interactions among the atmosphere, cryosphere, hydrosphere, biosphere, and even the lithosphere. This process serves as a critical link between Cryohydrology and other Cryospheric Science processes, marked by high sensitivity to climate change—a common trait among cryosphere elements. The hydrological processes of these elements exhibit diverse and complex response patterns. In specific basins, various cryospheric elements collectively influence river runoff, resulting in river runoff being very sensitive to climate change. Accurately predicting runoff changes in cryospheric basins, multi-source data, multiple experimental measures, multiple models, and multiple perspectives must be used together to project the future changes of different cryosphere elements, as well as their impact on hydrological processes.

Field observations and experiments are an important characteristic of Cryohydrology, although the harsh natural environment often complicates these activities, leading to a scarcity of experimental data. Cryohydrology not only demands an considering of the water cycle and water balance among different cryosphere elements at various scales but also necessitates analysis of energy and mass balances between them. Consequently, researchers in Cryohydrology must undertake numerous field observations and multi-measurement experiments to ensure the high quality of in-situ observations and experimental data.

The surplus effect is another significant characteristic of Cryohydrology. This discipline not only investigates the hydrological processes within the cryosphere but also considers the hydrological impacts on other regions outside the cryosphere. This surplus feature affects basins, regions, and even the world. The understanding of Cryohydrologic processes on the formation of water resources in related regions, the hydrological functions of different components, and the potential changes of different components under future climate change scenarios are closely related to short-term, medium-term, and long-term sustainable development. A key feature of Cryohydrology is its interdisciplinary nature, merging aspects of cryospheric science, hydrology, geography, and atmospheric science. The phase transition processes of water and related water chemical processes are closely related to traditional physics and chemistry. The study of the regional and global impact of the water cycle is also closely related to Social Science, such as sustainable development and economics.

The final characteristic of Cryohydrology is its varied temporal and broad spatial scales. Depending on the content of study, the temporal scale involved in Cryohydrology ranges from hours to hundreds of years (Fig. 1.6). For processes and mechanisms, these scales typically range from hours to hundreds of years, with daily or monthly scales commonly used. For instance, the duration of an extreme hydrological event, such as a moraine-dammed lake outburst flood, usually spans hours to days. In studies of basinal functions, the temporal scope often extends from one to hundreds years. For example, the intensity and frequency of extreme hydrological events probably changes under climate change, and its impact on the basin also varied. When examining the influence of the cryosphere's freshwater cycle on oceans, the time frame may expand to millennial scales or longer. Spatially, Cryohydrology studies can vary from localized points to global extents. Research on processes and

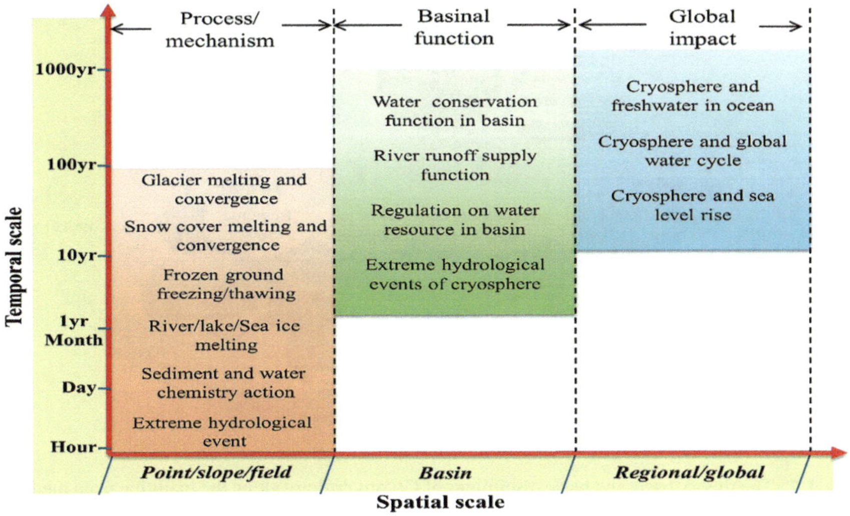

Fig. 1.6 Difference of main research contents of Cryohydrology at the temporal (hour, day, month, year, and more than 1000 years) and spatial (point/slope/field, basin, and region/global) scales (Ding et al. 2020)

mechanisms is typically conducted at point, slope, or field scales, hydrological functions are analyzed at the basinal scale, and impacts are assessed at regional or global scales (Fig. 1.6). Research has shown that temporal scales generally increase with the expansion of spatial scales.

1.3.3 Cryohydrology and Other Disciplines

The foundation of Cryohydrology discipline comes from other related specialties. Drawing on existing research foundation and scientific knowledge, energy balance and water balance form the physical basic of Cryohydrology, representing its most fundamental theoretical basis for development (Fig. 1.7). In terms of the energy balance theory, Cryohydrology is closely linked to Cryospheric Science and atmospheric science, whereas water balance connects more with hydrology and water resource science. Consequently, the basic disciplines for Cryohydrology are Cryospheric Science and hydrology (Fig. 1.7). Additionally, water resource science, geography, and atmospheric science are integral to the field, serving as its essential foundational disciplines (Fig. 1.7).

From this perspective, Cryohydrology considers various elements of the cryosphere—including glaciers, snow cover, frozen ground, river ice, lake ice, sea ice, ice sheets, ice shelves, and icebergs—as research subjects. Theoretical frameworks such as energy balance and water balance serve as the foundation for this

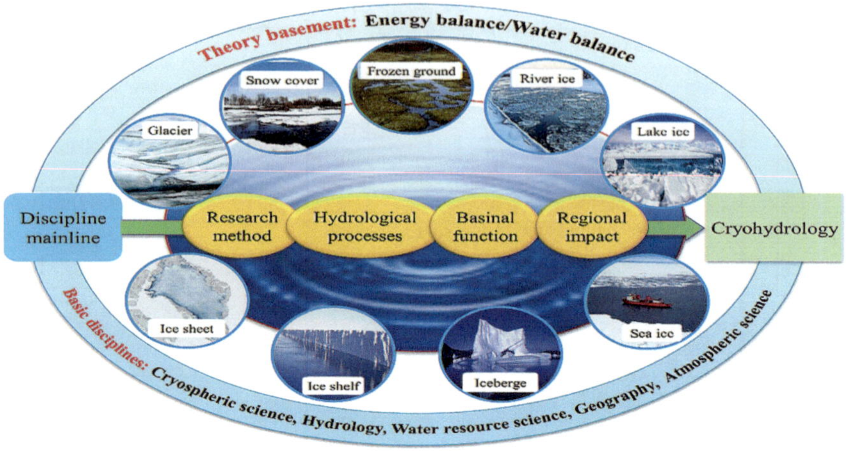

Fig. 1.7 The theory basis and basic disciplines of Cryohydrology along the mainline from method, processes, function to impact of different elements of cryosphere (Ding et al. 2020)

research. The main research approach follows the sequence of hydrological process, watershed effect, and regional impact. The objectives of Cryohydrology are to elucidate the hydrological processes, clarify the hydrological functions, and understand the hydrological impacts on society of the cryosphere.

References

Ding Y, Zhang S, Chen R, Han T, Han H, Wu J, Li X, Zhao Q, Shangguan D, Yang Y, Liu J, Wang S, Qin J, Chang Y (2020) Hydrological basis and discipline system of cryohydrology: from a perspective of cryospheric science. Front Earth Sci 8:574707. https://doi.org/10.3389/feart.2020.574707

Immerzeel WW, Lutz AF, Andrade M, Bahl A, Biemans H, Bolch T, Hyde S, Brumby S, Davies BJ, Elmore AC, Emmer A, Feng M, Fernández A, Haritashya U, Kargel JS, Koppes M, Kraaijenbrink PDA, Kulkarni AV, Mayewski P, Nepal S, Pacheco P, Painter TH, Pellicciotti F, Rajaram H, Rupper S, Sinisalo A, Shrestha AB, Viviroli D, Wada Y, Xiao C, Yao T, Baillie JEM (2019) Importance and vulnerability of the world's water towers. Nature 577(7790):364–369

IPCC (2021) Climate change 2021: The physical science basis. contribution of working group I to the sixth assessment report of the intergovernmental panel on climate change. In: Masson-Delmotte V, Zhai P, Pirani A, Connors SL, Péan C, Berger S, Caud N, Chen Y, Goldfarb L, Gomis MI, Huang M, Leitzell K, Lonnoy E, Matthews JBR, Maycock TK, Waterfield T, Yelekci O, Yu R, Zhou B (eds) Cambridge University Press, Cambridge, United Kingdom and New York, NY, USA, 2391 pp. https://doi.org/10.1017/9781009157896

Musselman KN, Lehner F, Ikeda K, Clark MP, Prein AF, Liu C, Barlage M, Rasmussen R (2018) Projected increases and shifts in rain-on-snow flood risk over western North America. Nat Clim Chang 8(9):808–812

Qin D, Ding Y, Xiao C, Kang S, Ren J, Yang J, Zhang S (2018) Cryospheric science: research framework and disciplinary system. Natl Sci Rev 5(2):255–268

Woo MK (2012) Permafrost hydrology. Springer Science & Business Media

Chapter 2
Research Methods of Cryohydrology

Haidong Han, Junfeng Liu, Qiudong Zhao, Xiangying Li, and Xiaobo He

Cryohydrology places significant emphasis on the integration of field experiments and laboratory analyses. Field observations and remote sensing data provide essential records and data for understanding the occurrence and progression of ice and snow hydrological processes. Experimental analyses not only offer fundamental information about the physical and chemical properties of ice, snow, soil, and other materials but also reveal changes and interconnections in the chemical composition of these samples. While field observations and experiments explain natural phenomena and their characteristics, model analyses link these observations to explore the processes, mechanisms, and trends of cryospheric hydrological changes. Hydrological models, which are based on a reasonable simplification of the objective world, mathematically represent the key processes of cryospheric hydrology through formulas and equations, thereby simulating actual changes. Consequently, hydrological models allow for a quantitative analysis of the mechanisms and influencing factors involved

H. Han (✉) · J. Liu · Q. Zhao · X. He
Northwest Institute of Eco-Environment and Resources, Chinese Academy of Sciences, Lanzhou, China
e-mail: hhd@lzb.ac.cn

J. Liu
e-mail: liujfzyou@lzb.ac.cn; liujfzyou@nwu.edu.cn

Q. Zhao
e-mail: dsslab@163.com

X. He
e-mail: hxb@lzb.ac.cn

X. Li
Shaanxi Key Laboratory of Earth Surface System and Environmental Carrying Capacity, Northwest University, Xi'an, China
e-mail: shaanxilxy@163.com

Y. Ding et al., *Cryohydrology*, https://doi.org/10.1007/978-981-96-7360-5_2

in hydrological processes. This chapter will introduce the main research methods in cryospheric hydrology, integrating traditional testing methods and instruments.

2.1 Field Observations

Field observations in cryospheric hydrology focus on the climatic environment, physical properties, and the temporal and spatial distribution of snow and ice, as well as the composition and changes of meltwater runoff. While there are commonalities in field observation methods across different cryospheric components, significant variations exist depending on the characteristics of each element and the specific research objectives (Qin et al. 2021). This section synthesizes the fieldwork associated with various cryospheric components, discussing the content, methods, and principles of field observations and experiments, highlighting both shared approaches and differences among these elements.

2.1.1 Surface Environment Observations

Unlike surfaces such as soil, rock, or grass, the surfaces of ice, snow, and frozen soil are inherently unstable. The melting, accumulation, and density changes of ice and snow result in continuous alterations of the ground surface. The generation and movement of meltwater runoff are closely linked to the nature of these surfaces. Therefore, observing near-surface meteorological conditions and surface characteristics is a crucial component of fieldwork in cryospheric studies.

1. Meteorological observations

The formation and evolution of the cryosphere are driven by specific meteorological and topographical conditions. To study the development and changes in ice and snow, it is essential to integrate local meteorological processes and climatic characteristics. As a result, meteorological observations are a fundamental aspect of cryospheric fieldwork, providing the essential data that underpin research in cryospheric hydrology. The installation of meteorological equipment and the collection of data should be conducted in accordance with the "Ground Meteorological Observation Regulations" issued by the China Meteorological Administration, as well as other relevant guidelines or manuals.

It is important to note that measuring precipitation in field meteorological observations presents significant challenges, particularly in cryospheric regions, due to the effects of wind and low temperatures. These areas are especially prone to strong surface winds, influenced by their unique geographical environments. Conventional precipitation measurement instruments may underestimate actual precipitation by more than 10% due to reduced capture efficiency caused by strong winds. To address

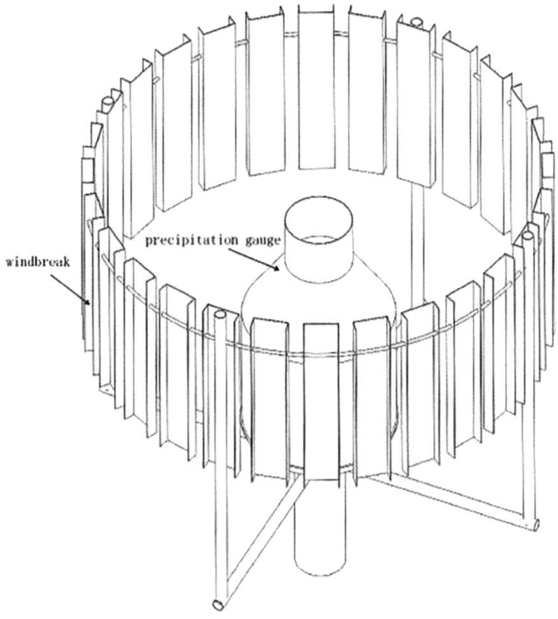

Fig. 2.1 Precipitation observation system with windbreak

this issue, it is necessary to establish a correction factor by comparing gauge-measured values with actual precipitation through experimental observations or by using precipitation instruments equipped with windbreaks (as shown in Fig. 2.1).

Low temperatures in cryospheric regions also increase the proportion of snowfall during precipitation events. Additionally, the freezing of liquid water inside precipitation instruments can lead to measurement errors or even damage the equipment. As a result, weighing-based methods for total precipitation measurement are commonly used in cryospheric regions, while tipping bucket and liquid level measuring instruments are less frequently employed.

2. Thickness survey

The thickness of snow, river ice, lake ice, and sea ice can be measured using a ruler (for snow) or stake measurements (for ice). For thick sea ice, ground-penetrating radar (GPR) is a convenient tool that provides a rapid assessment of regional thickness distribution. The basic guidelines for manual observation are as follows:

(1) **Representative Observation Points**: The chosen observation points should be representative of the larger area. The changes in snow and ice thickness at these points should reflect the overall changes across the region. For instance, snow observation points should avoid areas with wind drift, and river and lake ice measurements should be taken away from the riverbank.
(2) **Repeated Measurements**: Repeated measurements are generally necessary to reduce errors and improve accuracy.

Fig. 2.2 Radar measurement on the Chongce Glacier of West Kunlun Mountains

(3) **Consistent Timing**: The time interval between observations should correspond to the rate of change in ice and snow thickness. For periodic observations, measurements should be taken at the same time of day to ensure consistency in the data.

For automatic measurements, snow thickness can be observed using ultrasonic or infrared rangefinders, snow pillows, or timed photography. Ground-penetrating radar is also widely used for measuring the thickness of river, lake, and sea ice.

The measurement of glacier and frozen soil thickness is primarily carried out using geophysical prospecting techniques, such as ground-penetrating radar (GPR, as shown in Fig. 2.2), in combination with electrical instruments and drilling. Surveys are typically planned along multiple transect lines that are selected based on the topography or the distribution characteristics of the glacier and frozen soil, ensuring that the thickness measurements along these lines are representative of the area.

A key purpose of drilling in ice or frozen soil is to obtain direct thickness measurements, which can be used to calibrate and validate indirect measurements from methods like GPR. Additionally, drilling enables the retrieval of ice or soil samples, which are essential for analyzing the physical and chemical properties as well as the hydraulic characteristics of the ice or frozen ground.

3. Physical properties

Density is one of the fundamental physical properties of snow, ice, and frozen soil. The methods for manual density sampling and measurement are similar across these materials. Typically, a sampling box or drilling tool is used to collect undisturbed samples of snow, ice, or frozen soil, and the density or specific gravity is then determined by weighing the sample. Additionally, snow density can be measured in situ using a snow characteristic meter, which operates based on dielectric constant measurements (Fig. 2.3).

When conducting density measurements, it is crucial to ensure that the observation points are both representative and randomly selected to avoid bias. To reduce systematic errors, more than three effective measurements should be taken at each observation point, with the average value being used as the final result.

Fig. 2.3 Measuring snow density and water content using a snow characteristic meter

To accurately assess the hydrothermal parameters of ice and snow, it is essential to understand the composition and properties of surface materials. Surface roughness, impurities, and particulate matter can significantly impact surface albedo and heat flux, thereby affecting the melting rate of ice and snow. Common impurities include rock debris, dust, and black carbon. The impurity content in ice and snow can be determined using the thawing-filtration-drying method.

Soil porosity is categorized into capillary and non-capillary pores. Capillary pores indicate the soil's water-holding capacity, while non-capillary pores reflect its ability to aerate, permeate, and retain water. During the freeze–thaw cycles of frozen soil, porosity changes as ice forms and melts, directly influencing hydraulic parameters such as soil hydraulic conductivity and water potential. Capillary porosity is the percentage of capillary pores relative to the total soil volume and can be measured using the ring knife method or specific surface area meters. Non-capillary porosity represents the volume of pores excluding capillary pores within the total porosity. Total porosity can be calculated using soil bulk density and specific gravity, with bulk density referring to the weight of solid dry soil particles per unit volume compared to the weight of water in the same volume.

Snow water content represents the weight percentage of liquid water within snow and is typically measured using a snow characteristic meter. In contrast, the moisture content of frozen soil serves as an index for assessing the total moisture content within the soil. Common methods for measuring soil moisture content include the drying method, electrical resistance method, neutron method, and gamma ray method.

Unfrozen soil water content refers to the amount of liquid water present in frozen soil due to capillary action and adsorption on soil particle surfaces. Nuclear magnetic resonance (NMR) technology is a reliable method for determining unfrozen water content, operating by measuring the attenuation of the free magnetic induction of radio frequency signals as they pass through the soil sample. Additionally, time domain reflectometry (TDR) and frequency domain reflectometry (FDR) are straightforward and accurate methods for assessing soil moisture. TDR measures the difference in the dielectric constant of water compared to other media using time domain techniques, while FDR employs electromagnetic pulses to determine the apparent permittivity of the soil based on the propagation frequency of electromagnetic waves, thereby calculating the volumetric water content of the soil.

Lower unfrozen soil water content corresponds to higher soil water potential, which facilitates the migration and accumulation of moisture toward the freezing front. Soil water potential can be measured using soil water potential sensors or tensiometers. Soil water potential sensors operate on the principle of molar heat capacity, determining the matrix potential of the soil by measuring changes in heat capacity in a porous clay head. Tensiometers, on the other hand, are easy to install, cost-effective, and not affected by soil solute potential, though they can be cumbersome to use and are unsuitable for temperatures below 0 °C.

Hydraulic conductivity in soil and snow refers to the volume of water that passes through the medium per unit time under a unit water potential gradient, indicating the amount of water flowing through a unit area in a given time. Pore distribution characteristics significantly influence saturated hydraulic conductivity. Unsaturated hydraulic conductivity, defined as the water flux per unit water potential gradient in the unsaturated state, depends on both the water content of the medium and the water potential. Laboratory tests for hydraulic conductivity are based on Darcy's law, employing either the constant water head or falling water head methods to determine saturated hydraulic conductivity. The saturated-evaporation principle is used to measure unsaturated hydraulic conductivity. Additionally, saturated hydraulic conductivity can be assessed in the field using the disk permeameter method.

4. Thermal characteristics

Temperature variations in ice, snow, and frozen soil are key indicators of the energy exchange between the earth and the atmosphere. Ground temperature measurements can be performed using thermal resistance devices, thermocouples, or infrared temperature sensors to create detailed multi-level vertical profiles. For ice and snow, where the formation and movement of meltwater primarily occur at the surface, ground temperature monitoring focuses on the top 0–50 cm below the surface. Temperature sensors are typically positioned at depths of 0 cm (surface), 5 cm, and 10 cm, with subsequent intervals of 10 cm.

In frozen soil regions, temperature sensors should extend to the bottom of the active layer of permafrost or the maximum freezing depth of seasonally frozen soil. Soil temperature measurement sites are generally spaced at 10 cm intervals from the surface down to the vegetation root layer. Below this layer, sensors can be placed at 10 or 20 cm intervals, depending on soil conditions, until they reach the bottom of the active layer.

Heat flux, which refers to the rate of heat transfer per unit time per unit area, provides important insights into the heat budget of ice, snow, or soil. The direction and magnitude of heat flux can reveal the dynamics of energy exchange in these environments. Heat flux can be measured using soil heat flux plates, which should be positioned in alignment with ground temperature sensors or at specific characteristic levels, such as the ground surface, the bottom of the frozen soil active layer, the snow bottom, or the surface of the moraine layer.

Thermal conductivity is defined as the rate of heat transfer per unit area per unit time when the temperature gradient in a specified direction is 1 °C/m. The measurement of thermal conductivity typically involves two main techniques: the steady-state method and the transient method.

The steady-state method involves maintaining a constant temperature gradient across the medium and measuring the resulting heat flux through snow or soil. This method requires the temperature at both ends of the medium to be stable, allowing for precise measurement of the heat transfer.

In contrast, the transient method applies a pulse of heat to the medium and measures the resulting temperature change over time at a specific location. This method calculates thermal characteristics based on the observed temperature response. The transient method can be further subdivided into the single-needle and multi-needle methods.

The single-needle method combines the heating source and temperature sensor within a single probe, making it particularly useful and common in field applications due to its simplicity and effectiveness. The multi-needle method, although more complex, allows for more detailed thermal analysis by measuring temperature changes at multiple points within the medium.

2.1.2 Ablation Observations in the Cryosphere

The primary purpose of ablation observation is to monitor the processes of ablation and hydrothermal changes over a specific period. The data collected from these observations are essential for studying the characteristics of ablation across various cryospheric elements, understanding their relationships with climate factors, calibrating hydrological model parameters, and analyzing the physical and chemical composition of meltwater. Given the significant differences in surface characteristics among snow, ice, and frozen soil, the observation methods must be tailored to each type. This section will discuss these methods separately, detailing the specific approaches used for each cryospheric element.

1. Snow Melting Observation

The high porosity of snow means that snowmelt water must undergo a process of occurrence and migration within the snow layer before surface runoff can form. Therefore, to effectively observe snowmelt, it is essential to measure changes in snow thickness as well as the snow water equivalent (SWE). The SWE is defined as the depth of water that would result from the complete melting of snow over a unit area, representing the total amount of water stored in the snow.

Fig. 2.4 Single point integrated observation system for snow cover in river basin

Direct methods for measuring SWE include the snowboard measurement, lysimeter method, snow pillow method (Fig. 2.4), and gamma ray method (Fig. 2.5). Each of these methods provides valuable data for understanding snowmelt processes and their contribution to runoff.

2. Observation of Ice Surface Melting

Unlike snow, once the ice surface melts, the meltwater quickly flows along the surface, forming runoff. Therefore, monitoring ice surface ablation primarily involves tracking changes in the relative elevation of the ice surface. The stake (ablation rod) measurement is a commonly used manual observation method. This method involves

Fig. 2.5 Gamma ray method snow water equivalent observation system

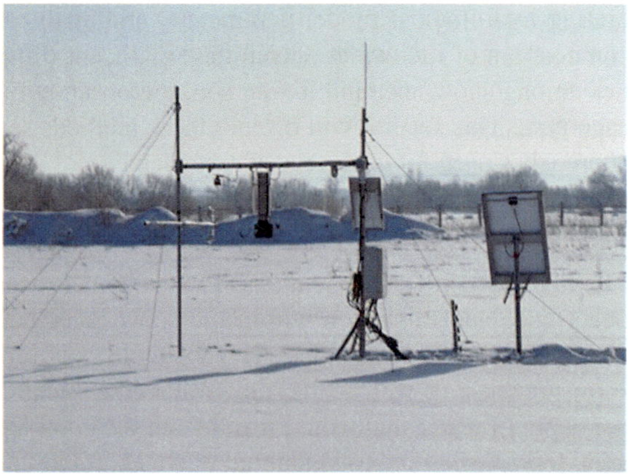

drilling a borehole using a hand-cranked or steam ice drill, inserting the stake, and then regularly measuring and recording the exposed height of the stake to track ablation.

In addition to manual methods, automated measurements based on ultrasonic distance measurement principles are increasingly utilized, especially in challenging environments such as river and lake ice or sea ice, where manual measurements may be difficult or unsafe.

In glaciated areas, the underlying surface topography and terrain properties often result in significant discrepancies between the surface ablation volume obtained from single-point observations and the actual volume of runoff. To accurately assess ice ablation, it is essential to establish a test runoff field specifically designed for slope ablation and runoff measurement. The runoff site should be located in a uniform and level area, typically arranged in a rectangular configuration of 2 m × (4–6 m) or 5 m × (10–20 m), with the longer side oriented along the direction of the maximum slope of the site.

To ensure accurate measurements, the runoff field should be enclosed with fences or intercepting ditches to prevent external meltwater from entering or leaving the area. A water outlet should be positioned at the lowest point of the terrain within the runoff field, allowing meltwater to flow into a storage tank for measurement. This setup enables precise monitoring of ablation and runoff, providing critical data for hydrological studies.

Ablation measurements on inclined ice surfaces, such as slopes and cliffs, are essential in the observation of glaciers and ice shelves. The stake method is commonly used for this purpose, where stakes are inserted vertically into the ice. However, to accurately calculate ablation and mass balance, the observed data must be adjusted to account for the incline, reflecting the direction perpendicular to the ice surface. This adjustment ensures the precision of the ablation measurement. The ablation amount can be calculated by:

$$m_p = m_v \cos\alpha \tag{2.1}$$

where m_p represents the amount of ablation in the direction perpendicular to the ice surface, m_v denotes the ablation observed by the stake, and α is the average slope of the ice surface. An ice cliff is an exposed ice surface characterized by a specific slope and aspect, typically found in debris covered areas. Given that the bare ice surface of an ice cliff often has a steep slope (>30°), measurements taken with distance-measuring sensors, such as ultrasonic devices, can yield significant errors. To mitigate this, a relatively stable and large moraine located behind the top of the ice cliff can be used as a fixed reference point. By regularly measuring the distance between this reference object and a specific point on the top of the ice cliff (Fig. 2.6), the backwasting of the ice cliff over time can be determined. The depth of ice cliff melting (R_c) is then calculated by:

$$R_c = \frac{\rho_i L_c \sin\beta}{\rho_w} \tag{2.2}$$

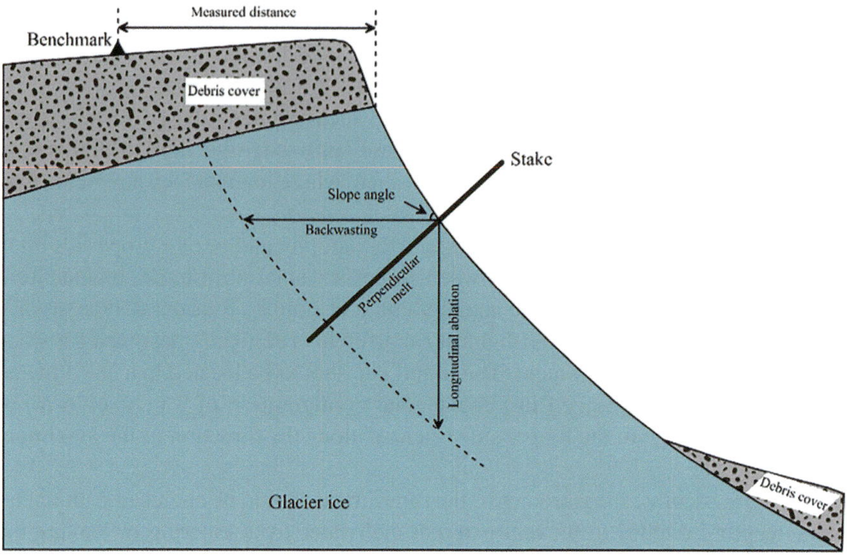

Fig. 2.6 Schematic diagram of the measurements of ice-cliff backwasting

where L_c is the ice-cliff backwasting; β is the mean slope of ice-cliff; ρ_i is the ice density; ρ_w is the density of water.

The stake method can also be employed to measure ablation in supraglacial debris-covered areas. In regions with thick debris, it may be necessary to excavate pits to facilitate the installation of stakes. Due to the minimal melting of the ice surface beneath the debris, the stakes do not need to be inserted deeply into the ice.

3. Glacier Mass Balance

Glacier mass balance observation aims to quantify the accumulation and ablation of glacier volume over a specific period. This can be achieved through direct and indirect observation methods. The direct method, also known as the glaciological method, involves establishing a measurement grid on the glacier (Fig. 2.7). By regularly measuring ablation and accumulation at each grid point, it is possible to comprehensively determine the glacier's mass balance and its components over the course of a year or during a specified time period.

In the ablation zone, the primary measurement point is a stake inserted into the ice. The amount of ice surface ablation observed over a specific period, when converted to water equivalent, represents the net balance at that point. In the accumulation area, where snow typically persists year-round, measurement points are primarily snow pits. The snow pit method is based on the principle that variations in temperature, precipitation, and atmospheric conditions cause changes in the density, composition, and color of the snow, leading to distinct layers or annual stratification. The snow water equivalent accumulated between these characteristic layers can be measured over a specified time interval. The depth of the manually excavated snow pit in

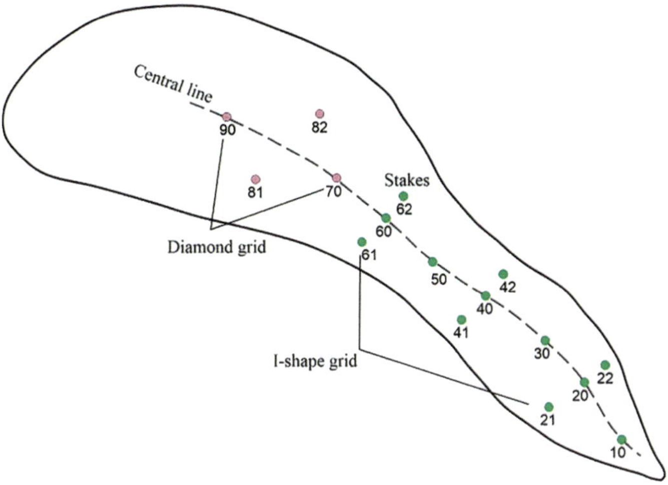

Fig. 2.7 Layout of mass balance measuring points

the accumulation area should exceed the annual snow thickness. The density and thickness of the snow are assessed layer by layer, with the snow water equivalent for each layer calculated and summed to determine the total annual accumulation. Accurate identification of annual layers is crucial for assessing mass balance. At the end of summer, significant ablation can cause dust, impurities, and other dark materials to accumulate on the snow, forming a polluted surface that serves as an important annual marker in the middle and lower parts of the accumulation area.

Once the balance data for each observation point on a glacier is obtained, the net balance value of the glacier over a specific period can be calculated. This is achieved by using the glacier area represented by each point and applying either the contour line method or the equal balance line method (Fig. 2.8).

Indirect methods for measuring glacier mass balance include repeated measurements, equilibrium line altitude (ELA) measurements, and climatic and hydrological calculations. The repeated measurement method, also known as the geodetic method, involves acquiring topographic data of the glacier through large-scale topographic mapping, ground photogrammetry, aerial photography, satellite remote sensing, and satellite altimetry. By comparing elevation differences on the ice surface, changes in glacier volume, area, and elevation can be assessed, enabling the calculation of the average net balance over the observation period.

The ELA measurement method is based on the principle that the shape of the mass balance gradient curve for a specific glacier remains relatively stable. Once the relationship between the glacier's ELA and its net balance is established (Fig. 2.9), annual mass balance data can be derived from ELA observations, even though the mass balance may fluctuate annually.

The climatic and hydrological calculation method relies on the principle of water balance. By using climate data and hydrological models, it is possible to calculate

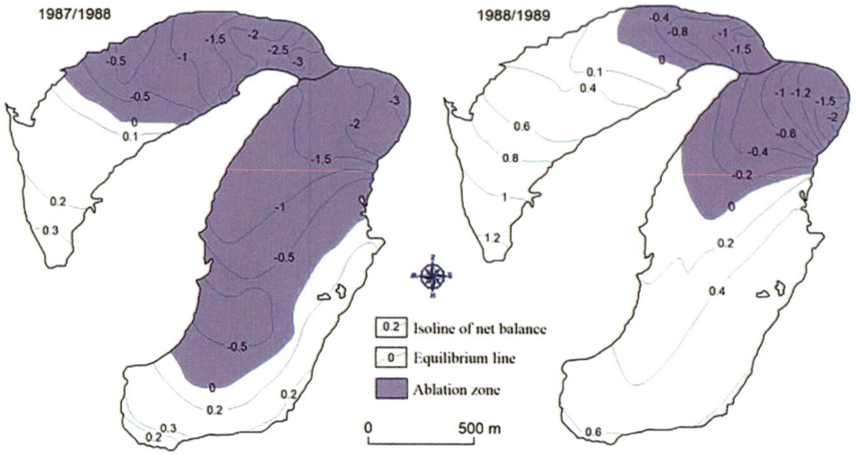

Fig. 2.8 Contour map of mass balance of Glacier No. 1 at the headwaters of Urumqi River (Based on the annual report of the Tianshan Glacier Observation and Experiment Station of the Chinese Academy of Sciences)

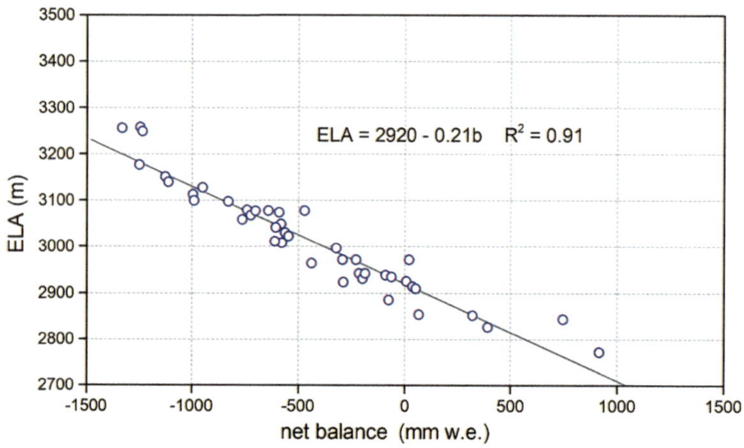

Fig. 2.9 The relationship between the ELA and annual mass balance of the Hintereisferner glacier in the Alps

accumulation, ablation, and net balance for any given area over a specified time frame. The net balance (bbb) per unit area of the glacier during a certain period is calculated using the following formula:

$$b = p - m - e \pm \Delta s \tag{2.3}$$

where p represents precipitation; m denotes ice ablation; e accounts for mass loss due to evaporation or sublimation; Δs includes other forms of mass accumulation or loss, such as snow replenishment, ice avalanches, condensation, or additional sublimation.

(4) Observation of Hydrothermal Processes in Frozen Soil

Observing hydrothermal processes in frozen soil differs significantly from monitoring ice and snow. While ice and snow undergo considerable thickness changes during accumulation and melting, the freezing and thawing processes in frozen soil primarily impact soil water content, soil flow, and the thickness of the active layer. Therefore, a systematic approach to understanding hydrothermal processes in frozen soil, coupled with automated near-surface meteorological observations, is essential. Key variables to observe include soil temperature, heat flux, thermal conductivity, soil water content, and soil water potential (Fig. 2.10).

Surface evapotranspiration in frozen soil can be measured using a lysimeter (Fig. 2.11), which involves directly weighing changes in the test soil column while accounting for precipitation input, water leakage, and drainage. Another method for directly measuring evapotranspiration is through an eddy covariance observation system (Fig. 2.12). This technique calculates turbulent transport volumes by measuring the covariance between fluctuations in temperature, water vapor, and vertical wind speed. It is widely regarded as a standard method for directly assessing energy and material exchange between the biosphere and the atmosphere.

The hydrological processes of frozen soil at the slope scale involve a complex interplay of physical factors, including the effects of precipitation, infiltration, and

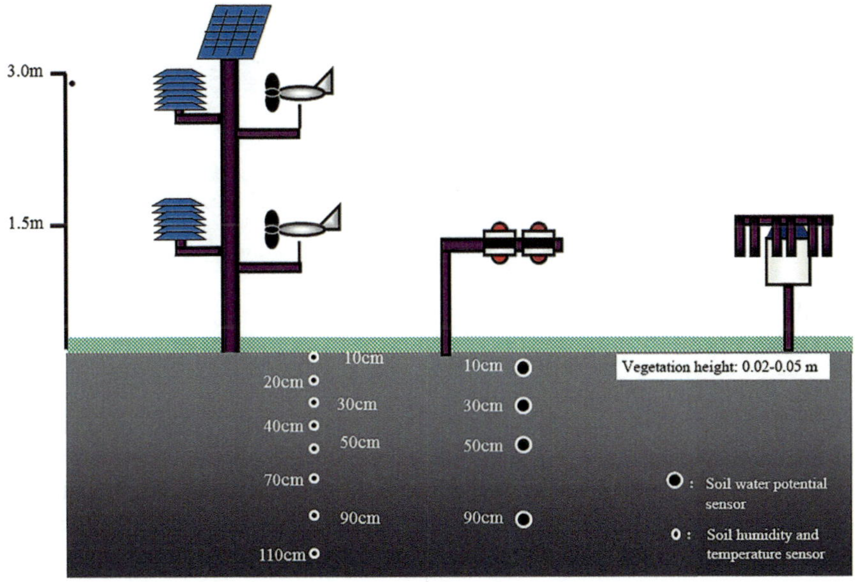

Fig. 2.10 Schematic diagram of field-based comprehensive observation of hydrothermal processes in frozen soil

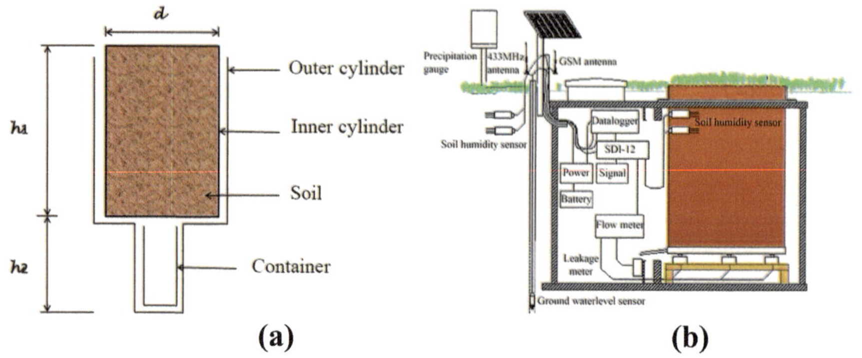

Fig. 2.11 Schematic diagram of soil lysimeter (**a** Small manual lysimeter, **b** Large automatic lysimeter)

Fig. 2.12 Open Eddy correlation observation system

surface runoff. Comprehensive testing is typically conducted by establishing slope runoff fields. Following established protocols, a relatively flat, straight slope on the hillside is selected, requiring a standard test field of 20 × 5 m. The runoff field includes a water catchment area, boundary markers, a water collection trough, and a drainage system (Fig. 2.13). Concurrent meteorological and surface observations should be conducted near the runoff field to provide comprehensive data.

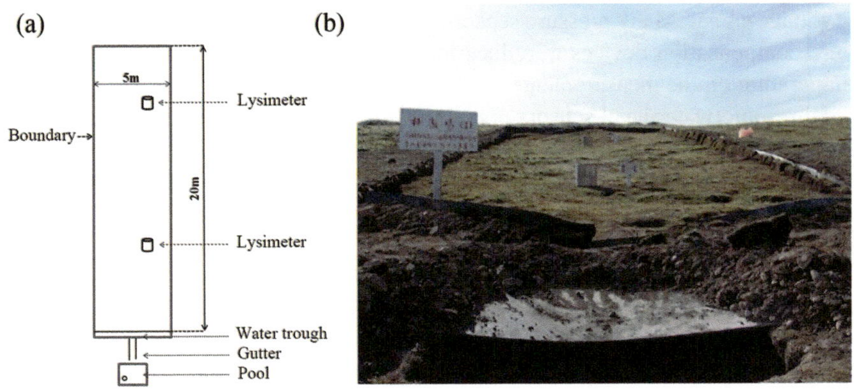

Fig. 2.13 Slope runoff field (**a** Schematic diagram of runoff field in frozen soil area; **b** Runoff field on the Yeniugou Slope in the Upper Reaches of the Heihe River)

2.1.3 Hydrological Section Observation

A hydrological section is a facility designed to regulate the surface runoff outlet of a catchment area or basin, with typical observation parameters including water level, discharge, and water quality. This section introduces the key principles and methods of hydrological section measurement, focusing on three main aspects: the establishment of the hydrological section, discharge measurement, and sediment and hydrochemical analysis.

The location of the hydrological section is closely tied to the river's topography and is ideally established in a flat, straight canyon area with steady water flow. In cryospheric watersheds, runoff is significantly influenced by temperature variations and exhibits a high interannual coefficient of variation. Therefore, when establishing a hydrological section, it is crucial to consider the historical flood line and maximum flow rate to prevent damage to observation facilities from sudden flooding. When the river channel is wide and the water volume is large, an observational well is typically installed near the riverbank to house the measuring equipment.

Water level measuring equipment includes water gauges, pressure water level sensors, and ultrasonic water level sensors. A water gauge is a specialized hydrological device used for manually observing water level changes, typically fixed to a pile or attached to the outer wall of a logging well. Electronic water gauges have a similar appearance and installation process but primarily measure water levels using the electrical conductivity of water or the electrical properties of materials. This is achieved by detecting electrical signals from distributed electrodes or by monitoring changes in capacitance and resistance.

The main component of a pressure water level sensor is a piezoelectric sensing device encased in a stainless steel shell, which converts water pressure into a current or voltage signal. Many pressure sensors are equipped with air pressure and temperature compensation devices to reduce the effects of fluctuations in air pressure or temperature on the measurements.

Ultrasonic water level sensors operate on the principle of sound pulse reflection from the water surface. By measuring the time difference between the transmission of the pulse and the reception of the echo, the distance between the sensor and the water surface is calculated based on the speed of sound in air, enabling accurate water level determination.

When discharge is low, a weir can be established to measure flow. A flow-measuring weir is a stable, artificial measurement section that is resistant to erosion by meltwater, offering advantages such as rapid measurement and maintenance-free operation. The weir flow formula can be applied according to the "Weir and Trough Flow Measurement Specification (SL24-91)." In large watersheds, where the river channel is often wide and braided, with water flow divided into multiple paths by floodplains or boulders, flow meters are necessary to measure flow in sections.

Flow meters used for this purpose include propeller, electromagnetic, and acoustic Doppler flow meters. The propeller flow meter operates by counting the signals generated by blade rotation within a specified time, which is then used to calculate the average flow rate during the measurement period. This type of flow meter is widely used in hydrological observations.

The electromagnetic flow meter uses water as a conductor, cutting through magnetic field lines to generate an electrical signal within a magnetic field. The voltage produced is proportional to the flow velocity, and by amplifying and converting this signal, the water flow velocity can be determined.

The segmented flow measurement method is the most commonly used approach for runoff observation. This technique divides the hydrological section into several equidistant subsections based on the width of the river channel and variations in flow velocity within the section. By calculating the average flow velocity and flow for each subsection, the total flow for the section is obtained.

To measure profile flow velocity, either single-point or multi-point measurements are taken, depending on the water depth, to determine the average velocity across the profile. For shallow waters ($h < 1.0$ m), single-point measurements can be used, with the flow velocity at 2/3h considered as the average profile velocity. For deeper waters, flow measurements can be taken at depths of 0.2h and 0.8h (two-point method) or at 0.2h, 0.6h, and 0.8h (three-point method).

In the two subsections near the riverbank, flow velocity decreases rapidly due to the drag effect of the riverbank. The water surface near the bank and the riverbed profile often form a triangular shape. Therefore, when calculating flow for these subsections, it is necessary to apply a bank coefficient (ranging from 0.5 to 0.9) to adjust the flow measurement accordingly.

During a hydrological year, flow measurements should be conducted at various water levels, with particular emphasis on capturing the highest and lowest levels for cross-sectional flow analysis. By correlating the obtained flow rates with their corresponding water levels, a water level-flow curve (Fig. 2.14) can be constructed, from which a fitting formula can be derived. This formula allows for the conversion

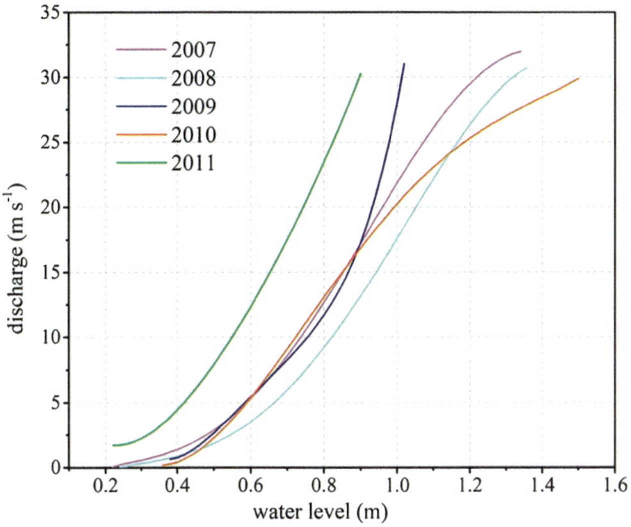

Fig. 2.14 The relationship curve between water level and discharge in the Koxkar Glacier Basin in Tianshan Mountains

of continuous water level records into continuous flow data. In natural river sections, where section morphology changes frequently, it is crucial to establish a water level-discharge curve and corresponding fitting formula at annual or intra-annual intervals, depending on the extent of the morphological changes.

Field water quality observations primarily include parameters such as pH, conductivity, oxidation–reduction potential (ORP), salinity, turbidity, suspended solids, dissolved oxygen, and total dissolved solids (TDS), all of which can be measured using online automatic equipment. The chemical composition of water mainly comprises inorganic matter (primarily soluble ions and trace components), organic matter, microorganisms, and gases present in the runoff. Component analysis requires stringent conditions for both the testing environment and instruments, typically necessitating the collection of water samples for laboratory analysis. An overview of the indoor analysis of samples will be provided in Sect. 2.2.

2.1.4 *Small Watershed Integrated Observations*

Integrated observations in a small cryospheric watershed involve a comprehensive approach to field observations, encompassing the measurement of hydrology, meteorology, and surface attributes, along with the integration and application of spatial data across the watershed. At the basin scale, cryosphere elements such as ice, snow, and frozen soil are not isolated; they coexist with rocks, grasslands, forests, and other non-cryospheric surfaces, forming a cohesive hydrological system. For example, in a glacial basin, the accumulation area is perpetually covered with snow, while the ablation area consists of a thick layer of glacial ice. Surrounding the glacier are rocks, permafrost, and grasslands, accompanied by seasonal snow and frozen soil, which are widely distributed and often occur simultaneously. Consequently, hydrological observations at the watershed scale must align with the characteristics of runoff generation and concentration, as well as the research objectives related to different underlying surface types within the watershed.

1. Spatial Survey of Cryosphere Elements

The spatial survey of hydrological research in cryospheric basins involves several critical components: basin topography, the distribution and thickness of snow, ice, and frozen soil, vegetation types and coverage, soil types and structure, and the distribution of key hydrological indicators such as characteristic lines or points marking changes in hydrological processes. Additionally, it includes the distribution of geomorphic units and structures that significantly influence runoff patterns. Conducting repeated measurements over time enables the detection of temporal and spatial changes in these variables.

Advancements in satellite remote sensing technology have significantly improved our ability to capture various parameters—such as surface elevation, land cover types and extents, surface temperature, and the thickness of snow and ice—across different temporal and spatial scales. These technological developments have greatly enhanced the capacity to monitor regional changes. An overview of the application of remote sensing technology in cryospheric hydrological research will be discussed in Sect. 2.3. This section will specifically focus on the methods and techniques used in field investigations and experiments for collecting spatial information within watersheds.

In field observations, information such as topography, landforms, and surface types can be obtained through optical image analysis using photogrammetry or UAV (drone) topographic measurement techniques. Photogrammetry involves capturing fixed photographs of the target area using either a measurement camera or a calibrated non-measurement camera. These images are then corrected through orthographic projection, enabling the extraction of positional or area information of the target (see Fig. 2.15). For small watersheds, a digital elevation model (DEM) can be generated using close-range photogrammetry or drone-based measurement and control techniques. Both methods rely on image pairs taken from different perspectives with an overlap greater than 60% to analyze the spatial position of the target, thereby creating a three-dimensional model or a detailed topographic map.

Three-dimensional laser scanning technology, which emerged in the 1990s, has rapidly found applications across various fields. In pulsed ranging, the distance to a target is determined by calculating the time difference between the transmission and reception of laser pulse signals, making it particularly suitable for large-area topographic measurements. On the other hand, phase difference measurement involves modulating a laser beam with radio waves to determine the phase delay during one round trip of the modulated light along the measurement line. This phase delay is then converted into distance based on the wavelength of the modulated light. Although

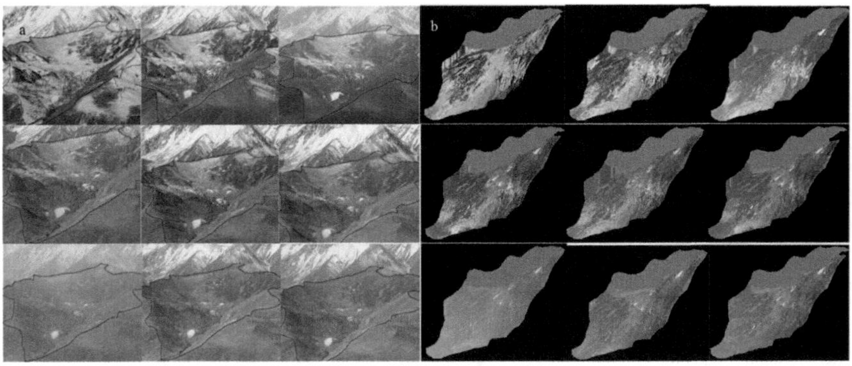

Fig. 2.15 Photogrammetry of the snow melting process in a small watershed of the first tributary of Hulugou, Qilian Mountains. **a** Progression of a snowfall, **b** Corresponding orthographic projection image

this method has a shorter measurement range, it offers high measurement accuracy, making it ideal for three-dimensional measurements of complex targets at close distances.

During field investigations, special attention should be paid to characteristic lines or points that indicate changes in hydrological processes. For example, in a glacier basin, the accumulation area lies above the equilibrium line, while the melting area is located below it (excluding the superimposed ice region). An upward shift in the equilibrium line indicates an expansion of the glacier melting area and a reduction in the accumulation area, leading to a negative mass balance compared to the previous year and an increase in meltwater runoff; conversely, a downward shift would suggest the opposite. At the end of the ablation season, the lower boundary of snow distribution often coincides with or is close to the equilibrium line, allowing for the acquisition of data on changes in the equilibrium line through photogrammetry.

2. Surface Observation System

Surface observation systems for studying watershed hydrological processes are designed according to specific research objectives. Generally, these systems should include hydrological section observations within the watershed, multi-element meteorological gradient observation systems, networks for observing underlying surface attributes, ice and snow ablation observation networks, and experimental observation fields (Fig. 2.16).

For watersheds dominated by ice and snow, the ablation process is directly linked to the energy balance of these components. Single-point meteorological observations typically measure variables such as temperature, air humidity, wind speed, wind direction, radiation, and precipitation. In frozen soil basins, the interactions between water and heat transfer, along with evapotranspiration, play a significant role in the water balance. Therefore, it is crucial to enhance observations of ground temperature, soil moisture content, and evapotranspiration.

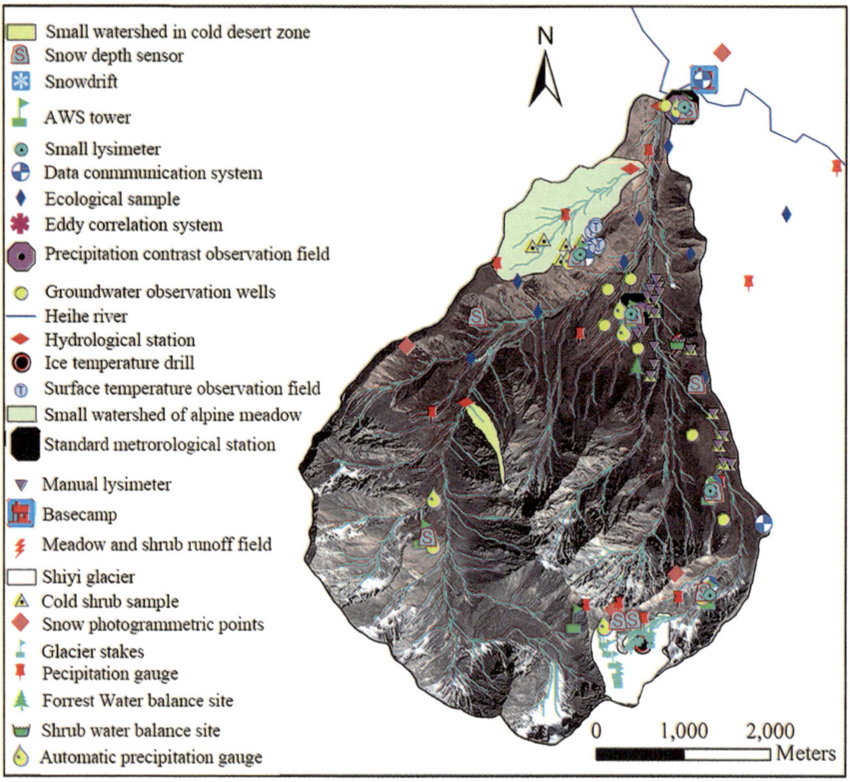

Fig. 2.16 Monitoring of the Hulugou Small Watershed in the Upper Heihe River, Qilian Mountains

In snow-covered areas, artificial observation networks should be established using ablation stakes, strategically placed based on the basin's size and elevation changes, to regularly measure snow and ice ablation rates. In permafrost and other non-ice and snow regions, the processes of water replenishment, loss, and migration within the soil are more complex. Sample plots can be selected to create observation fields for monitoring slope runoff and energy-water balance, enabling the observation of runoff processes and energy balance across various underlying surfaces. Furthermore, decentralized ground observation facilities should be closely aligned with meteorological observation systems at different altitudes to ensure comprehensive data analysis.

The observation of the Tanggula Mountain glacier watershed, located in the source region of the Yangtze River, is vital for advancing our understanding of the cryosphere and its interactions with other Earth systems. Monitoring the Dongkemadi Watershed, situated in the upper reaches of the Yangtze River within the Tanggula Mountains, represents a pioneering effort in alpine hydrology. This watershed, characterized by its high-altitude environment, glaciers, and permafrost, serves as a unique natural laboratory for researchers to investigate the complexities of hydrological processes under extreme conditions. Spanning a wide range of altitudes and featuring diverse landscapes, the Dongkemadi Watershed encapsulates the ecological and hydrological dynamics of the Tibetan Plateau. With an average elevation of 5,418 m above sea level, the terrain presents both challenges and significant scientific opportunities.

Field observations in the Tanggula Mountain glacier watershed are essential for collecting data on meteorological and hydrological conditions (see Fig. 2.17). Techniques such as drilling and pitting, combined with geophysical methods like electromagnetism and ground-penetrating radar, enable scientists to monitor the status of glaciers and permafrost. Detailed assessments of cryospheric components, including glacier mass balance, snow depth, and permafrost temperature, offer valuable insights into the current state and historical changes of these systems.

Remote sensing technology has significantly advanced cryospheric research in the Tanggula Mountain region. Aerial and satellite imagery, employing visible light, near-infrared, thermal infrared, microwave, laser, radio, and gravity technologies, facilitate the efficient collection of geometric, mass, and energy parameters at large scales and with high resolution. This capability extends the spatial coverage of observations beyond the limitations of manual methods, which are often restricted to accessible areas due to the remoteness of many cryospheric environments. Recently, the integration of remote sensing with field observations, coupled with advancements in automatic field measurements—such as automatic weather stations, eddy covariance systems, and automated photography—has significantly improved the accuracy

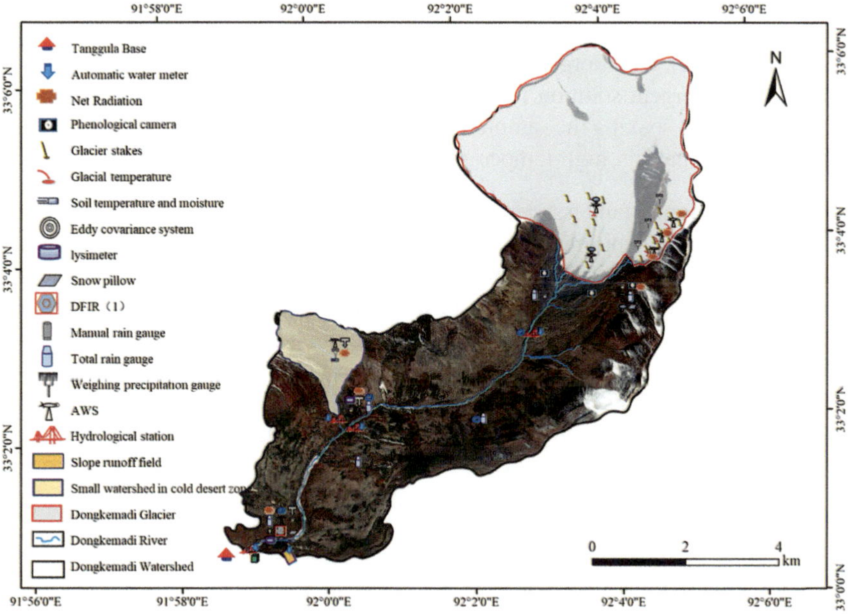

Fig. 2.17 Monitoring of the Dongkemadi Watershed in the Upper Reaches of the Yangtze River within the Tanggula Mountains

and efficiency of cryospheric monitoring. These technologies enhance the Tanggula Mountain glacier watershed's capacity for continuous and precise observations of mass and energy processes.

2.2 Water Chemical Analysis

Water chemical analysis, combining field monitoring with experimental analysis, is essential for understanding the evolutionary patterns of hydrological processes in the cryosphere. It reveals the internal connections between hydrological and hydro-chemical processes and assesses their impacts on biogeochemical cycles. This section focuses on the unique aspects of cryospheric hydrological research, primarily introducing the sampling and experimental methods for snow and ice, meltwater, precipitation, groundwater, and river water.

2.2.1 Water Sample Preparation

In cryospheric hydrological research, sample collection methods vary depending on the indicators being analyzed (Fig. 2.18). To prevent contamination, sampling bottles and tools must be thoroughly cleaned before collection, with cleaning procedures tailored to specific test requirements. For instance, when analyzing pH, electrical conductivity (EC), and inorganic ions, it is common practice to soak the equipment in an acidic detergent solution followed by rinsing with ultrapure water. For trace element analysis, soaking the samples in an acid solution is necessary. In the case of organic carbon analysis, high-temperature combustion in a muffle furnace is required.

(1) **Precipitation**: Stainless steel or plastic basins are typically used to collect precipitation. To minimize contamination, a sterile plastic bag should be placed over the sampling basin. Alternatively, a specialized precipitation collection bucket with a larger diameter can be used, which is particularly useful for small precipitation events. The interior of the sampling bucket should also be lined with a replaceable sterile plastic bag to prevent contamination.
(2) **Snow and ice**: This includes snow, snow pits, and glacial ice. Samplers should wear clean suits and gloves, using a wide-mouth bottle positioned downwind of the sampling site for direct snow collection. Glacial ice can be gathered using ice axes or ice picks. For larger sample volumes, initial samples should be collected in a sterile sampling bag and transferred to a sampling bottle after they have naturally melted.
(3) **Meltwater, River Water and Groundwater**: Narrow-mouth bottles are used to directly collect meltwater, river water, and groundwater. Before collection, the bottle and cap should be rinsed multiple times with the target water source to avoid contamination. When collecting meltwater or river water, the

Fig. 2.18 Collection of typical samples in cryospheric areas

sampling bottle should be positioned facing the direction of the flow, and it is recommended to sample from areas with higher water velocity.

(4) **Sediment**: Glacial sediments should be collected from representative locations at the front of the glacier. Samples should be placed in ziplock bags, labeled, and stored appropriately.

After sampling, all samples should be labeled with the collection location, geographic coordinates, collection time, sample name, a brief description, and the name of the personnel involved. This information is essential for the proper organization and analysis of the samples.

Once the sample enters the laboratory, it must be filtered to separate solid particles from the liquid phase (suspended load and bed load). The filtrate is used for the detection and analysis of soluble components, while the solids retained on the filter membrane can be frozen or dried for subsequent particle analysis (see Fig. 2.19). It is crucial to use specific filter materials and pore sizes tailored to the type of analysis. For soluble ion and trace element testing, polysulfone filters and 0.45 μm nitrocellulose membranes are typically employed. For organic carbon analysis, all-glass filters and GF/F glass cellulose membrane filters with a pore size of 0.7 μm are commonly used.

2.2.2 *Test and Analysis*

Laboratory water chemistry analysis typically includes the measurement of pH, electrical conductivity (EC), major ions, trace elements, organic carbon, nutrient

Fig. 2.19 Sample filtration, suspension drying, and weighing

elements, and stable isotopes. This section provides a brief overview of the testing methods and principles for analyzing the primary chemical constituents.

(1) **pH and EC**: pH measures the acidity or alkalinity of a solution, while electrical conductivity (EC) indicates the solution's ability to conduct electrical current, providing an estimate of the total concentration of charged substances in the water. A pH meter and a conductivity meter are typically used to measure pH and EC, respectively. It is essential to regularly calibrate both instruments with standard solutions, whether in the field or laboratory.
(2) **Inorganic ion**: Common inorganic ions in meltwater include Ca^{2+}, Mg^{2+}, Na^{+}, K^{+}, SO_4^{2-}, NO_3^{-}, Cl^{-}, HCO_3^{-}, and PO_4^{3-}. These ions are used to study chemical weathering processes in watersheds, measure chemical erosion rates and solute fluxes, and trace confluence processes. Ion chromatography is typically employed to determine the concentration of soluble ions. This instrument is designed for isocratic elution and is often used with an online electrolytic eluent generator, enabling gradient elution capabilities.
(3) **Trace elements**: Trace elements, such as iron, aluminum, zinc, lithium, copper, manganese, mercury, lead, titanium, cobalt, and nickel, are found in relatively low concentrations in meltwater. Despite their low concentrations, the toxicity, bioactivity, and bioconstraints of some elements can significantly affect downstream water environments and ecosystems. Inductively coupled plasma mass

spectrometry (ICP-MS) is commonly used to determine the concentration of trace elements. For example, in analyzing soluble iron (DFe), the filtered sample must be acidified to pH < 2 with nitric acid before storage in a low-temperature environment. For particulate iron (SSFe), the concentration is determined by measuring the suspended mass attached to the filter membrane.

(4) **Organic carbon**: Organic carbon refers to the total amount of carbon in soluble and suspended organic matter in water, serving as a key indicator of organic pollution. A total organic carbon (TOC) analyzer is typically used to measure the concentration of dissolved organic carbon (DOC), while an elemental analyzer is used to measure particulate organic carbon (POC). For POC analysis, the sample is first dried and then combusted in a combustion tube. Oxygen is introduced to oxidize the organic matter into CO_2, H_2O, N_2, and nitrogen oxides. The nitrogen oxides are reduced to N_2 in a reduction tube, and the gases are separated using a trap adsorption column or gas chromatography column before being detected by a thermal conductivity detector to determine the POC concentration.

(5) **Nutrient elements**: Nutrient elements are essential for plant growth and development. As climate change accelerates, nutrients exported from the cryosphere may influence downstream biogeochemical cycles. These nutrients include: ① **Nitrogen**: For inorganic nitrogen, concentrations of NO_3^- and NO_2^- are typically measured using ion chromatography or automatic analyzers, while salicylate spectrophotometry or colorimetry is used for NH_4^+ concentrations. For organic nitrogen, particulate organic nitrogen (PON) is measured with an elemental analyzer, and total dissolved nitrogen (TDN) is determined using a TOC meter. The concentration of dissolved organic nitrogen (DON) is calculated using a specific formula. ② **Phosphorus**: Total dissolved phosphorus (TDP) concentrations are measured with a flow injection analyzer. Soluble inorganic phosphorus (DIP) concentrations are determined using colorimetry, spectrophotometry, or standard wet chemical methods, allowing for the calculation of dissolved organic phosphorus (DOP) concentrations. ③ **Silicon**: The concentration of silicon is measured using a flow injection analyzer. The molybdate colorimetric method is applied for determining dissolved silicon (DSi), while amorphous silicon (ASi) is measured using the molybdic acid colorimetric method and alkaline digestion method.

(6) **Stable isotope**: Isotope analysis is a crucial aspect of cryospheric hydrology research. By utilizing the principles of stable isotope fractionation and radioactive decay, researchers can trace and time water cycle processes, revealing the sources, causes, conditions, and dynamics of ice, snow, and water bodies. The most significant stable isotopes studied are ΔD and $\delta^{18}O$ in ice, snow, and precipitation. Other important isotopes include $\delta^{13}C$, $\delta^{17}O$, and deuterium excess, all of which can be analyzed using isotope mass spectrometry.

2.3 Remote Sensing and Geographic Information Applications

Remote sensing technology provides essential tools for cryospheric hydrology, including ground surface identification, extraction of surface water and heat parameters, inversion of water cycle elements, and monitoring regional water storage changes. Geographic information science plays a crucial role in supporting the integration of hydrological models with geographic information systems, as well as in spatial data management, retrieval, and visualization. For comprehensive theoretical methods, data processing techniques, and applications related to cryosphere remote sensing, please refer to the "Cryosphere Remote Sensing" section of this series. This section will focus exclusively on aspects directly related to cryospheric hydrology.

2.3.1 Application of Remote Sensing in Hydrology

The rapid advancement of multi-source remote sensing data and retrieval technologies has significantly enhanced the qualitative and quantitative assessment of land use, land cover, water cycle observation parameters, and surface water reserves. These advancements also present new opportunities for the multi-method verification of hydrological models. While remote sensing technology enables the retrieval of surface parameter information at unprecedented temporal and spatial scales, its inversion algorithms still heavily rely on ground observation data. The accuracy of these algorithms is highly dependent on the quantity and quality of ground reference data used during data training. Below is a brief summary of some commonly used remote sensing products in cryospheric hydrology research.

Precipitation: Remote sensing images from spaceborne visible light, infrared sensors, passive microwave sensors, and active microwave sensors can all be utilized for precipitation retrieval. Visible and infrared image analysis estimates precipitation indirectly by establishing relationships between cloud top temperature and the probability and intensity of precipitation. Passive microwave sensors use microwave radiometers to capture the radiation intensity emitted by raindrops, which helps calculate rainfall and precipitation intensity. In contrast, active microwave sensors emit electromagnetic waves at specific frequency bands and derive vertical precipitation profiles through echo analysis, enabling precipitation inversion.

Currently, remote sensing products based on the integration of multiple satellites and sensors are essential for measuring precipitation in mountainous regions. Various global precipitation datasets are available for cryospheric hydrology research. Notable products include the Tropical Rainfall Measuring Mission (TRMM) dataset, CPC MORPHing technique (CMORPH) dataset, Climate Hazards Group Infrared Precipitation with Station Data (CHIRPS), Global Satellite Mapping of Precipitation (GSMaP), and the Global Precipitation Climatology Project (GPCP). Each dataset has its advantages and limitations due to differences in data sources and inversion algorithms. Generally, the accuracy of precipitation measurements in mountainous areas obtained through remote sensing is lower compared to ground observations. Therefore, the use of multi-satellite (both geostationary and low-Earth orbit), multi-channel (visible, infrared, and microwave), and multi-mode (passive and active) approaches for high temporal and spatial resolution precipitation observations is an increasingly promising field for research and application.

Ice and snow: Monitoring the spatial distribution of ice and snow is crucial in cryospheric hydrology research. Due to the challenging topography and climatic conditions, manual observations are often impractical, making it difficult to obtain accurate surface information. However, visible light and infrared remote sensing images can quickly delineate the distribution and changes in snow and ice, as well as efficiently retrieve the temporal and spatial distributions of snow and ice albedo and surface temperature. Additionally, microwave remote sensing can be utilized to assess snow particle size, thickness, and snow water equivalent.

Currently, widely available global snow products include the IMS snow area products, MODIS snow products (which encompass snow area and albedo), NASA and ESA passive microwave snow water equivalent products, and GLASS albedo products. In China, notable snow products include daily cloud-free snow cover data for the Qinghai-Tibet Plateau based on MODIS, daily passive microwave snow depth products, and snow data retrieved from satellite remote sensing by the China Meteorological Administration.

Glacier-related products primarily consist of the GLIMS global glacier database, GLAS/ICESat ice surface elevation data, and glacier distribution data from two glacier catalogs in China. Additionally, the use of GRACE (Gravity Recovery and Climate Experiment) satellites to detect regional gravity changes for measuring the mass balance of glaciers in concentrated areas is an emerging method in this field.

Frozen soil: Unlike the relatively straightforward detection of precipitation, ice, and snow, frozen soil develops beneath the ground surface, making remote sensing detection significantly more challenging. The Earth Observation System (EOS), Cold Region Scientific Experiments (CLPX), and various international research initiatives have identified remote sensing of frozen soil as a key objective. NASA's Hydro satellite, launched in 2014, focused on soil moisture and the freezing and thawing of soil as its two primary goals.

Remote sensing applications for frozen soil primarily involve detecting its distribution and measuring the active layer thickness using optical and infrared remote sensing data. Additionally, microwave remote sensing is employed to monitor the freezing and thawing processes of frozen soil.

2.3.2 *Geographic Information Method*

Watershed-scale hydrological analysis is central to cryospheric hydrology research. Geographic Information Systems (GIS) provide a powerful platform for integrating watershed data, conducting spatial analysis, applying models, and visualizing results. This approach enables comprehensive assessments and deepens the understanding of hydrological processes within watersheds.

Data integration: Using GIS interactive graphics processing and mapping tools, topography, landforms, hydrological environments, river systems, hydrometeorological data, glaciers, snow, and frozen soil within cryospheric basins are transformed into three-dimensional representations of points, lines, surfaces, and volumes for input into a spatial database. This cryosphere data can also be integrated with other datasets, such as foundational natural and socio-economic information, within the database.

Spatial analysis: GIS facilitates the processing of multi-source data, overlaying diverse information, and manipulating spatial attribute data. This capability allows for the acquisition, management, analysis, simulation, and visualization of data with temporal characteristics, thereby improving the efficiency of hydrological information management and application. GIS can establish different thematic layers based on various influencing factors within the spatial database and overlay these layers to enable rapid and informed evaluation and planning of cryospheric water resources.

Model ensemble: GIS provides essential input data for distributed hydrological models through effective data management and spatial analysis. It offers a data conversion interface for models with different projections and spatial scales, as well as graphical displays of simulation results. GIS plays a crucial role in hydrological modeling in the following key areas: ① **Spatial Data Management**: GIS integrates extensive spatial and attribute data related to hydrological models, offering tools for data querying, retrieval, updating, and maintenance. ② **Layer Generation**: GIS can generate layers from various data sources; for example, terrain data can be used to calculate slope, aspect, and confluence paths, while hydrological data can be employed to define river networks. ③ **Automatic Model Parameter Acquisition**: GIS can automatically acquire model parameters, such as using land use maps extracted from remote sensing data to determine roughness coefficients for each calculation grid. ④ **Facilitation of Hydrological Modeling**: Many hydrological models rely on numerical solutions like finite difference and finite element methods, which require the research area to be divided into regular or irregular grids. GIS supports this process. ⑤ **Visualization of Analysis and Results**: GIS enhances the visual representation of the analysis and calculation processes. Its spatial display capabilities create an advantageous modeling environment, enabling model visualization and computation while allowing analysts to interactively adjust model parameters.

2.4 Mathematical Statistics and Models

The analysis, exploration of mechanisms, and estimation of trends in hydrological changes within the cryosphere rely heavily on quantitative mathematical methods. As a result, mathematical statistics and hydrological simulations are fundamental techniques for understanding and analyzing hydrological processes in cryospheric environments.

2.4.1 Mathematical Statistics

Mathematical statistical methods are based on the statistical relationships between different variables. These methods often focus on statistical correlations rather than the detailed processes involved, thereby simplifying complex physical processes into univariate or multivariate statistical models. For example, in cryospheric basins, the ablation of ice, snow, and frozen soil is positively correlated with temperature changes, while precipitation has a significant impact on basin runoff. Consequently, temperature and precipitation are often the primary parameters used in runoff estimation methods for cryospheric basins.

For instance, in the Yarkant River Basin, which has high glacier coverage, the relationship between the annual runoff (Y) and the average summer temperature (T) at the Kaqun Hydrological Station, as well as the precipitation (P) from July to August, can be expressed as follows:

$$Y = 9.088T - 0.049P - 70.77 \tag{2.4}$$

This statistical relationship effectively captures the annual runoff patterns of the Yarkant River Basin over the past few decades (see Fig. 2.20).

It is important to note that the selection of variables in the statistical relationship varies significantly depending on the hydrological characteristics of the river basin, the degree of influence of each factor, data availability, and the research period. For example, in glacial basins with low rainfall, such as those on the Qiangtang Plateau, modeling based solely on temperature may yield more accurate results. Conversely, in the source area of the Yellow River, where permafrost is widespread and glacier replenishment rates are low, runoff is primarily driven by precipitation. Additionally, the impact of soil evapotranspiration on water loss cannot be overlooked. Therefore, a comprehensive consideration of multiple influencing factors is essential for accurate analysis.

Mathematical statistical methods offer several advantages, including fewer parameters, simple calculations, and high accuracy within specific temporal and spatial ranges, making them widely used in cryospheric hydrology. However, they also have

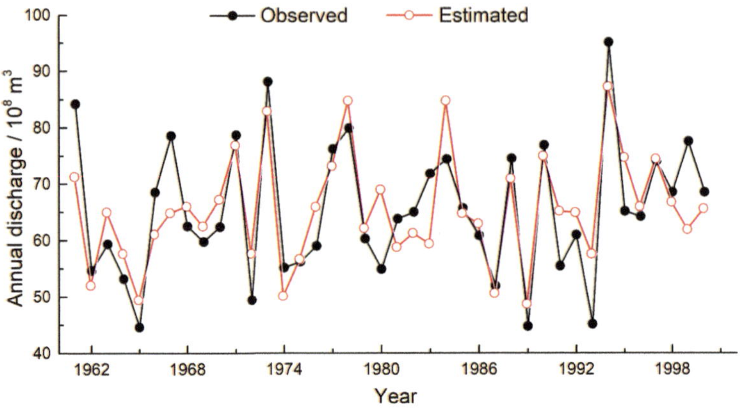

Fig. 2.20 Comparison of observed runoff and simulated values at Kaqun Station, Yarkant River Basin (Liu et al. 2008)

notable limitations. First, establishing reliable statistical relationships requires long-term observational data. Longer observation periods tend to stabilize the relationships between parameters and target variables, minimizing the influence of secondary factors. Conversely, shorter observation periods increase susceptibility to fluctuations from other variables, leading to larger calculation errors.

Second, statistical models focus primarily on the relationships between selected parameters and target variables, treating the influence of other factors as constants through fixed coefficients. While this approach may yield small errors when secondary factors remain stable, significant changes in these factors can result in substantial prediction errors, especially if they become primary influences.

Third, due to variations in climate and topography across different basins, the relationships between runoff and related elements display significant temporal and spatial heterogeneity. As a result, a statistical relationship that performs well in one basin may not produce the expected results when applied to another, demonstrating the limited regional adaptability of statistical methods.

Therefore, when using mathematical statistical methods to study hydrological processes in the cryosphere, it is essential to consider the length of the data series, the primary and secondary relationships among variables, and the specific requirements and conditions of the analysis.

Currently, two primary methods are used to estimate future runoff in river basins using mathematical statistics: (1) **Direct Prediction Based on Historical Data**: This method relies on historical runoff characteristics and trends, employing various mathematical techniques—such as GDP, periodic variance, neural networks, and R/S gray prediction—to forecast short-term future runoff. However, because future climate conditions will not exactly mirror past conditions, there is significant uncertainty in extrapolating future runoff changes from historical data. (2) **Statistical Modeling with Meteorological Data**: This approach uses historical runoff data to develop statistical models that incorporate meteorological observations (e.g., temperature, precipitation, evaporation) to predict future runoff under different climate change scenarios. This method assumes that the mathematical relationship between runoff and meteorological variables remains constant over time. However, in high mountainous basins, changes in glaciers, snow, and frozen soil due to climate change are likely to alter this relationship significantly, leading to considerable deviations in future runoff predictions. Both methods face challenges in accurately forecasting future runoff in the context of ongoing climate change.

2.4.2 Hydrological Model of Cryosphere Basin

Mathematical statistical models often overlook the processes connecting selected parameters to the target variable. However, a detailed understanding of these processes is essential for comprehending the underlying mechanisms of change. This understanding enables the mathematical representation of the processes driving changes in the target variable, leading to the development of hydrological models. This section begins by discussing the specific needs and characteristics of hydrological simulation in cryospheric basins. It then provides an overview of the main types of hydrological models, highlighting their respective advantages and disadvantages.

1. Types and Characteristics of Hydrological Models

River runoff in cryospheric basins results from a complex interplay of various interrelated processes. For example, in a small frozen soil basin, runoff is influenced by multiple factors, including precipitation, infiltration, soil freezing and thawing, vegetation interception, evapotranspiration, and changes in soil water content. These processes can be mathematically described using two main approaches: the empirical relationship method and the physical process method. Models developed using these approaches are referred to as empirical models and physical models, respectively.

Empirical models primarily rely on statistical relationships to describe different processes, yielding final results through integration. In contrast, physical models are constructed using a framework of mathematical equations that have clear physical significance. In some cases, limitations in process understanding and data availability may prevent a model from fully representing all subprocesses using either empirical or physical methods. Consequently, some processes in empirical models may be described using physical principles, while physical models may incorporate empirical relationships. Such configurations are termed hybrid models, though they are still categorized as either empirical or physical models based on their predominant characteristics.

Empirical models are characterized by a simple structure, fewer parameters, fast calculations, and, to some extent, higher accuracy. However, like mathematical statistical methods, empirical models have certain drawbacks, including a greater reliance on observational data, limited extrapolation capabilities, and weak regional adaptability. Therefore, empirical models are primarily suited for runoff analysis and medium- to short-term trend forecasting in large-scale watersheds. The abundance of observation sites and long-term data series in these watersheds allows for the establishment of robust statistical relationships, while the reduced number of parameters facilitates their application over large temporal and spatial scales.

In contrast, physical models provide a more detailed representation of the various aspects of hydrological processes within a basin, effectively revealing the mechanisms behind runoff generation and confluence. This makes them particularly suitable for analyzing hydrological processes and studying the underlying mechanisms of change. However, the increased number of parameters in physical models necessitates a strong observational foundation within the study basin, which can significantly limit their applicability.

Over the past 30 years, advancements in cryospheric observations and remote sensing technology have substantially increased the availability, variety, and volume of data. Consequently, the application of physical models has expanded considerably.

In addition to differences in process description methods, hydrological models can be categorized based on spatial representation into two types: lumped models (or conceptual models) and distributed models. Lumped models treat the entire watershed as a single hydrological unit, ignoring spatial differences in model parameters. This approach results in relatively simple and fast calculations. In contrast, distributed models account for spatial variability by dividing the watershed into multiple relatively homogeneous hydrological units, each calculated separately, with the results subsequently aggregated for output.

Distributed models provide a more accurate representation of basin conditions but require more parameters and involve more complex calculations. Typically, they segment the watershed into numerous computation units of uniform size, forming a calculation grid. The size of these units defines the model's spatial resolution: the more grids there are, the better the model approximates actual processes, though this also results in exponentially increased computational demands.

To balance the need for detailed descriptions with computational efficiency, some models utilize irregular grid segmentation methods. In these semi-distributed hydrological models, larger homogeneous areas are divided into coarse grids, while smaller but significant areas are segmented into finer grids. This approach captures actual spatial characteristics more effectively but places higher demands on the simulation environment.

The essence of a distributed model lies in the spatial distribution of model parameters. Therefore, both empirical and physical models can employ a distributed calculation approach. Physical models often utilize more detailed distributed representations, which is why they are commonly referred to as distributed physical models, or simply distributed models. Some conceptual models can also be considered distributed if they use sub-watersheds or grids in their calculations. Consequently, it is important for readers to distinguish between different model types based on their specific descriptions and methodologies.

In watershed hydrology, there is a category of models known as land surface hydrological models or large-scale hydrological models. These models are designed for application at larger regional scales and are based on physical processes that detail the moisture and energy balance among the atmosphere, vegetation, and soil in the vertical dimension. By integrating these processes with a confluence model, they provide a comprehensive mathematical representation of the mechanisms involved in runoff generation within the basin. Since land surface hydrological models can typically be coupled with large- and medium-scale climate models and are intended for large-scale hydrological simulations, they are often referred to as land-surface models. However, their fundamental nature remains that of a distributed physical hydrological model.

Watershed-scale hydrological simulation is a crucial method for quantitatively predicting future hydrological changes and is a significant focus of international research. The prevailing approach for predicting regional or watershed hydrological processes under climate change involves using General Circulation Model (GCM) climate scenario data to drive hydrological models. Initially, GCMs provide projections of global temperature, precipitation, radiation, and other variables under various future scenarios. These projections are then refined using statistical downscaling or dynamic downscaling (Regional Climate Models, RCMs) to generate high-resolution climate data compatible with the spatial scale of the hydrological model. This high-resolution data is essential for driving the model and predicting future hydrological changes in the region or watershed (Fig. 2.21).

In this process, the accuracy of climate-driven data is influenced by future climate change scenarios, the simulation accuracy of General Circulation Models (GCMs), and the spatial downscaling algorithms used. Hydrologists play a crucial role in

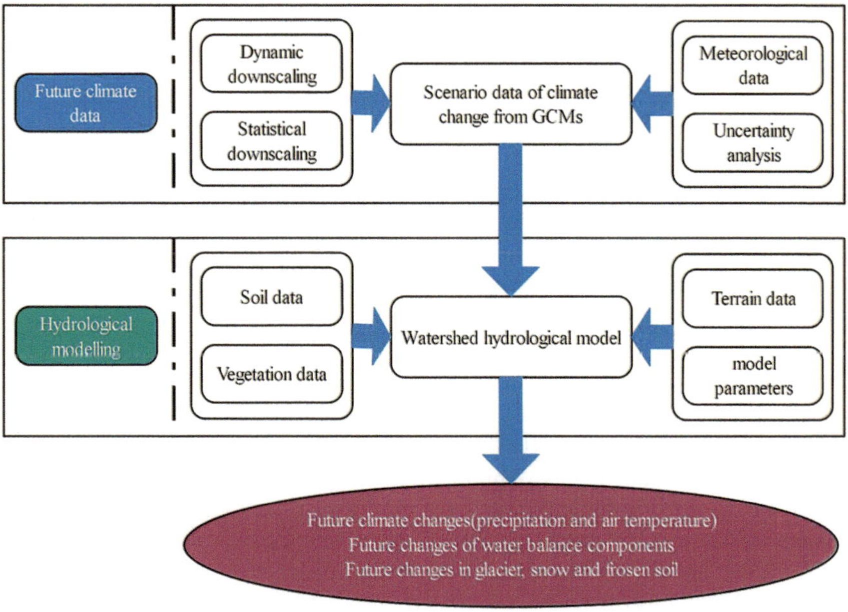

Fig. 2.21 Schematic diagram for the estimation method of watershed hydrological processes

selecting appropriate downscaling methods. The dynamic downscaling method, commonly referred to as the regional climate model (RCM) approach, uses RCMs coupled with GCMs to project future climate change scenarios at a regional level. This approach depends on the availability of observational data and requires substantial computational resources, as it is constrained by the boundary conditions provided by GCMs.

While dynamic downscaling is theoretically not limited by spatial resolution, it can be constrained by the principles of the model and the imposed boundary conditions. Excessive resolution may fail to accurately capture local climatic characteristics, potentially leading to significant systematic errors in forecasts of temperature, precipitation, and other climate variables.

The statistical downscaling method uses years of observational data to establish statistical relationships between large-scale climate conditions (primarily atmospheric circulation) and regional climate elements. This relationship is validated using independent observational data and then applied to the large-scale climate information generated by GCMs, enabling the estimation of future climate change scenarios for the region. Essentially, this process involves creating a statistical functional relationship between large-scale climate forecast factors and regional climate forecast variables.

$$Y = F(X) \tag{2.5}$$

where X represents the large-scale climate predictors, Y denotes the regional climate forecast variables, and FFF characterizes the functional relationship between these large-scale climate factors and the regional climate variables. Typically, F is unknown and must be determined through dynamic methods (such as regional climate model simulations) or statistical methods (derived from observational data). In statistical downscaling, X encompasses large-scale climate states, while F includes regional or local factors such as topography, land-sea distribution, and land types. The statistical downscaling method establishes relationships among climate elements across different scales for a specific area by extensively analyzing climate variables. This approach can correct some systematic errors in GCMs and is favored for its relatively low computational cost and other advantages, leading to its widespread application.

2. Some Hydrological Models with Cryospheric Elements

Different parameterization schemes and calculation strategies have led to the development of various hydrological models. Over the past century, these models have evolved alongside advancements in hydrology and space technology, resulting in more than 70 applications today. Table 2.1 lists some commonly used hydrological models in cryospheric hydrological research. Most hydrological models currently account for snow cover processes. For instance, lumped hydrological models like HBV, PRSM, and SRM utilize degree-day factor algorithms, while models such as DHSVM, DWHC, WEB-DHM, CLM, and VIC employ energy-balance snowmelt algorithms. Frozen soil, as a unique underlying surface characteristic, impedes the infiltration of snowmelt and rainfall, thereby increasing watershed flow rates. The

Table 2.1 Description of cryosphere elements in common hydrological models

Model type	Abbreviation	Snow module	Frozen soil module	Glacier module
Lumped/conceptual hydrological model	HBV	Yes	No	Yes
	SRM	Yes	No	No
	TAC-D	Yes	No	Yes
	UBC	Yes	No	No
	WASMOD	Yes	No	No
Distributed hydrological model	CBHM	Yes	Yes	Yes
	CRHM	Yes	Yes	No
	DHSVM	Yes	No	No
	DWHC	Yes	Yes	Yes
	IHDM	Yes	No	No
	MIKE-SHE	Yes	No	No
	SWAT	Yes	No	No
	SWAT-Luo	Yes	Yes	Yes
	GBEHM	Yes	Yes	Yes
	SPHY	Yes	No	Yes
	TOPMODEL	Yes	No	No
	WEB-DHM	Yes	Yes	No
	WaSiM	Yes	No	Yes
Land surface hydrological model	CLM	Yes	Yes	No
	CoLM	Yes	Yes	No
	ITP-SIB2	Yes	Yes	No
	LSM	Yes	Yes	No
	SVAT	Yes	Yes	No
	VIC-CAS	Yes	Yes	Yes

water storage capacity, infiltration strength, and evaporation potential in frozen soil areas differ significantly from those in non-frozen soil regions. To address these differences, projects like PPILPS and GEWEX-GCIP have significantly advanced the understanding of frozen soil land surface processes. Recently developed distributed physical hydrological models and land surface models, such as CRHM, WEB-DHM, CLM, VIC-CAS, and ITPSIB2, now incorporate the hydrological processes associated with frozen soil. However, few hydrological models adequately account for glacier hydrological processes, and those that do often provide a simplified representation. Most of these models, including HBV, SPHY, and WaSiM, rely on degree-day factor approaches and do not consider glacier movement or confluence processes. Overall, while the number of hydrological models that comprehensively include cryosphere elements remains limited and their descriptions relatively simple, the field is advancing rapidly.

The CBHM model is currently recognized as a comprehensive distributed hydrological model suitable for cold watersheds. Additionally, the GBEHM model, developed as part of the Heihe Project, integrates hydrological and ecological processes in cryospheric mountainous regions. It is a systematic and complete distributed hydrological model that accounts for hydrological processes, ecological dynamics, and slope hydrological characteristics within the cryosphere. Below, we will briefly introduce the structure and features of the CBHM model.

The CBHM model effectively accommodates spatial interpolation methods for meteorological factors across different time scales, separates solid and liquid precipitation, and corrects observational errors. It incorporates typical vegetation interception and evapotranspiration processes found in alpine regions, along with a straightforward algorithm for estimating glacier area, volume, and meltwater runoff. Additionally, the model simulates wind and snow processes, snowmelt dynamics, and the water-heat coupling processes in frozen soils, as well as methods for estimating frozen soil areas (Fig. 2.22). Given the limited observational data available in China's alpine regions, the model is designed to require fewer input variables—primarily precipitation, temperature, and wind speed—along with conventional data such as soil, vegetation, and glacier distribution. Most parameters are supported by relevant observational data. The CBHM model can output changes in cryosphere elements and water balance components. It also utilizes parallel computing, enabling it to run efficiently on standard desktop computers.

3. Simulation of Ice and Snow Melting Processes

Ice and snow melting calculations can generally be categorized into two types: statistical models based on temperature and energy balance models grounded in physical processes.

Air temperature serves as a comprehensive indicator of the radiation balance and is highly correlated with the rate of snow and ice melting. The degree-day factor model is one of the most widely used statistical models based on temperature. It is founded on the linear relationship between temperature and snowmelt, particularly in relation to positive accumulated temperature. Its general form is:

$$M = DDF \times (T_a - TT) \tag{2.6}$$

where M is the melted snow in water equivalent over a certain period of time, DDF is the degree-day factor, Ta is the air temperature (°C), and TT is the threshold temperature for melting, usually set at 0 °C.

In practice, the degree-day factor is typically treated as a fixed value; however, it actually encompasses various influences beyond temperature. Changes in snow and ice albedo significantly affect the degree-day factor, and terrain features such as slope and aspect also play a role. As a result, numerous studies have sought to integrate additional variables into the model to enhance simulation accuracy. For example, the degree-day model can be updated to include radiation variables, which can generally be expressed as follows:

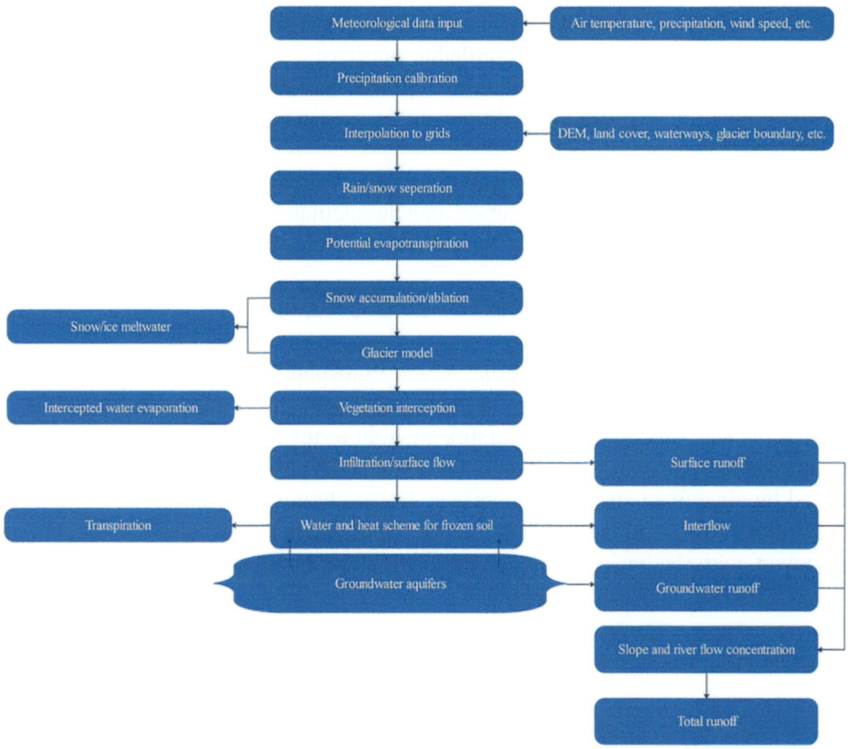

Fig. 2.22 The framework of CBHM hydrological model

$$M = DDF \times (T_a - TT) + \alpha R \quad (2.7)$$

where α is the radiation adjustment factor, and R represents solar shortwave radiation or net radiation. The degree-day factor model primarily relies on temperature as its main variable, simplifying data acquisition and facilitating distributed calculations. These characteristics have contributed to the model's widespread application in ice and snow ablation studies and runoff simulations.

The energy balance principle is a fundamental theoretical foundation for cryospheric hydrological research and serves as the cornerstone of physical hydrological models. This principle adheres to the law of conservation of energy, analyzing the sources and sinks of energy at the Earth's surface or other surfaces to examine the hydrothermal processes occurring at the interface. The energy balance equation for the surface of ice and snow can generally be expressed as follows:

$$R + H + LE + G + Q_r + Q_M = 0 \quad (2.8)$$

where R is the net radiation, H is the sensible heat flux, LE is the latent heat flux, G is the underground/under-ice heat flux, Q_r is the heat transferred by precipitation, and Q_M is the melting heat of ice and snow. Energy input at the surface is considered positive, while energy output is regarded as negative.

For various ice and snow interfaces, the fundamental calculation principles and methodologies are largely similar, despite differences in heat flux calculation schemes and specific parameter values.

Net radiation is defined as the algebraic sum of shortwave and longwave radiation received at the surface of ice and snow. It can be calculated using the surface radiation balance equation:

$$R = R_s(1 - \alpha) + LW \downarrow - LW \uparrow \tag{2.9}$$

where R_s is solar shortwave radiation, $LW\downarrow$ is atmospheric longwave radiation, LW↑ is longwave emission from the ice surface, and α is the surface albedo.

Sensible heat flux and latent heat flux are both forms of turbulent heat flux, representing heat transfer and the heat exchange associated with water vapor phase changes caused by air movement over the surface. In simulating turbulence on the surfaces of ice and snow, methodologies based on aerodynamic principles are commonly employed, with the basic expression generally being:

$$H = \rho c_p c_h (\theta_z - \theta_s) u_z \tag{2.10}$$

$$LE = \rho c_e L (q_z - q_s) u_z \tag{2.11}$$

where c_p is the heat capacity of the atmosphere, L is the latent heat of vaporization or sublimation, c_h and c_e are the turbulent transport coefficients for sensible heat and water vapor, respectively, and u_z, θ_z, and q_z are the wind speed, potential temperature, and specific humidity at height z, respectively. θs and q_s are the potential temperature and specific humidity at the ground surface.

The change in the energy balance of a glacier's surface due to precipitation consists of two components: first, the sensible heat transported by warm rain falling on the ice and snow surface; and second, the heat released by freezing rainwater that falls below 0 °C onto the surface. The sensible heat transport from precipitation (Q_{rh}) can be expressed as follows:

$$Q_{rh} = P_i \rho_w c_w (T_r - T_s) \tag{2.12}$$

where P_i is the intensity of precipitation, ρ_w is the density of water, c_w is the heat capacity of water, T_r is the temperature of the rainwater, and T_s is the temperature of the ice and snow surface. The heat released by rainwater refreezing (Q_{rf}) can be expressed as:

$$Q_{rf} = P_i \rho_w L_f \tag{2.13}$$

where L_f is the latent heat of phase change between water and ice.

Ice/underground heat flux (G) refers to the heat loss or increase in the ice/snow layer due to the vertical temperature gradient, which can be calculated according to Fourier's law:

$$G = k \cdot dT/dz \tag{2.14}$$

where k is the thermal conductivity of ice/snow, z is the depth of the ice and snow layer, and T is the temperature of the ice and snow.

Snow layering can be classified into single-layer and multi-layer types. The single-layer model treats snow as a homogeneous entity, calculating the total snowmelt for the entire snowpack. These models are commonly used in hydrological simulations, focusing primarily on the quantity of snowmelt water while often neglecting detailed energy processes such as the freezing and thawing of the snow surface and radiation influx. Most estimated variables pertain to the outflow of snowmelt water, with models employing the degree-day factor method fitting within this category.

In contrast, the multi-layer model implements distinct layering schemes tailored to specific research objectives. Simpler models, like DHSVM, divide snow into two layers: a surface layer and a lower layer. More detailed models, such as SNOWPACK, facilitate in-depth vertical analysis of snow stratification. Fundamentally, whether considering snow as a unified mass or through detailed stratification, snowmelt follows the principle of "heating, reaching maximum water holding capacity, and then snowmelt outflow." Advances in remote sensing technology enable comparisons of the spatial dynamics of snow cover with model-simulated dynamics.

After obtaining the energy components mentioned above, the melting heat (Q_M) of ice and snow can be determined, allowing for the calculation of the amount of ablation:

$$M = Q_M/(\rho_w L_f) \tag{2.15}$$

In a watershed or hydrological unit, after the formation of meltwater from ice and snow, runoff is typically not fully realized, as some of the water is lost through various processes. This change in water volume within the unit is represented by the water consumption balance model, which can be expressed as follows:

$$\Delta S - M + P + W - E - I - Q \tag{2.16}$$

where ΔS is the change in water volume within the calculation unit, M is the amount of melting/freezing of ice and snow, P is the amount of rainfall, W is the amount of groundwater replenishment, E is the amount of evaporation or sublimation, I is the amount of water infiltrated into the soil or bedrock, and Q is the unit discharge flow. The water balance model follows the law of conservation of mass and accounts for all water inputs and outputs within the unit. In practical applications, the model can be adjusted based on the specific water components present in the unit, or certain secondary variables can be ignored to simplify the model.

4. Frozen Soil Hydrological Modeling

The principles of energy balance and water balance are also relevant to hydrological simulations in frozen soil research areas. However, due to differences in the underlying surface, the components of the balance equations experience some modifications. In the energy balance Eq. (2.9), when there is no snow on the surface, the meltwater from snow and ice (Q_M) is absent, and geothermal flux (G) assumes a more significant role. In the water balance Eq. (2.17), there may be an increase in surface interception and evaporation due to vegetation. Moreover, unlike the ice-snow zone, which emphasizes the energy and water balance at the ground surface, the frozen soil zone focuses on the energy and water change processes within the active layer (permafrost) or the freezing depth (seasonally frozen soil). Consequently, in models addressing frozen soil hydrology, the transfer of heat and water within the soil takes on critical significance.

The coupling of soil heat and water transmission is described by the classic soil energy transmission process equation and the unsaturated soil water movement equation:

$$C_s\frac{\partial T}{\partial t} - \rho_i L_f\frac{\partial \theta_i}{\partial t} = \frac{\partial}{\partial z}\left(k_h\frac{\partial T}{\partial t}\right) - \rho_l c_l\frac{\partial q_l T}{\partial z} - L\left(\frac{\partial q_v}{\partial z} + \frac{\partial \rho_v}{\partial t}\right) \tag{2.17}$$

$$\frac{\partial \theta_l}{\partial t} + \frac{\rho_i}{\rho_l}\frac{\partial \theta_i}{\partial t} = \frac{\partial}{\partial z}\left[k_w\left(\frac{\partial \psi}{\partial z} + 1\right)\right] + \frac{1}{\rho_l}\frac{\partial q_v}{\partial z} + U \tag{2.18}$$

where C_s and T are the soil volumetric heat capacity and soil temperature, respectively, ρ_i is the density of ice, L_f is the latent heat of freezing and thawing, θ_i and θ_l are the volumetric water content of ice and liquid water, respectively, k_h is the thermal conductivity of the soil, and k_w is the unsaturated soil hydraulic conductivity. Additionally, ρ_l is the density of water, c_l is the heat capacity of liquid water, q_l is the liquid water flux, q_v is the water vapor flux, L is the latent heat of evaporation, ρ_v is the water vapor density in the soil, ψ is the soil water potential, and U is the soil water flux sink/source term.

The following section will take the SHAW model as an example to introduce the analysis of its main parameters.

The thermal conductivity k_h is fundamental for calculating the hydrothermal processes in frozen soil. The SHAW model treats the air, water, ice, and soil particles in the soil as a conceptual whole, with the calculation as follows:

$$k_h = \sum m_j k_j \theta_j / \sum m_j \theta_j \tag{2.19}$$

where m_j, k_j, and θ_j refer to the weighting factors, thermal conductivity, and volume content of various components in the soil (such as sand, silt, clay, organic matter, water, ice, and air).

There is a specific relationship between the hydraulic conductivity kwk_wkw of unsaturated soil and the relative saturation degree of soil water content. The calculation scheme in the SHAW model is as follows:

$$k_w = k_{mat}(\psi_a/\psi)^{(2+3\lambda)} \tag{2.20}$$

where k_{mat} is the saturated matrix potential conductivity, ψ_a is the inlet pressure, ψ is the water potential, and λ is the pore size distribution index.

The SHAW model assumes that the hydraulic conductivity of frozen soil is similar to that of unsaturated soil and that when the porosity is lower than 0.13, the hydraulic conductivity of frozen soil is considered zero. The soil water potential in ice-containing soils is influenced by the saturated water vapor pressure on the surface of the ice within the soil. The SHAW model uses the freezing point depression equation of frozen soil to describe and determine the unfrozen water content in frozen soil:

$$\theta_l = \theta_s\left[\frac{1}{g\psi_a}\left(\frac{L_f T}{T+273.15}\right) + \frac{cR(T+273.15)}{g}\right]^{-\lambda} \tag{2.21}$$

where g is the acceleration of gravity, c is the solute concentration, and R is the universal gas constant.

Surface water infiltration is the primary mechanism for converting surface water into soil water and groundwater. Infiltration during non-freezing periods can be evaluated using conventional methods, such as the Horton and Kostiakov infiltration equations. For the freezing period, surface infiltration volume can be calculated using empirical formulas that relate surface water volume to soil moisture content, or the Green-Ampt infiltration model may be employed. The Green-Ampt model operates on the principle of infiltration rate (INF) in dry and homogeneous soil under a thin and stable hydraulic head:

$$INF = k_s[1 + (\theta_s + \theta_{ini})\psi_w/z] \tag{2.22}$$

where k_s is the saturated hydraulic conductivity in unfrozen conditions, θ_s and θ_i are the saturated water content and initial water content, respectively, ψ_w is the water potential at the wetting front, and z is the infiltration depth.

The previous discussion on water and heat transfer in frozen soil provides only the foundational equations, offering the reader a preliminary understanding of the model's key components. These mathematical expressions are grounded in the description of physical processes. It is evident that the physical model involves numerous complex factors and parameters. Therefore, reasonable assumptions and simplifications are often necessary to effectively manage specific processes.

2.4.3 Application of Machine Learning in Glacier Runoff Simulation

Artificial Intelligence (AI), which emerged in the 1950s, is designed to perform tasks that typically require human intelligence. A core area of AI is machine learning (ML), which enables computers to automatically learn from data and extract meaningful insights. ML uses algorithms and models to analyze and process large datasets, uncovering hidden patterns, regularities, and relationships. Unlike traditional methods, ML models do not require domain-specific knowledge, allowing them to address complex tasks across various fields and capture intricate nonlinear relationships within data.

In recent years, ML has achieved remarkable results and gained significant recognition in hydrology and water resources. Techniques such as Random Forest (RF), Support Vector Machine (SVM), Extreme Gradient Boosting, Multilayer Perceptron (MLP), Convolutional Neural Network (CNN), and Recurrent Neural Network (RNN) have been effectively applied to simulate hydrological processes in glacier basins. These approaches can be broadly categorized into two types: single ML model approaches that use meteorological elements as inputs and methods that couple hydrological models with ML models.

Chen et al. (2022) studied Glacier No. 1 in the Urumqi River Source in Western China, utilizing the advantages of Recurrent Neural Network (RNN) models for capturing temporal features. They proposed a model called the Exogenous Regularization Network (ERNet), which emphasizes the relationships between exogenous variables (such as temperature and rainfall) and endogenous variables (like runoff). The model employs a parallel dual-branch structure, where exogenous variables are processed in separate branches to extract temporal features. A fully connected layer then merges the outputs from both branches, enabling simultaneous predictions of temperature, rainfall, and runoff.

In comparisons with the FLEXG model—which incorporates terrain discretization, meteorological drivers, snow and glacier components, and a rainfall-runoff module for non-glacial areas—ERNet not only accurately simulates runoff and peak flows but also demonstrates superior generalization performance (Fig. 2.23).

However, in the context of climate change, future glacier shrinkage and shifts in precipitation patterns may significantly alter the relationships between runoff and meteorological variables. As a result, single ML models that rely solely on meteorological data may be inadequate for long-term simulations and projections. To address this, some researchers have sought to couple hydrological models with ML models to enhance the spatiotemporal extrapolation of glacier runoff simulations.

For instance, Yang et al. (2023) integrated the SWAT+ model, which accounts for glacial hydrological processes, with Gated Recurrent Unit (GRU) neural networks to create a hybrid model (GRU-GSWAT+) aimed at improving glacier runoff simulation at the basin scale. In runoff simulation experiments conducted in two glacial basins within the Tarim River Basin (TRB) in Northwest China—the Aksu River Basin (ARB) and the Weigan River Basin (WRB)—the study used daily maximum

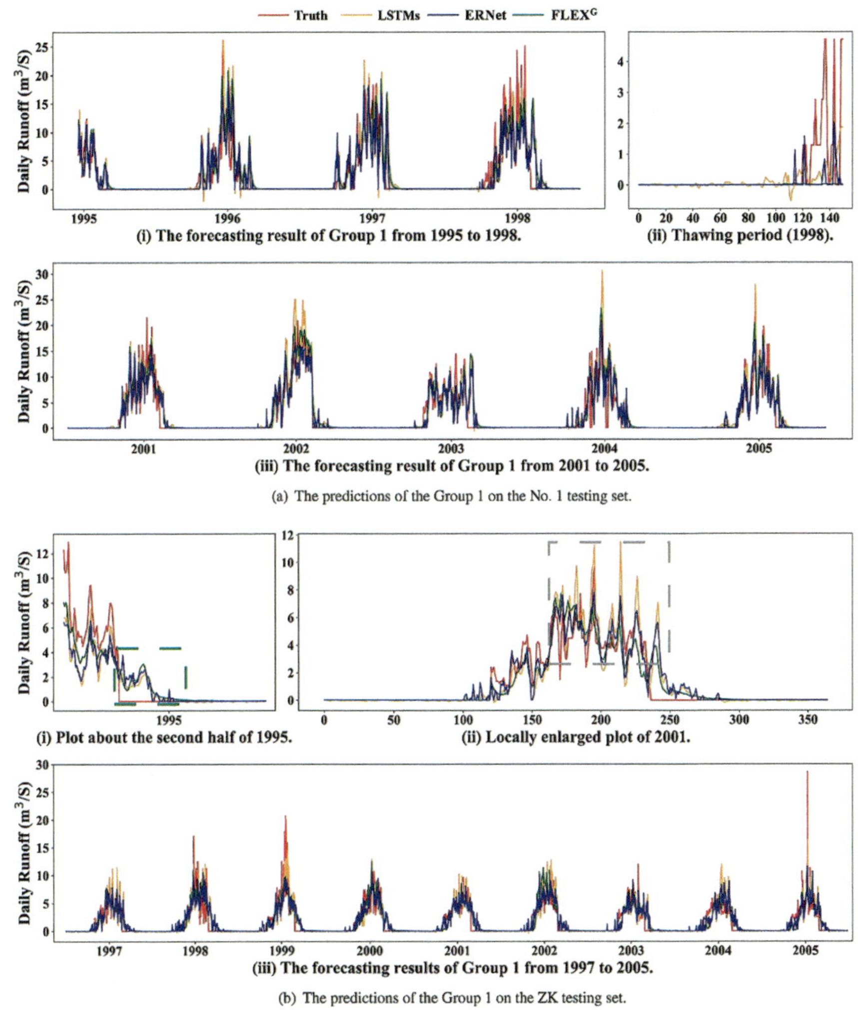

Fig. 2.23 The forecasting results on the two catchments. Both catchments have missing data for some years, we deleted them: **a** No. 1 catchment, 1999 and 2000; **b** ZK catchment, 1996

and minimum temperatures, total precipitation, solar radiation, and wind speed as input data for the GSWAT+, GRU, and GRU-GSWAT+ models. The results indicated that GRU-GSWAT+ outperformed the other two models in both simulation and prediction, achieving the most accurate and robust runoff simulations at four hydrological stations within the sub-basins, particularly in modeling high and low flows (Fig. 2.24).

In summary, compared to traditional physically-based hydrological models, machine learning (ML) algorithm-based models—often referred to as "black box" models—struggle to explain hydrological phenomena and processes. However, they

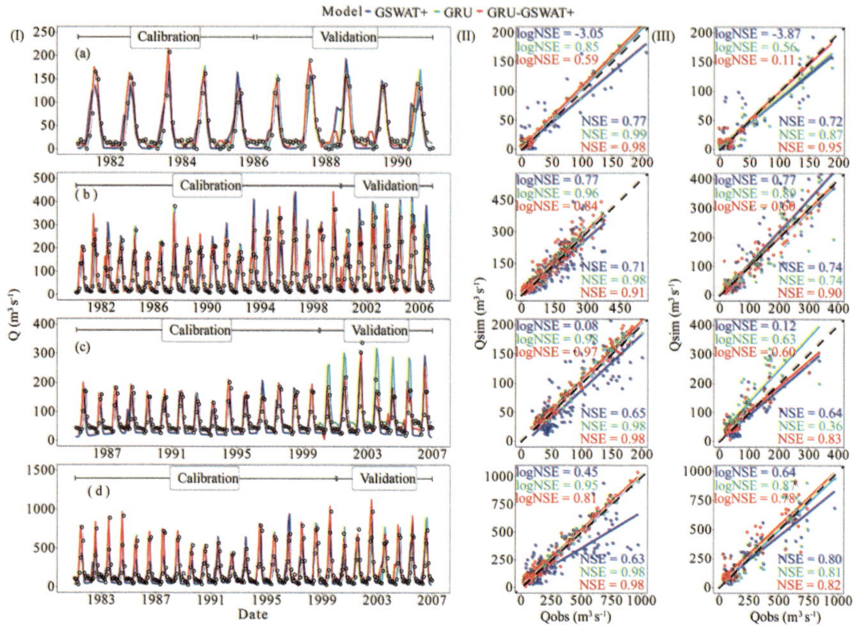

Fig. 2.24 Monthly simulated streamflow comparison using the GRU, GSWAT+, and GRU-GSWAT+ models. Black dots represent observed streamflow, while colored lines correspond to simulated streamflow using the different models. Subfigure (I) shows the comparison of observed monthly streamflow versus the simulation results from the models for the entire period, while subfigures (II) and (III) present the comparison during the calibration (training) and validation (testing) periods, respectively. Each row corresponds to a specific hydrological station: **a** Qianfodong, **b** Shaliguilanke, **c** Tuokexun, and **d** Xidaqiao

can achieve higher simulation accuracy with fewer input data. Recently, many researchers have sought to integrate physical hydrological processes into ML models to better explain these phenomena. This integration offers valuable insights for the development of future hydrological models. Overall, ML-based hydrological models show significant promise in simulating glacier runoff at both individual glacier and watershed scales.

Questions

1. From which aspects are the observations of the cryosphere's surface environment carried out?
2. What are the basic requirements for the establishment of a hydrological section? What are the main observation items?

3. For a small watershed with permafrost, how to establish a field observation system?
4. What are the main analytical indicators in the study of cryosphere hydroculture?
5. What are the main ways to perform cryosphere hydrological simulations? What are the main features?

References

Chen X, Wang S, Gao H et al (2022) Comparison of deep learning models and a typical process-based model in glacio-hydrology simulation. J Hydrol 615:128562

Liu T, Yang Q, Qin R et al (2008) Climate change towards warming-wetting trend and its effects on runoff at the headwater region of the Yarkant River in Xinjiang. J Arid Land Resour Environ 22(9):49–53

Qin D, Yao T, Ding Y, Ren J (2021) Field observations and measurements for Cryospheric science. In: Qin D, Yao T, Ding Y, Ren J (eds) Introduction to Cryospheric science. Springer Geography. Springer, Singapore. https://doi.org/10.1007/978-981-16-6425-0_10

Yang C, Xu M, Kang S et al (2023) Improvement of streamflow simulation by combining physically hydrological model with deep learning methods in data-scarce glacial river basin. J Hydrol 625:129990

Extended Reading

Ding Y et al (2017) Introduction to hydrology in cold regions. Science Press, Beijing

Zhang W, Yuan J (2007) Principles and methods of atmospheric sounding. China Meteorological Press, Beijing

Chapter 3
Cryosphere Ablation, Runoff Generation and Flow Routing Processes

Hongkai Gao, Donghui Shangguan, and Tianding Han

3.1 Overview of Runoff Generation and Flow Routing of Cryosphere

Cryohydrology examines the hydrological processes, impacts, and interactions involving cryospheric elements (Ding et al. 2020). A defining characteristic of cryohydrology is the storage of water in its solid phase and its subsequent release as a liquid. This discipline focuses on the interconversion between ice and water and its implications for both resources and the environment. The primary research subjects include the hydrological responses of snow cover, glaciers, permafrost, sea ice, river ice, and lake ice, as well as the associated hydrological processes of mountainous streamflow, lakes, and wetlands. Investigations into runoff generation, flow routing, and their temporal changes are key components of cryohydrology.

Runoff generation occurs when rainfall reaches the land surface and produces a volume of water that can be collected at the watershed's outlet through processes such as interception, infiltration, and evaporation. Flow routing refers to the movement and convergence of various runoff components, generated by rainfall with specific spatiotemporal distributions, to the watershed's outlet via surface runoff, subsurface stormflow, or groundwater discharge. The variations in runoff generation and flow

D. Shangguan · T. Han
Northwest Institute of Eco-Environment and Resources, Chinese Academy of Sciences, Lanzhou, China
e-mail: dhguan@lzb.ac.cn

T. Han
e-mail: tdhan@lzb.ac.cn

H. Gao (✉)
School of Geographic Sciences, East China Normal University, Shanghai, China
e-mail: hkgao@geo.ecnu.edu.cn

Y. Ding et al., *Cryohydrology*, https://doi.org/10.1007/978-981-96-7360-5_3

routing between regions are influenced by the complexity of rainfall conditions and underlying surface characteristics. The cryosphere, as a distinct type of landscape surface, further complicates these processes. Factors such as snow cover extent, snow area, and mountainous terrain influence snowmelt. Similarly, glacier area, glacier characteristics, and glacier type impact not only ice melting and subglacial channel confluence but also ice surface morphology, debris cover extent, and crevasse development. Permafrost affects the interaction between soil water and groundwater, with the melting of ground ice serving as a source of groundwater recharge, thereby influencing runoff generation. Additionally, meltwater from river ice, lake ice, and sea ice, which connects with water bodies, cannot directly contribute to runoff generation or flow routing. As a result, this chapter will focus on the runoff generation and flow routing of snow cover, glaciers, and permafrost, while briefly discussing the melting of river ice, lake ice, and sea ice.

3.2 Snow Ablation and Snowmelt Runoff Generation and Routing

Snow ablation, along with the runoff generation and flow routing of snowmelt, is a fundamental area of study in snowmelt hydrology. Compared to rainfall, the processes of runoff generation and flow routing from snowmelt are more complex. Snow undergoes intricate water-heat interactions, including accumulation, melting, infiltration, and routing, before generating runoff. This section focuses on the physical characteristics of snow cover, the processes of snowmelt runoff generation, and the routing of this runoff.

Upon deposition on the land surface, solid precipitation, known as snowfall, accumulates to form snow cover. Snow cover is a crucial and distinct hydrological component of the terrestrial system. With its high albedo and low thermal conductivity, snow cover plays a pivotal role in regulating weather and climatic systems, making it an essential factor in these processes. Additionally, snow cover serves as a significant water resource in high-latitude regions, where snowmelt runoff accounts for approximately 50–70% of the annual surface runoff. This contribution is also notable in lower-latitude mountainous regions, where snowmelt significantly contributes to runoff (Qi et al. 2020).

Once snow falls on the land surface, its physical properties undergo changes driven by factors such as temperature, pressure, thermal gradients, and the movement of interstitial water. These dynamics lead to modifications in snow density, thermal conductivity, and hydraulic conductivity, which in turn affect the energy and water balance within the snowpack and influence the release of snowmelt.

The ablation of snow cover, along with runoff generation and flow routing, are fundamental components of snow hydrology research. Compared to the hydrological processes associated with precipitation, the dynamics of snowmelt are more complex. Once deposited on the ground, snow undergoes intricate hydrothermal processes

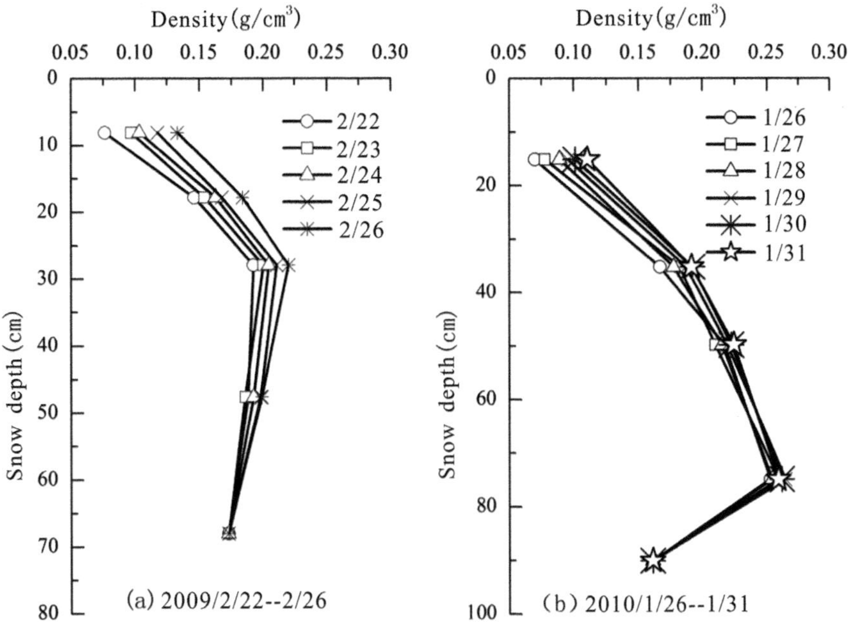

Fig. 3.1 Variation of snow density, snow cover duration, and snow depth at the Tianshan Avalanche Station (Lu et al. 2011)

before contributing to runoff, including the accumulation and melting of snow cover, as well as the infiltration and concentration of snowmelt.

3.2.1 Physical Characteristics of Snow

1. **Snow Density and Compaction**

Snow density generally increases as a result of snow metamorphism, the accumulation of liquid water within the snow cover, and the reduction of porosity (Eq. 3.1). This gradual increase in snow density is referred to as snow compaction.

$$\rho_s = \rho_i(1 - \alpha) + \rho_w \varphi S_w \tag{3.1}$$

where, ρ_s is the density of snow cover (kg/m^3); ρ_i is the ice density (917 kg/m^3); ρ_w is the density of water (10^3 kg/m^3); S_w is the saturated volume water content in pores (m^3 m^{-3}) and φ is the porosity in snow cover (m^3 m^{-3}).

The initial density of new snow is primarily determined by the type and quantity of snow crystals, with typical values ranging from 40 to 100 kg/m^3. During the late ablation period, snow density can increase to as much as 500 kg/m^3.

Figure 3.1 shows the variations in snow cover density at the Tianshan Avalanche Observation Station in China, as a function of time and depth within the vertical profile. As the accumulation period progresses, the density of the snowpack tends to increase. Vertical profile density measurements reveal that the surface layer of the snowpack, composed of fresh snow, typically has a lower density than the base layer. However, during the stabilization phase of the snowpack, the highest density is often observed in the mid-profile, while the surface and base layers exhibit lower densities, influenced by various factors.

2. **Heat Conduction of Snow Cover**

Heat transport within the snow cover along temperature gradients involves three primary processes: solid heat conduction through the ice crystals in the snow, air heat conduction within the pores of the snow cover, and latent heat conduction resulting from the diffusion of water vapor molecules.

The thermal conductivity of ice crystals is approximately 100 times greater than that of air. As a result, the thermal conductivity of snow is primarily determined by the degree of adhesion between ice crystals and the overall structure of the snow. Generally, air heat conduction within the snow pores can be neglected. However, if the latent heat generated by condensation or sublimation in the snow pores drives water vapor transport, heat conduction can increase by up to 50% (Sturm et al. 1997). Water vapor that vaporizes or sublimates from the surface of higher-temperature ice crystal particles diffuses into the atmosphere through pores in the snow layer or condenses on the surface of cooler ice crystals, generating latent heat conduction along the temperature gradient.

The thermal conductivity of snow typically refers to a combination of solid heat conduction and latent heat transfer. While the conductivity of ice generally decreases as temperature rises, latent heat transfer exhibits the opposite trend. As a result, the overall thermal conductivity of the snowpack tends to increase with rising snow temperatures.

$$q = -K_{eff} dT/dZ \tag{3.2}$$

where, q is heat conduction in snowpack (W m^{-2}), K_{eff} is effective thermal conductivity of snow layer (W m^{-1} K^{-1}), dT/dZ is temperature gradient in the snow layer (K m^{-1}).

For snow with high intergranular ice adhesion and wind compaction, the correlation between thermal conductivity (K_{eff}) and snow density (ρ_s) is strong. However, for heavy frost with low intergranular ice adhesion, the correlation between Keff and ρ_s is weak or nonexistent. The K_{eff} of heavy frost is only 1/4 to 1/2 that of other thin snow with the same density but low ice crystal adhesion. In such cases, K_{eff} is often represented by a simple statistical value. The thermal conductivity (K_{eff}) of stable, uniform snow cover with high intergranular ice adhesion can be calculated using the following equation (Sturm et al. 1997):

$$K_{eff} = \begin{cases} 0.138 - 1.01\rho_s + 3.233\rho_s^2 & 0.156 \le \rho_s \le 0.6 \\ 0.023 + 0.234\rho_s & \rho_s < 0.156 \end{cases} \quad (3.3)$$

When snow density reaches 0.6 kg/m^3, the effective thermal conductivity of snow remains lower than that of solid ice. At lower snow densities, the effective thermal conductivity becomes comparable to that of air (0.024 W/m K).

3. **Snow Temperature**

The temperature distribution within the snowpack is influenced by the energy exchange processes at the snow-atmosphere interface, including solar radiation, sensible heat flux, latent heat flux, longwave radiation from the snow surface, atmospheric longwave radiation, and heat flux from the underlying surface. Due to the complex energy transfer mechanisms within the snowpack, its temperature exhibits both diurnal and seasonal variations.

Figure 3.2 illustrates the temporal evolution of snowpack temperature observed at the Western Tianshan Avalanche Station from December 15th to April 1st in 2009. During the winter months, the temperature within the snowpack remains predominantly below 0 °C. Due to the low thermal conductivity of the snow cover, heat cannot rapidly conduct from the surface to deeper layers, nor can it quickly dissipate from deeper layers to the surface. As a result, significant temperature gradients develop near the snow surface, delaying the propagation of daily temperature fluctuations and rapidly dampening diurnal temperature variations within the deeper snow layers. Consequently, the diurnal amplitude of snow temperature gradually decreases with depth. As air temperatures rise, the temperature profiles of the snowpack exhibit a wavelike upward trend. Temperatures at depths exceeding 30 cm are notably influenced by air temperature, exhibiting pronounced fluctuations, especially near the snow surface. In contrast, temperatures below 30 cm show minimal response to air temperature fluctuations, remaining relatively stable, typically ranging between −3 and 0 °C, particularly at the snow-soil interface.

4. **Liquid Water Content in Snow Cover**

Surface snowmelt and rainfall infiltrate into the upper layers of the snowpack, initially satisfying the snow's cold storage capacity and gradually reaching its water-holding capacity. Typically, excess liquid water within the snowpack percolates downward to the soil surface, contributing to either infiltration excess or saturation excess runoff. The liquid water content of the snowpack can be quantified using either volumetric water content (θ_v) or mass water content (θ_m), which are related by the following equation:

$$\theta_v = \theta_m(\rho_s/\rho_w) \quad (3.4)$$

where ρ_s is the density of the snowpack (kg/m^3), and ρ_w is the density of water (1000 kg/m^3).

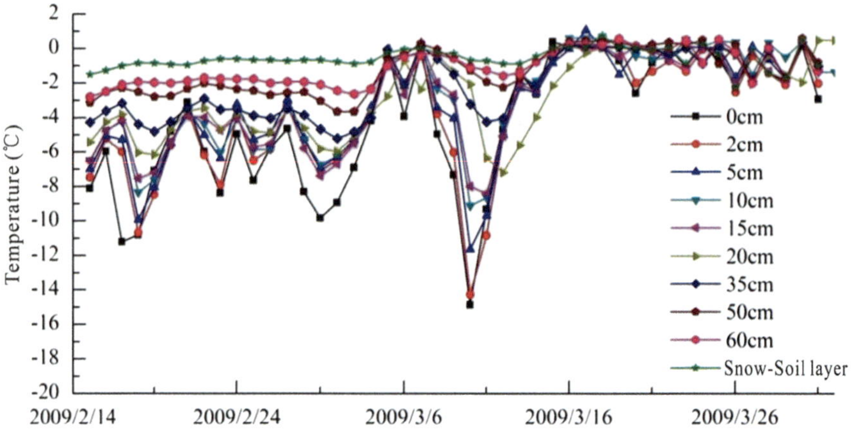

Fig. 3.2 Daily temperature of different snow layers at the Western Tianshan Mountains Avalanche Station (Gao et al. 2012)

The liquid water content within the snowpack fluctuates over time due to precipitation or snowmelt events at the snow surface. Minimal changes in water content are observed during nighttime, while significant variations occur during the daytime, when extensive snowmelt leads to the infiltration of meltwater, causing fluctuations in the mass water content of the snowpack by up to 30% (Heilig et al. 2015).

The liquid water content within the snowpack significantly influences snowmelt runoff. Higher liquid water content leads to the release of more water per unit of energy input. When calculating snowmelt, it is essential to consider both the liquid water content within the snowpack and the effect of cold storage. This can be achieved using a dimensionless parameter called the "heat ratio" (B), defined as the ratio of the energy required to melt a unit mass of snow to the energy consumed in melting a unit mass of ice at 0 °C.

The energy required to melt snow per unit mass includes: (1) the energy needed to raise the snow temperature to 0 °C and (2) the energy required to melt the solid ice crystals in the snow. The heat ratio, B, can be calculated using the following equation:

$$B = \left[(1 - \theta_m)L_f + C_i T\right] / L_f \tag{3.5}$$

where, θ_m is the liquid water content of the snowpack, L_f is the latent heat of fusion of ice at 0 °C (0.334×10^6 J kg^{-1}), C_i is the specific heat capacity of ice (2.1×10^3 J kg^{-1}°C^{-1}), T is the deviation of the snow temperature from 0 °C.

When the snow temperature reaches 0 °C and contains a certain amount of liquid water, the heat ratio is less than 1. When the snow temperature is below the freezing point of 0 °C, B is greater than 1. For snow containing liquid water at 0 °C, and neglecting the influence of air in the snowpack, the heat ratio B is equivalent to the mass ratio of ice within the snowpack.

3.2.2 Accumulation and Ablation of Snow

The accumulation and ablation processes of snow typically occur within an annual cycle, beginning with the formation of winter snow and continuing until the complete melting of summer snow, constituting a full accumulation and ablation year. In most cases, snow completely melts in summer, and even in high-altitude areas where snowfall may occur, it is challenging to retain snow for extended periods as it melts quickly. The only locations where summer snow accumulation can occur are glacier accumulation zones, where the snow formed contributes to the glacier mass balance rather than forming a typical snow cover.

Snow cover, formed from surface snowfall, is primarily controlled by climatic conditions and typically persists for less than a year, making it a seasonal phenomenon. Surface snow undergoes various physical processes such as metamorphism, compaction, wind drifting, avalanches, evaporation, sublimation, melting, and refreezing during the accumulation and ablation phases, leading to changes in snow properties and mass. In the warm season, snow may melt quickly upon contact with the surface or within a few days; this type of snow is called transient snow. Transient snow cover tends to occur at lower elevations but can also occur at higher elevations during melting seasons. In cold seasons, snowfall events are frequent, and snow cover can last for several weeks or even months. Due to differences in the physical characteristics of new and old snow, the snowpack often develops multiple layers. The initial characteristics of each snow layer are determined by the weather conditions at the time of snowfall, while its subsequent evolution is controlled by metamorphism within the snow layer. Snow thickness (snow depth) is primarily determined by the amount of snowfall, but factors such as melting, refreezing, and wind-blown snow can also lead to changes in snow depth.

Snow ablation refers to the overall loss of snow mass, encompassing processes such as melting, sublimation, evaporation, and snow drift. Snowmelt can be categorized into two types: radiative melting and convective melting. (1) **Radiative Melting**: Snowmelt primarily arises from solar radiation, with air temperature being the most direct factor influencing it. The reflectivity of snow determines the amount of solar shortwave radiation absorbed by the snow surface. The high reflectivity of winter snow and low solar elevation angles limit the absorption of solar shortwave radiation, thereby preserving the snow. In spring, the metamorphosis and compaction of snow reduce surface reflectivity, enhancing solar irradiance and initiating snowmelt. (2) **Convective Melting**: When strong warm, moist air masses arrive over snow-covered areas, they transfer energy to the snow through various mechanisms, including longwave radiation emitted from the cloud base, heat from rainfall, and turbulent heat flux caused by strong winds passing over rough terrain. Strong turbulence can disrupt the stable near-surface atmospheric conditions (neutral or inversion) formed over the 0 °C snow surface. The presence of forests further enhances convective melting. Although the heat content of rainfall is generally on the order of 40–60 J/g, significantly less than the latent heat of ablation (335 J/g), it can still contribute to snowmelt, albeit limited, with approximately 0.2 g of snow being melted per gram of rainfall. However, the more significant impact of rainfall-induced snowmelt lies in its reduction of snow surface reflectivity.

Influencing Factors

Temperature and precipitation are the two most direct factors influencing snow accumulation and ablation. Precipitation directly determines the mass of snow, while temperature controls the processes of snowfall and snow ablation. In addition to these two factors, topography (including elevation, slope, and aspect), vegetation cover, and wind also influence snow accumulation and ablation to varying degrees.

Temperature: Temperature primarily influences snow accumulation by determining the form of precipitation—whether it falls as snow, rain, or sleet. However, the temperature thresholds distinguishing between these forms of precipitation are not fixed but are closely related to altitude and relative humidity. The temperature ranges for rainfall and snowfall vary across different regions. For instance, analysis of data from over 1,000 observation stations in the United States reveals that both rainfall and snowfall events can occur within the temperature range of 0–4 °C (Auer 1974). Similarly, observations from 19 stations in Sweden report a temperature range for rainfall and snowfall events from −3 to 5 °C (Feiccabrino and Lundberg 2008). Data from 643 stations in China, analyzed by Han et al. (2010a, 2010b, 2010c), indicate that the threshold temperature for distinguishing between rainfall and snowfall varies widely, ranging from 1 to 6 °C. Therefore, in many cases, 0 °C cannot be directly used as the temperature threshold to distinguish between rainfall and snowfall. The impact of temperature on snowmelt involves processes such as direct heating and the transfer of latent and sensible heat. When the temperature is above freezing, heat from the air transfers to the snow surface, initiating melting. As temperatures rise further, the rate of melting accelerates. Higher temperatures result in faster snowmelt because a significant amount of latent heat is required to transform solid snow into liquid water. Sensible heat transfer also increases with rising air temperatures, further warming the snowpack and accelerating the melting process. Additionally, diurnal temperature variations cause repeated cycles of melting and refreezing, leading to a looser snow structure that is more susceptible to melting during each subsequent cycle.

Precipitation: In the winter of high-latitude regions in the Northern Hemisphere and high-altitude mountain areas, most of the precipitation occurs as snowfall. Snowfall is the primary mass source of snow cover, and elevation gradient is the key geographical factor affecting the phase of precipitation. Due to the variation of precipitation with altitude, there are significant differences in precipitation across different zones, elevations, slope aspects, seasons, and between dry and wet years. These variations result in distinct zonal distribution patterns of precipitation processes and snow cover in mountainous areas. The elevation distribution of precipitation is the primary reason for the elevation effect on snow distribution. Compared to other factors such as terrain, forest vegetation, and others, elevation has the most significant impact on the distribution of snow depth, especially in years with abundant snow cover, where the effect of elevation on snow depth distribution is even more pronounced (Lehning et al.

2011). The impact of precipitation on snowmelt depends on the type, temperature, and intensity of the precipitation, as well as the climatic conditions during the precipitation event. When rainfall occurs at temperatures above freezing, it directly contacts and melts the snow. Warm rain provides additional heat, accelerating the snowmelt process. If the temperature of the precipitation is near or below freezing, the rate of snowmelt will be slower and may even result in an increase in snowpack. Heavy rain events can lead to rapid snowmelt due to the large amount of warm rainwater quickly melting the snow. Frequent rainfall keeps the snow wet, reducing its density and making it easier to melt. Temperature changes during precipitation significantly impact snowmelt; rising temperatures accompanying precipitation result in faster melting, while falling temperatures slow down or even halt the snowmelt process.

Vegetation Conditions: Vegetation, particularly forests, plays a significant role in both the accumulation and ablation processes of snow. In regions with a continental climate and extensive distribution of forests and shrubs, the interception, sublimation, and evaporation of snow by forests, shrubs, grasslands, and other vegetative covers are crucial factors affecting snow accumulation. Additionally, surface vegetation characteristics influence snow redistribution through factors such as differences in snow cover and thermal conductivity. Research in the Qinling Mountains region indicates that forest type and snowfall amount are the primary factors influencing the spatial distribution of surface snow in forested areas. The type of snowfall also has an impact. Different forest types exhibit interception rates for snowfall ranging between 19.22 and 48.60%, with coniferous forests showing superior interception effectiveness compared to broadleaf forests (Dang and Wu 1991). Snow interception experiments conducted in the Lesser Khingan Mountains region of China revealed that snow interception rates by forests are mainly controlled by canopy closure and tree species composition. Among these, spruce-fir and Korean pine forests demonstrate the highest interception effectiveness, with rates reaching 39.7%, approximately five times higher than in secondary birch forests and 2.5 times higher than in artificial deciduous forests. Additionally, snowfall characteristics significantly influence interception rates in forests with the same canopy closure and tree species composition. Spruce-fir and Korean pine forests, as well as Korean pine plantations, exhibit the strongest interception effects during both heavy and light snowfall, while secondary birch forests and larch plantations demonstrate better interception effectiveness during light snowfall (Liu et al. 2012).

Blowing Snow and Topography: Wind is the primary driving force for snow redistribution, and variations in topography (such as valleys and ridges) and land cover (including bare soil, vegetation, and scrub) significantly influence wind-blown snow. In some snow-abundant mountainous areas, wind-blown snow is the main factor controlling the development of surface snow cover and the heterogeneity of its spatial distribution. Wind speed and the age of snow, particularly the characteristics of the surface snow cover, determine the extent of wind-blown snow. The minimum wind speed required for snow to be blown by the wind is known as the critical wind speed for wind-blown snow. For relatively dry and loose new snow, the critical wind speed ranges between 0.07 and 0.25 m/s. As the snow accumulation process progresses, snow compaction enhances the interaction between snow particles, causing the critical wind speed to gradually increase to 0.25 to 1.0 m/s (Pomeroy and Jones 1996). Additionally, the liquid water content in the snow layer increases the cohesive force between snow particles, further raising the critical wind speed.

Different terrain and surface cover conditions significantly influence the distribution of snow cover. First, topography affects the direction and speed of wind and its spatial distribution characteristics to a certain extent. Various geomorphic features alter the aerodynamic characteristics near the ground by changing the surface roughness. Additionally, different topography and surface conditions directly impact local snow storage and seasonal variations (Pomeroy et al. 1997). For example, in the Arctic, there is a 10–25% difference in snow depth between regions with varying levels of shrub development (size and coverage) (Sturm et al. 2001). By the end of winter, wind-blown snow can cause snow cover in the Arctic region to range between 54 and 419% of the original snowfall. Global warming is leading to the expansion of shrubland and grassland into alpine areas, and changes in continental surface conditions will inevitably alter the spatial distribution of regional snow cover, thereby affecting the hydrological processes associated with snow cover.

Snow sublimation caused by wind-blown snow is a significant component of snow ablation. A wind-blown snow model revealed that in 2008, sublimation of snow due to wind-blown snow accounted for 41.5% of the total snowpack sublimation in the Binggou watershed of the Qilian Mountains (Li et al. 2012). In the Canadian Rockies, sublimation of snow is estimated to be about 20–32% of the total snowfall, with sublimation losses from wind-blown snow accounting for approximately 17–19% (MacDonald et al. 2010). Compared to the Qilian Mountains and the Rocky Mountains, snow sublimation in the Swiss Alps is relatively low. The regional average sublimation of wind-blown snow, calculated using the Alpine3D model, is only 0.1% of the precipitation during the same period (Groot Zwaaftink et al. 2013).

3.2.3 *Processes of Runoff Generation and Flow Routing of Snowmelt*

1. **Process of Runoff Generation**

The water balance of snow is determined by the accumulation and ablation of snow. Over a specific period, the change in the water balance of snow cover at a single point can be expressed as follows:

$$\Delta SWE = Q_c - Q_a \tag{3.6}$$

where, $\Delta\ SWE$ is the change of snow water equivalent; Q_c is the net accumulation of snow; Q_a is the net ablation of snow.

For a complete snow accumulation-ablation cycle, the water balance of the snow cover is zero.

$$Q_c = Q_a \tag{3.7}$$

where, P_{snow} is the most important mass source of snow, and the amount of snow directly determines the snow water equivalent. Additionally, snow accumulation includes the amount of P_{rain} that freezes into ice due to rainfall and the amount of Q_f that condenses into ice from water vapor in the air on the snow surface. Therefore, snow accumulation can be expressed as:

$$Q_c = P_{snow} + \mathrm{P}_{rain} + Q_f \tag{3.8}$$

The melting of snow cover mainly includes two components: snow sublimation loss and snowmelt. During the snowmelt period, snowmelt is the most significant form of snow cover ablation. In some areas, the maximum daily snowmelt rate can reach 7–8 cm per day, and the hourly snowmelt rate can reach 0.1–0.3 cm per hour. However, over the entire snow accumulation period, snow sublimation plays a crucial role in the water balance of snow. Additionally, processes such as redistribution by trees, shrubs, and wind-blown snow affect the mass balance of snow. Therefore, the amount of snowmelt loss can be expressed as:

$$Q_a = Q_{is} + Q_s + Q_e + Q_m \tag{3.9}$$

where, Q_{is} is canopy interception sublimation amount; Q_s is sublimation of snow cover; Q_e is evaporation of liquid water in the snow layer; Q_m is the snowmelt.

Therefore, the mass balance equation for the complete snow accumulation-ablation cycle can be expressed as follows:

$$P_{snow} + \mathrm{P}_{rain} + Q_f - Q_a \pm \Delta Q = 0 \tag{3.10}$$

where Q represents snow displacement due to avalanches, wind-blown snow, and other factors. The units of all terms are kg m^{-2} from formula (3.6) to (3.10).

It should be noted that Q_m represents the amount of meltwater flowing out of the snow cover. If only a phase transition occurs but the water remains in the snow layer, the liquid water content of the snow layer increases, while the snow layer's mass remains unchanged. The changing process of snow mass balance is also the process of snowmelt runoff production.

$$Q_m = P_{snow} + \mathrm{P}_{rain} + Q_f - Q_{is} - Q_s - Q_e \pm \Delta Q \tag{3.11}$$

2. **Process of Flow Routing**

Compared to the runoff generation and flow routing associated with rainfall, those involving snow are more complex. Once snowfall reaches the ground, it undergoes intricate hydrothermal processes before contributing to runoff; these processes primarily include the storage and melting of snow, as well as the infiltration and concentration of meltwater. Prior to the completion of snowmelt, the upper soil layers

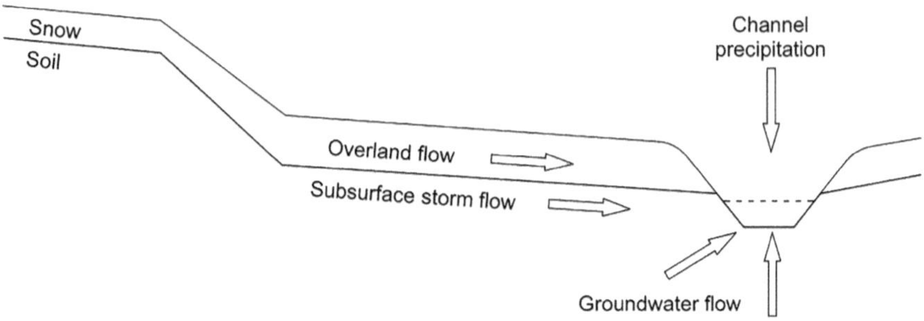

Fig. 3.3 Schematic diagram of snowmelt (rainfall) flow routing (DeWalle and Rango 2008)

often remain frozen, limiting the infiltration capacity of meltwater within the soil. As soon as meltwater reaches the ground surface, it rapidly converges at the base of the snow layer, forming surface runoff. After snowmelt water reaches the ground surface, it can enter watercourses through four primary runoff pathways: direct runoff, overland flow, interflow, and groundwater flow. A comprehensive illustration of the runoff process is presented in Fig. 3.3. The following sections will detail each of these flow routing pathways.

Direct runoff generation occurs when precipitation—whether solid, liquid, or mixed—directly falls into a river channel, making it the most direct, effective, and fastest way to form runoff. In cold regions, lakes or wetlands may undergo freeze–thaw cycles. When the water surface is frozen, winter snow is stored on the ice in the form of snow until it melts and the channel thaws, at which point the snow and meltwater directly contribute to runoff. In basins where river ice develops, a large amount of snow accumulates on the surface of river ice during winter. In the early stages of snowmelt, river ice blocks the path of meltwater, causing a large amount of snowmelt water to form runoff on the surface of the river ice. As runoff increases, the snowmelt water directly carries the snow on the surface of the river ice, forming mixed snow-water runoff. In this process, solid snow directly participates in runoff production and convergence as part of the runoff. For example, in the Kaiertes River basin, a primary tributary of the Irtysh River, the thickness of river ice in winter exceeds 1 m. Before the snowmelt in 2014, the snow thickness on the surface of river ice was generally greater than that on the riverbanks. On March 19, 2014, a large amount of snowmelt water converged on the surface of the river ice, resulting in snowmelt runoff formed by snowmelt water combined with snow on the river ice (Fig. 3.4).

Overland flow refers to the form of runoff generation and flow routing that occurs directly on the surface after rain or snowmelt water touches the ground. During snowmelt, the soil layer beneath the snow is generally frozen, with only a thin surface layer of soil melting due to the latent heat input from the snowmelt. Because frozen soil has relatively low water permeability (low hydraulic conductivity), it is difficult for snowmelt (or rainwater) to infiltrate and recharge the soil layer. Consequently, a large amount of snowmelt water (or rainwater) flows directly to the riverbed by

(a) (b) (c) (d)

Fig. 3.4 Process diagram of ice melt runoff in the preformative stage (**a**), early stage (**b**), clear snow period (**c**), and no ice period (**d**) in the Kayertes River basin. *Note* Images **a**, **b**, and **c** were taken in March 2014; photo **d** was taken in September 2013

concentrating on the slope. An observation experiment during the spring snowmelt period of 2014 in the source area of the Ertysi River showed that it is very common for snowmelt to form river runoff directly through slope confluence. Additionally, due to the melting of the surface soil, surface gullies formed by slope flow are widespread in areas with relatively steep terrain, making it easy for small snowmelt debris flows to develop during snowmelt (Fig. 3.5).

Interflow refers to the hydrological process in which snowmelt or rainwater, upon reaching the ground surface, infiltrates into the soil and rapidly converges to contribute to runoff. Under conditions of sufficient snowmelt and rainfall, when deeper soil layers are frozen, infiltrating water can quickly reach riverbeds via flow channels in the topsoil, thereby generating runoff. The primary stages of this process are as follows.

In the initial phase of snowmelt, the meltwater first replenishes the topsoil moisture. The pre-existing soil water stored in the soil layers is discharged into rivers under the influence of meltwater. As the snowmelt intensifies or is accompanied by precipitation, a substantial volume of water enters the topsoil, causing a sharp increase in the moisture content within this layer, which continues to supply rivers through interflow. In arid and semi-arid regions, the moisture content of the upper soil layers often reaches its annual peak during the spring snowmelt period.

Fig. 3.5 The process of overland flow routing during snowmelt in the Cailts River basin. *Note* The image was taken during the snowmelt in March 2014

Figure 3.6 illustrates the variation in soil moisture at a depth of 5 cm as observed at the Kuwei Snow Station of the Northwest Institute of Eco-Environment and Resources, Chinese Academy of Sciences, located in the middle section of the Altai Mountains within the Kax River Basin, a source area of the Irtysh River. The results indicate that the regional snowmelt process commenced on March 19, and the moisture content in the topsoil simultaneously reached its maximum value for the year. Comprehensive observations of soil moisture and snowmelt demonstrate that due to the prolonged duration of the snowmelt process, which can provide stable replenishment to surface soil water, the contribution of snowmelt to interflow is substantial and can even exceed that supplied by precipitation.

Unlike interflow, groundwater is fed through rock voids by gravity and external pressure. The mechanisms of groundwater flow direction and velocity are the same in permafrost and seasonal tundra, and the theoretical basis for its calculation is Darcy's law:

$$V_g = -K_w(dh/dx) \quad (3.12)$$

where, V_g is velocity of groundwater flow, K_w is saturated hydraulic conductivity (m/s), *dh/dt* is hydraulic gradient.

K_w is influenced by porosity, pore size, and the degree of soil fragmentation. In frozen soil, the pores occupied by ice crystals also have a significant impact on K_w. Additionally, the formation of ground ice can block the flow of groundwater, while the degree of soil compaction and the chemical concentration of soil water greatly affect the formation of soil ice crystals. If cracks in the soil contain ice, the

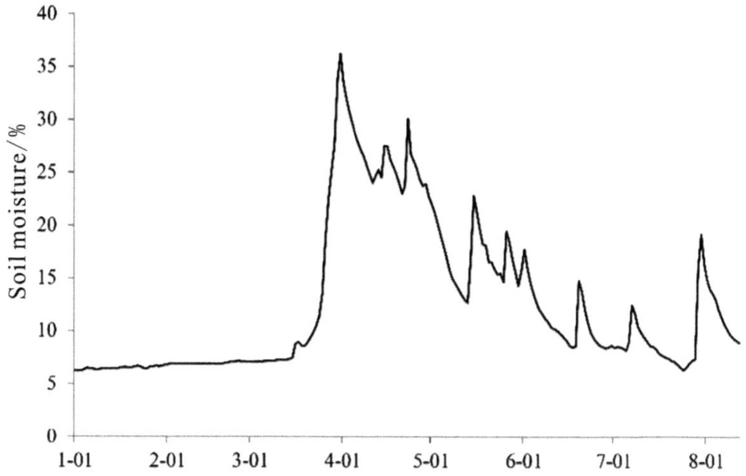

Fig. 3.6 Evolution of daily soil moisture at a depth of 5 cm observed at the Kuwei Snow Observation Station in the Kayertes River Basin in 2014

permafrost is often considered an impermeable layer or a weakly permeable layer within the subsurface flow system.

Influence of Frozen Ground on Snowmelt Runoff

The interaction between snow cover and permafrost is bidirectional. On the one hand, the thermal insulating properties of snow cover significantly influence the freeze–thaw dynamics of permafrost. On the other hand, the low hydraulic conductivity of the permafrost layer markedly alters the generation and convergence processes of snowmelt runoff. The depth of the snowpack is often used as a critical index for assessing the thermal insulation effect of snow. When the snowpack is thin, its high albedo prevents most solar radiation from reaching the ground, but its insulating effect is minimal, failing to effectively hinder the dissipation of thermal energy from the soil to the atmosphere, resulting in a colder soil layer. As snow depth increases, the insulating effects of the snow gradually become more dominant, favoring soil thermal retention (Gao et al. 2022). However, the relationship is not linear; as the snow depth continues to increase, the duration of the snow cover is significantly extended, potentially leading to perennial snowpack or glaciation. At this stage, the impact of snow on soil thermal conditions becomes more complex (Zhang et al. 2005).

Based on these theoretical foundations, the threshold snow depth emerges as a valid indicator for judging the insulating capacity of snow. Research from Russia suggests that a snow depth of 40 cm provides optimal thermal protection for the soil. Observational data spanning five decades from the Altai region of Xinjiang, encompassing meteorological conditions, snow cover, soil temperature, and frost penetration depth, indicate that snowpacks thinner than 20 cm do not significantly insulate the soil from freezing. A snow depth of 20 cm represents the lower limit of snow influence on underlying soil freezing, beyond which the maximum frost penetration depth is unaffected. When snow depth exceeds 20–40 cm, the snow can reduce the soil frost penetration depth by 15–50 cm, exerting a certain insulating effect. Once the snow depth surpasses 40 cm, the temperature's influence on subnivean soil freezing stabilizes, and beyond 70 cm, the snow's impact on soil freezing further alters, with monitoring results varying by location, showing either an increase or decrease in frost penetration depth (Wang et al. 2012).

Observations from the Xidatan region reveal that during cold seasons with ground and air temperatures around −10 °C, snow cover less than 10 cm thick has a negligible effect on ground temperatures. However, during warm seasons, when snow depth exceeds 10 cm and remains for over 10 days, the insulating effect of snow becomes significant (Sun et al. 2010). At the Qilian Mountain Bingou Station, simulation results from the CoupModel show that snow depths of 0–20 cm facilitate the freezing process of permafrost, while depths of 20–70 cm provide considerable thermal insulation (Zhang et al. 2013).

In cryohydrological research, snow hydrological processes in permafrost regions are altered due to the presence of permafrost. When significant snowmelt occurs over a short period, the meltwater cannot infiltrate the permafrost layer, leading to surface water accumulation and subsequent hillslope runoff, thereby generating substantial snowmelt runoff in a brief span.

In a typical snow-covered catchment in Vermont, USA, differences in permafrost freezing depth led to significant variations in snow hydrological processes over two consecutive years (Fig. 3.7). In 1993, due to greater soil freezing depth, when snowmelt began, the underlying soil remained frozen, and rapid melting of a substantial amount of snow generated slope convergence, reaching the river channel in a shorter time, resulting in a significant snowmelt runoff event from March 26 to April 2, 1993. In contrast, in 1994, due to a thinner underlying permafrost layer, snowmelt water showed a progressive increase without the snowmelt flood observed in 1993 (Shanley and Chalmers 1999). Therefore, the freezing depth of permafrost influences the recharge of groundwater systems that rely on snowmelt to some extent (e.g., the headwaters of the Irtysh River in the Altai Mountains).

In the typical hydrological treatment of meltwater infiltration, permafrost is conventionally regarded as an impermeable layer, while seasonally frozen ground exhibits certain permeable characteristics. Theoretically, the permeability of frozen soil is primarily determined by the soil texture before freezing and the water content (or ice content) during the freezing process. Generally, soils with larger particle sizes, such as sandy soils, exhibit enhanced permeability after freezing due to their larger interstitial spaces between particles. In contrast, soils with higher water content tend to form more ice crystals during the freezing process, which consequently reduces the soil's permeability. Therefore, soils with larger particle sizes and lower water content (or ice content) demonstrate greater permeability upon freezing, evidenced by higher infiltration coefficients, thereby facilitating the infiltration of meltwater and recharging groundwater systems.

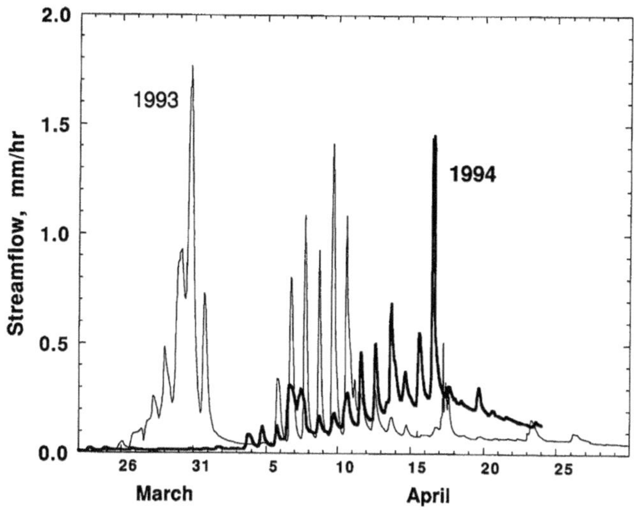

Fig. 3.7 Snow hydrological processes at different frozen depths (390 mm and 180 mm, respectively) in 1993 and 1994. The hydrograph pattern, supported by isotopic evidence, suggests that ground frost in 1993 promoted direct runoff of meltwater while preventing groundwater recharge. In 1994, under low-frost conditions, most meltwater infiltrated to recharge groundwater, which in turn sustained a very gradual flow recession after the main melt period

3.3 Glacier Melt, Runoff Generation and Flow Routing

Glaciers are perennially preserved, dynamically active large bodies of natural ice on land, formed through the compaction and metamorphic evolution of accumulated snow. They exhibit distinct shapes and persist over extended periods in Earth's colder regions. Temperature, precipitation, and topography are the three essential conditions for the formation and development of glaciers. Glaciers can develop in any region on Earth where the annual mean temperature remains below 0 °C, and snow can persist over several years. Therefore, glaciers are a product of cold climates and fluctuate with climatic variations.

During the accumulation phase, the glacier surface gains mass through rainfall, snowfall, deposition of hoarfrost, rain refreezing, and the redistribution of windblown and avalanche-deposited snow due to wind and gravitational forces. During the ablation phase, mass loss occurs through processes such as meltwater runoff, evaporation, sublimation, disintegration of the ice mass, and the removal of windblown snow and avalanches beyond the glacier margins. The presence of a vertical temperature gradient (the phenomenon of a linear decrease in temperature with increasing elevation, generally measured as 0.6 °C per 100 m) results in slower or negligible melting at the glacier's upper reaches and more intense melting at its lower reaches. Over the course of a year, these melting processes lead to the formation of distinct accumulation and ablation zones, demarcated by the equilibrium line altitude. The hydrological processes on and within glaciers—particularly runoff generation, evaporation, sublimation, and the convergence of flows on the surface, within the interior, and at the base—constitute the foundation of glacial hydrology.

3.3.1 *Physical Characteristics of Glacier*

Glacier Mass Balance

The mass balance of a glacier refers to the algebraic sum of mass input, predominantly in the form of solid precipitation (accumulation), and mass output, primarily through glacial melting (ablation), within a unit of time, expressed in terms of water equivalent. The term "glacial mass" here specifically denotes solid water bodies, excluding rock debris, atmospheric deposition, or other substances. Hence, the mass balance can also be termed the ice-mass balance. The mass balance is determined by the energy budget conditions within the glacial zone, where the energy balance dictates the surplus or deficit of glacial accumulation and ablation. The process of receiving solid water in the glacial system is known as glacier accumulation or replenishment, including precipitation, icefalls, avalanches, condensation and deposition of water vapor, rain refreezing, windblown snow, and more. Conversely, the expenditure of solid water within the glacier system, known as glacial ablation or loss, encompasses processes such as meltwater runoff, meltwater percolation, evaporation of water vapor, sublimation of snow and ice, and the removal of snow and ice due to wind and gravity. Typically, glacier accumulation primarily stems from precipitation, while glacial loss is mainly due to surface melting of snow and ice. The glacier undergoes a net positive or negative mass balance over a certain time scale (usually annually) through the processes of accumulation and ablation.

Based on the characteristics of meltwater production and refreezing, glaciers can be divided into accumulation and ablation zones. These zones are demarcated by the equilibrium line, which represents the threshold where the annual glacial mass input equals the output—that is, where annual accumulation equals annual ablation, also referred to as the glacier equilibrium line altitude (Duan et al. 2022). Accumulation mainly occurs in the basins above the equilibrium line, known as the accumulation zone, while below it is termed the ablation zone. The ablation zone also receives mass input, typically in the form of seasonal snow accumulation, but the rate of glacial melting always exceeds that of accumulation. Even in the accumulation zone above the equilibrium line, some melting occurs, but it diminishes with increasing altitude. Near the equilibrium line within the accumulation zone, snowmelt can generate meltwater runoff and form a layer of refrozen ice at the base of the snowpack, leading to the designation of this area as the additional ice zone. The threshold where melting occurs but does not produce meltwater runoff is known as the firn line. Located between the equilibrium line and the firn line, the superimposed ice zone is where, even if snow melts, it only creates percolation and refreezing within the snowpack, without any loss of meltwater; thus, the ablation in this region is zero (no mass loss) (Figs. 3.8 and 3.9).

Equation of glacier mass balance:

$$GMB = C - B \tag{3.15}$$

where, C is net equation accumulation; B is net ablation.

For a glacier, when B < 0, ablation exceeds accumulation, resulting in a negative balance. Conversely, when B > 0, the glacier is in positive equilibrium. When B = 0, the glacier is in equilibrium. A negative balance over a certain period leads to glacier retreat, while a long-term positive balance results in glacier expansion,

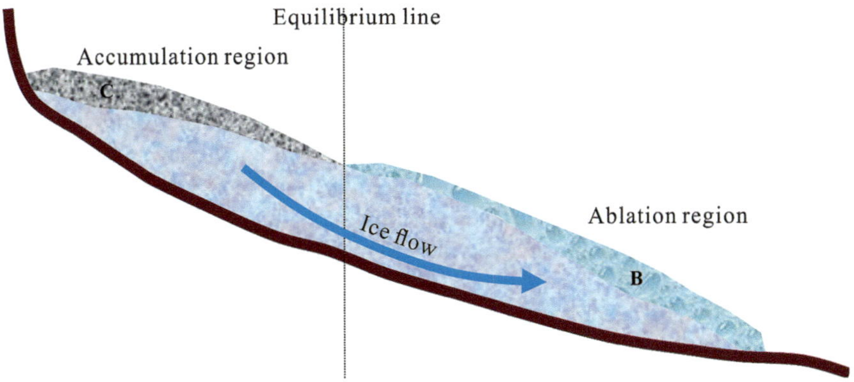

Fig. 3.8 Schematic diagram of glacier mass balance (Ding 2017)

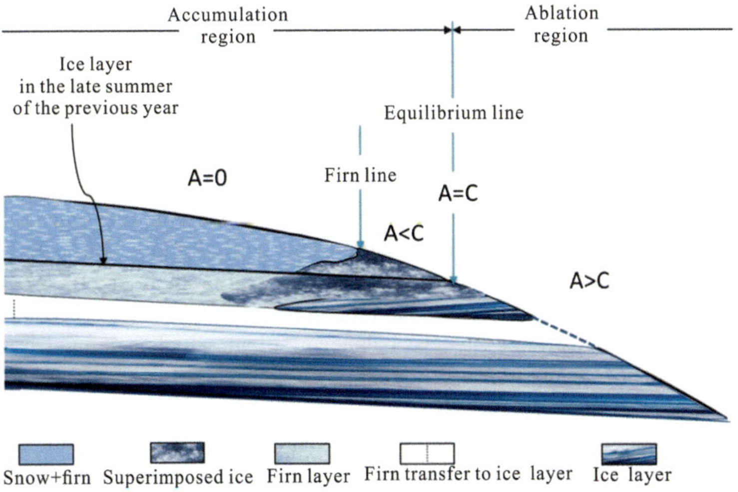

Fig. 3.9 Schematic diagram of the surface zoning of a glacier (ice sheet) (Ding 2020)

with an equilibrium glacier remaining in a stable state. For glaciers in continuous negative equilibrium, the large amount of meltwater stored on the glacier causes a significant increase in yield and discharge over a certain period, leading to higher glacier meltwater runoff. Conversely, for a glacier in long-term positive equilibrium, the glacier's meltwater decreases over time. For a glacier in a stable state, the yield reflects the average value over a certain period. A negative or positive balance in a given time.

Net Balance and Balance Year

The net change in a glacier's mass budget between the same date in two consecutive years is called the annual mass balance, also referred to as the annual balance or net balance. The equilibrium year begins with the date of least glacier mass at the end of summer and ends with the date of least glacier mass in the following year. The equilibrium year accurately defines the period during which the glacier mass budget changes, providing clear physical significance. However, due to climate fluctuations, the onset and duration of the equilibrium year vary for each glacier, with significant differences between glaciers. This variability makes it difficult to compare mass balance data and to analyze and simulate with other data, such as hydrometeorological data. For this reason, a fixed-date scheme based on the hydrological year was developed. In the Northern Hemisphere, the hydrological year runs from October 1 to September 30 of the following year. In China, glaciers generally stop melting at the end of August, so in Chinese glacier research, the mass balance year is typically from September 1 to August 31 of the next year. For mid- to low-latitude marine glaciers and some continental glaciers, the mass balance year can also coincide with the hydrological year, as the melting period may extend until the end of September.

The annual mass balance of glaciers is the sum of the total accumulation and total loss of glacial mass during the year. In mid- to low-latitude glacier regions with significant seasonal variations in snowfall and temperature, the annual balance is usually divided into a winter balance and a summer balance based on the date of maximum glacier mass in the equilibrium year. This division allows for the separate quantification of the contributions of winter snowfall and summer melting to the glacier's mass balance.

Mass Balance Elevation Gradients

In the context of most continental glaciers, both glacier accumulation and ablation typically show an increasing or decreasing trend with altitude. Thus, the rates of change in accumulation and ablation along the vertical gradient can be represented by the accumulation gradient and the ablation gradient, respectively. The summation of these gradients constitutes the mass balance gradient, which is defined as the rate of change of the net balance with altitude. Figure 3.3 presents a selection of common glacial mass balance gradient curves.

The mass balance gradient intuitively reflects the impact of local climatic conditions on the glacial mass balance. A larger mass balance gradient indicates a condition where snowfall is abundant in the upper reaches of the glacier accumulation area, while ablation at the glacier tongue is exceptionally intense. Glaciers exhibiting such characteristics typically develop in mid-latitude regions with plentiful precipitation and lower elevations at the glacier terminus, such as in southern Alaska, New Zealand, and western Norway. Conversely, a smaller mass balance gradient reflects minor variations in accumulation and ablation along the vertical profile of the glacier and a lower total mass turnover rate, indicative of slower renewal of glacier mass. Glaciers in cold and dry regions often exhibit this type of mass balance gradient, such as the extreme continental glaciers in the hinterlands of the Qinghai-Tibet Plateau in China, as well as some ice caps and glaciers in northern North America.

In the ablation zone of glaciers, air temperature generally decreases linearly with increasing altitude, with a temperature gradient (lapse rate) of approximately 0.4–0.9 °C per 100 m. Due to the significant role that atmospheric longwave radiation and sensible heat flux—both influenced by temperature—play in snow and ice melting, the amount of glacier ablation also shows an approximately linear variation with altitude. This is manifested as a straight line in the lower portion of the mass balance gradient curve, where the slope represents the ablation gradient. The higher the average air temperature in the glacier's ablation zone, the greater the ablation gradient, resulting in a steeper curve.

For large valley glaciers in regions such as the Himalayas, Tianshan, and Karakoram in China, ablation with elevation often shows a nonlinear variation (as in the Koxkar Glacier depicted in Fig. 3.10) due to continuous debris cover on the glacier tongues. This debris cover principally originates from rock debris entering the glacier from avalanches in the upper and middle parts of the glacier, which then accumulates as the glacier flows and melts at the glacier tongue, or from rock debris at the glacier base that is squeezed out along weak layers in the ice during glacial motion. The thickness of the debris decreases from the terminus of the glacier upwards. A thick debris layer significantly inhibits the melting of the glacier surface, resulting in an increased rate of glacier ablation with altitude at the glacier tongue. As the debris thickness diminishes, its insulating effect weakens, exposing the ice surface in the middle to upper parts of the glacier tongue. Only near the equilibrium line does the ablation rate begin to show a decreasing trend with increased altitude.

Unlike the melting gradient, which is controlled by temperature, the variation in the glacier accumulation gradient is primarily governed by the longitudinal change in snowfall. For many valley glaciers and cirque glaciers, snowfall increases approximately linearly with elevation, leading to a linear variation in glacier accumulation. However, for many continental glaciers, due to the high altitudes of their sources and low temperatures, most ascending water vapor condenses and precipitates before reaching the highest elevations, creating a zone of maximum precipitation height. Consequently, snowfall in the accumulation area exhibits a nonlinear change with altitude: it initially increases with elevation from the equilibrium line and, upon reaching a certain height, gradually decreases. This precipitation distribution characteristic is more common in continental glaciers located in high-altitude regions such as the Himalayas, Karakoram, and Tianshan mountains. Additionally, a decrease in snowfall with increasing altitude is also commonly observed in the ice caps and large ice fields of the polar regions. This phenomenon is due to the glacier termini

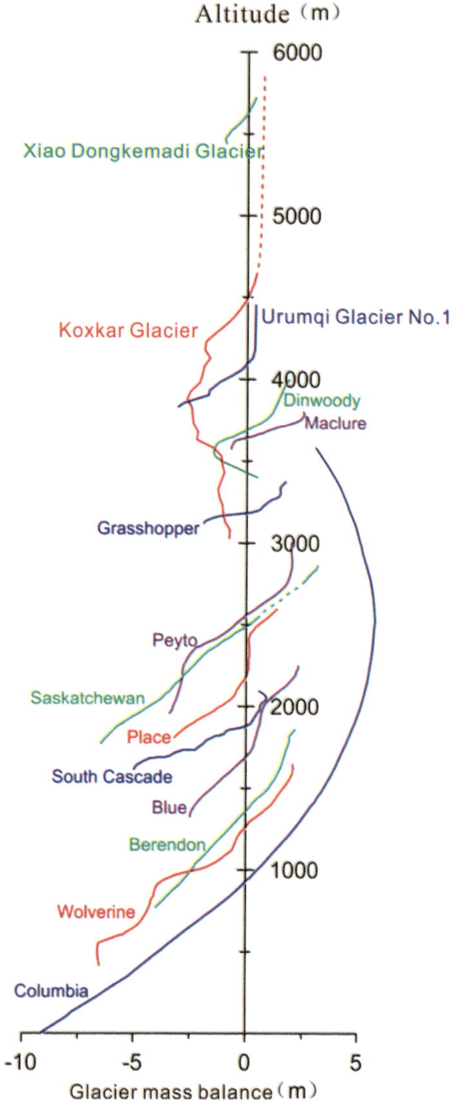

Fig. 3.10 Annual mass balance gradient curve of High Asian and North American glaciers (Mayo 1984)

being near moisture-rich oceans, while the polar interiors are far from the coasts and exceptionally cold, resulting in minimal moisture transport.

Mass Balance Ratio

Due to variances in climatic and topographic conditions across different glacial regions, substantial differences in their mass balance profiles are observed. However, to facilitate comparative analysis of the relative changes in accumulation and ablation across diverse glaciers, the concept of the mass balance ratio has been introduced.

Defined as the ratio of the ablation gradient to the accumulation gradient, the mass balance ratio reveals the comparative rates of ablation and accumulation within a glacier. For instance, some maritime glaciers in North America, specifically in Alaska and the Pacific Northwest, have mass balance ratios ranging between 1.5 and 2.2, whereas the continental glaciers in the High Asian regions exhibit mass balance ratios approximately between 4 and 12. Glaciers in low-latitude regions, which undergo ablation year-round, demonstrate higher mass balance ratios; an example includes the Zongo Glacier in Bolivia, which has a mass balance ratio of 25 (Francou et al. 1995).

Typically, glaciers with smaller mass balance ratios possess larger ablation zones, a characteristic common to many glaciers at mid- to high-latitudes. Conversely, glaciers with large mass balance ratios may have comparatively smaller ablation areas, such as those found in low-latitude tropical and subtropical regions, where the equilibrium line is much closer to the glacier terminus than to the source. For debris-covered glaciers, due to the nonlinear variation of the ablation gradient, utilizing the mass balance ratio to determine the relationship between ablation and accumulation is not substantively meaningful.

Equilibrium Line Altitudes

Glacial accumulation areas receive snow and ice mass, yet during the summer months, some melting of snow and ice also occurs in these regions due to the elevation of the 0 °C isotherm. Similarly, although the primary mass loss of glaciers occurs within the ablation zone, snowfall within the ablation area during winter also contributes to the glacier's accumulation. Therefore, there may exist a boundary between the accumulation and ablation zones where the amount of accumulation equals the amount of ablation within a balance year; this boundary is termed the equilibrium line, and its altitude is known as the equilibrium line altitude (ELA). Variations in the ELA are closely associated with the climatic conditions of the glacial region (particularly precipitation and temperature) and directly reflect glacier-climate interactions. A sustained increase in average temperature could lead to intensified glacier melting, placing the glacier in a state of persistent negative mass balance and causing a continuous rise in the ELA. Conversely, an increase in snowfall could result in a positive mass balance with an increase in accumulation, thus lowering the ELA.

The ELA in a state of zero balance (where the amount of accumulation equals the amount of ablation) is known as the steady-state equilibrium line altitude, indicating that under specific climatic conditions, the glacier is in a state of equilibrium between mass input and output. Due to constant climate fluctuations, it is rare for glaciers to achieve exact parity between ablation and accumulation. Hence, the steady-state ELA serves as an idealized metric to gauge the deviation of the current mass balance state from equilibrium: the greater the difference between the ELA and the steady-state ELA, the further the glacier's mass balance state is from equilibrium, with positive values indicating accelerated shrinkage and negative values indicating growth. The steady-state ELA can be obtained through regression analysis using long-term series of net balance data and ELA measurements. It reflects the average state of the current climate, and when there is a fundamental change in the climate system, the steady-state ELA will also shift accordingly. When the steady-state ELA is near or exceeds the altitude of the glacier's source region, it suggests that the glacier has lost the basic climatic conditions for its development and will soon disappear or will do so due to a persistent negative mass balance.

3.3.2 Accumulation and Ablation of Glacier

Accumulation of Glacier

Snowfall in glacier regions is the primary source of glacial accumulation (Fig. 3.11). The snow that falls in glacier accumulation areas undergoes various processes, such as the deformation of snow crystals, densification of the snowpack, and ice formation, which transform it into glacial ice. The duration required for snow to transform into ice varies significantly depending on the mechanism of ice formation. For instance, isothermal metamorphism, which leads to ice formation, often occurs in the accumulation zones of polar ice caps or inland glaciers on the Tibetan Plateau, where temperatures are extremely low. Snow crystals gradually grow and thicken through rounding processes and under the force of gravity, causing the distance between grains to diminish, and thus gradually transform into glacial ice. These processes occur under consistently negative temperature conditions, making the formation of ice exceedingly slow. On alpine glaciers, this type of ice formation can take decades, while in polar glaciers, the process can extend for hundreds or even thousands of years. Conversely, when the snow surface temperature rises above zero degrees, causing melting, or when rainfall occurs, liquid water enters the snowpack and initiates melt-freeze metamorphism, which can transform the snowpack into glacial ice within minutes to hours.

The amount of snowfall in glacier regions exhibits considerable spatial variability due to local climatic characteristics. Generally, glaciers closer to the ocean, where there is an ample supply of moisture, receive larger annual snowfall, such as the maritime glaciers found in Alaska, southern Iceland, southern Argentina, western New Zealand, and the southeastern part of the Himalayas in China (Wu et al. 2020; Black and Kurtz 2023; Phillips et al. 2014). It is reported that snowfall on the Patagonian Ice Fields in Argentina can reach up to 8 m per year (Warren and Sugden 1993). Conversely, in inland areas, the long pathways of moisture migration and lower temperatures result in a sharp decrease in snowfall, as in the interior of Antarctica, where annual precipitation is less than 50 mm.

Blowing snow (a multiphase flow of dispersed snow particles carried near the ground by air currents, or snow transported by wind) is another important means of accumulation for continental glaciers (Fig. 3.11). Snow accumulating on the slopes and summits adjacent to the accumulation area can be transported to the glacier surface by wind action. In regions with relatively low annual precipitation, such as the western Tien Shan, Kunlun Mountains, Karakoram, and Qiangtang Plateau

Fig. 3.11 Glacier snowfall recharge, wind-blown snow and avalanche recharge

in China, blowing snow can expand the accumulation area of glaciers and plays a significant role in the mass balance of glaciers.

Glacier Ablation

Glacial ablation is the collective term for all processes that result in the loss of ice and snow from glaciers, signifying the expenditure of solid water from glaciers. This includes the formation of runoff from ice and snow melt, evaporation, sublimation, the disintegration of ice bodies, snow blown beyond the glacier margins, and avalanches. Sublimation of snow and the evaporation of meltwater are significant aspects of glacial ablation. Both sublimation and evaporation are manifestations of a glacier's absorption of latent heat and are related to near surface wind speed, air temperature, snow (water) temperature, and underlying surface characteristics. An increase in air temperature and wind speed both favor the absorption of latent heat. Due to the relatively low temperature of glacial meltwater (0–3 °C), the evaporation of meltwater is minor compared to that of conventional water bodies. It has been estimated that sublimation of snow in mountainous regions of the Northern Hemisphere can account for more than 20% of the annual snowfall (Zhang et al. 2008), thereby directly impacting the glacial mass balance. For glacier terrains such as the edges of polar ice caps, ice caps, and plateau glaciers elevated above the surrounding topography, the loss of glacier mass due to blown snow is exceedingly important. For example, in coastal regions of the Antarctic continent, cold inland air moving toward the coast creates strong katabatic winds that strip significant amounts of snow from the glaciers. It is estimated that mass losses due to blown snow in eastern Antarctica are approximately 6% of the annual accumulation of the glacier (Van Den Broeke and Bintanja 1995).

Ice avalanches are not only a common form of glacier accumulation but also a significant pathway of glacier ablation (Alean 1985). At the margins of the Arctic and Antarctic and in other coastal regions, glacier termini often extend into the ocean, where they disintegrate due to marine tidal erosion, forming various sizes of floating ice bodies, or icebergs. For the Antarctic ice sheet, for instance, because of the low air temperatures, surface melting is minimal, and the primary loss of ice mass occurs through the calving of ice shelves extending into the ocean and melting at the interface of the ice shelf with the sea. Glaciers that lose mass predominantly through calving are widespread in the ice-rich polar regions and their peripheries. Calving also impacts continental glaciers to some extent (Benn and Åström 2018). For example, in the Himalayas and the Karakoram ranges in China, sizable ice-dammed lakes or glacier-dammed lakes often form at the terminus and flanks of large valley glaciers. The erosive action of lake water on the ice and the exertion of stress can cause the glacier to calve, leading to substantial mass losses in a very short time. Additionally, surges caused by massive ice collapses and the increase in the volume of ice-dammed lakes can threaten the stability of moraine dams or ice dams and may even trigger glacial lake outburst floods.

For hanging glaciers, cirque glaciers, and the transitional types between them (cirque-hanging glaciers) that develop in mountainous regions, the glacier tongues often terminate on steep slopes. Under the influence of gravity and climate, portions of the terminal ice are prone to collapse and detach from the glacier. Ice avalanches can contribute to the accumulation of other glaciers developing in the valley below, or they can form deposits in areas with gentler slopes. If the rate of ice ablation is slower than the rate of replenishment from collapse, the ice volume continuously grows, thereby forming regenerative glaciers. In essence, the upper source of the glacier and the lower regenerative body constitute the two parts of a glacier. However, compared to ordinary glaciers, the connection between their accumulation zones and ablation zones transitions from continuous ice replenishment to intermittent mass supplementation.

Glacier Melting Mode

Bare Ice Ablation. For most glaciers, the relatively flat and exposed ice surfaces serve as the main ablation zones and are the primary sources of a glacier's meltwater yield (Di Mauro et al. 2020; Naegeli and Huss 2017). Factors influencing the intensity of bare ice ablation include air temperature, direct solar radiation, atmospheric longwave radiation, surface albedo, and the degree of topographic shading, among others (Hock 2005; Gao et al. 2017).

Snowpack Ablation. Within the ablation zone, meltwater from the snowpack primarily consists of two components: first, the melting of snowfall from the cold season upon the arrival of the melting period; second, the melting of snowfall during the summer melt season. In regions such as Europe, North America, South America, and New Zealand, influenced by warm and humid air masses from the ocean, winter snowfall in glacier areas is abundant (Shakoor and Ejaz 2019). When spring arrives, the snow stored in the glacier's ablation zone begins to melt first, forming the main part of the glacier's meltwater during the spring season. The meltwater from summer snowmelt also represents a significant proportion of the glacier's total runoff, particularly for glaciers that accumulate during the summer season. For instance, the Kokshal Glacier in the Tianshan range receives only 11% of its annual precipitation during winter, with the majority occurring in summer, predominantly as snow in the upper parts of the glacier. In this glacier's runoff, meltwater from the summer snow accounts for approximately 20% of the total runoff from the glacier-covered area.

Ice Cliff Ablation. Ice cliff ablation is a distinctive form of ablation. Steep ice slopes or ice ramps formed due to ice fracturing, collapse, or differential ablation are referred to as ice cliffs (Fig. 3.12). As ice cliffs are situated in the lower parts of the ablation zones, their exposed ice surfaces undergo exceptionally strong ablation. For example, the average annual ablation intensity of the ice cliffs on the Kokshal Glacier in the Tianshan mountain range is 2.2 times that of the bare ice zone. Meanwhile, on the Lirung Glacier on the southern slope of the Himalayas, the ice cliff area constitutes only 1.8% of the total glacier area, yet it contributes 69% of the glacier's meltwater (Sakai et al. 1998).

Englacial and Subglacial Ablation. Due to the vertical structural heterogeneity of glaciers, as well as the erosive impact of glacial movement, hydraulics, and thermodynamics, intricate drainage channels commonly develop within and at the base of glaciers. Surface meltwater runoff enters the glacier through crevasses and moulins (Fig. 3.13) and migrates downstream via englacial drainage pathways. The meltwater erodes the walls and base of these channels through dynamic scouring and thermal

Fig. 3.12 Ice cliffs developing in the surface moraine area of the Koqikal Glacier in the Tianshan Mountains

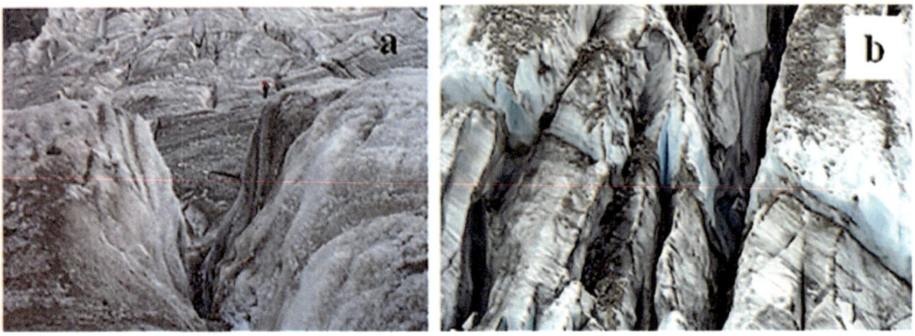

Fig. 3.13 Examples of crevasses in continental glaciers (**a**) and temperate glaciers (**b**)

erosion, leading to partial melting of the ice and its incorporation into the flow, consequently enlarging the englacial conduits. Similar processes occur at the base of the glacier when substantial volumes of meltwater are present. Additionally, heat from the underlying bedrock can promote basal melting of the glacier. Compared to the total volume of glacial meltwater runoff, the amount of englacial and subglacial meltwater constitutes only a minor fraction and is typically considered negligible in glacier meltwater runoff models.

Influencing Factors of Glacier Melting

Temperature. In glacial regions, characterized by capricious weather, total solar radiation is significantly influenced by cloud cover, resulting in a general decrease in total radiation with an increase in altitude. Regarding albedo, fresh snow exhibits the highest reflectivity, followed by granular snow, with bare ice having the lowest (Skiles et al. 2018). Consequently, the net shortwave radiation reaching the glacier surface typically diminishes with rising altitude. Downwelling longwave radiation is positively correlated with air temperature, while upwelling longwave radiation is primarily associated with surface temperature. Both air and ice surface temperatures tend to decrease with increasing altitude, thereby reducing net radiation in the glacier region. The net radiation reaching the ground is primarily partitioned into sensible heat, latent heat, and the heat of melting. Latent heat is closely related to turbulence, vapor pressure, and relative humidity, while sensible heat depends on various environmental conditions (Hock 2005).

Precipitation. Precipitation releases a portion of heat that contributes to glacial melting; the latent heat released by water vapor condensation also aids in glacial melt. However, the thermal flux generated by heat conduction within the glacier consumes some of the heat of melting, thereby slowing the rate of surface ablation. Intense rainfall can severely erode the snow and other loose material on the glacier surface, accelerating surface melt (Jouberton et al. 2022).

Light Absorbing Substances. Continental glacier surfaces and interiors are not composed of pure ice but contain light-absorbing substances such as black carbon, dust, and cryoconite. During periods of intense melting, the surface of continental glaciers melts, and numerous dark rocks form a polluted layer that absorbs more heat, thereby accelerating the glacial melting process (Kang et al. 2020). Furthermore, black carbon has a significant impact on glacier melting. Simulations from Global Climate Models (GCM) indicate that black carbon aerosols from East and South Asia contribute approximately 0.6 °C to the warming of the mid and lower stratosphere, comparable to the contribution from greenhouse gases (Meehl et al. 2008).

Glacier Surface Features. Glacier surface morphology and snow cover conditions influence the process of glacial melt. The presence of debris cover, ice lakes, and ice cliffs also significantly affect surface melting (Rounce et al. 2021). For exposed ice surfaces, over 90% of the heat contributes to surface melting. For instance, on the debris-covered surface of the Rongbuk Glacier near Mount Everest, only 33% of the total heat is used for surface melting. A 10 cm thick layer of debris can reduce melting by about 10%, while a 20 cm thick layer can generally reduce it by approximately 56% (Han et al. 2010a, 2010b, 2010c).

Glacier Types. affects glacier variations through its impact on mass balance, though these changes are not entirely uniform. The rate of mass balance variation depends on the size of the glacier; the larger the glacier, the longer the lag time in its response (Huss et al. 2010; Gao et al. 2021a, 2021b). Additionally, the thermal properties of glaciers are decisive factors. Temperate glaciers, which exhibit basal sliding, have a noticeably faster movement speed and a greater magnitude of advance and retreat compared to cold glaciers under similar mass balance changes (Han et al. 2023). These factors mean that within the same region, glaciers of different sizes, or glaciers of the same size in different regions, have varied responses to climate change (Huss and Hock 2018).

3.3.3 Runoff Generation and Flow Routing in Glacier Regions

1. **Runoff Generation**

The runoff from glaciers is derived from the algebraic sum of the total amount of melting or loss (A) and the equivalent amount of snow and ice removed by evaporation, sublimation, and wind and gravity.

$$RI = (AW + AS + AF + AP + AI + AD) - (EI + WF + FF) \tag{3.16}$$

where, RI is glacier melting runoff; AW is the amount of seasonal glacier ablation in the winter half year, AS is the amount of snow melt in winter; AF is the amount of additional ice zone ablation; AP is solid and liquid precipitation in summer; AI is the amount of glacial ice loss in the ablation area; including the ice surface, intra-ice and sub-ice loss; AD is the amount of snow that breaks away from the glacier in the form of collapse or wind-blown snow; EI evaporation of ice surface and sublimation of snow surface; WF is refreezing amount for winter snowmelt; FF is the amount of meltwater refreezing in the additional ice belt. In the formula, the items in the first bracket represent the total melt of the glacier, and the items in the second bracket represent the amount that did not form runoff after melting.

Winter Ablation on Oceanic Glaciers (*AW*): Ablation during the winter season predominantly occurs on oceanic glaciers, affecting both snow cover and ice. This process can generate runoff. In contrast, winter ablation is relatively rare on continental glaciers, where typically no meltwater runoff is observed.

Ablation of accumulated snow cover in winter (*AS*): The accumulation of snow on glacier surfaces throughout the winter begins to undergo ablation in the spring. The snowmelt in the ablation zone, save for a minor proportion that sublimates, contributes to runoff. Within the accumulation zone, only a fraction of the melting snow contributes to runoff in the superimposed ice zone, while the meltwater at elevations above the firn line does not run off but rather refreezes, forming ice lenses.

Ablation in the superimposed ice zone (*AF*): As aforementioned, a certain volume of snowmelt in the superimposed ice zone contributes to runoff, with the remaining quantity refreezing to form the superimposed ice that constitutes the glacial structure.

Summer Precipitation on Glaciers (*AP*): Glaciers experience both solid and liquid precipitation during the summer, both of which contribute to the meltwater runoff that drains over the glacier surfaces.

Ablation of glacier ice (*AI*): Melting of glacier ice predominantly takes place in the ablation zone, following the disappearance of the winter snow cover. This melting process affects the ice at the surface, within the glacier, and beneath the glacier. Surface ice melting is highly dependent on solar radiation, whereas the melting of internal and subglacial ice is largely driven by the heat from liquid water, subsequently producing runoff.

Avalanche Losses in the Ablation Zone (*AD*): Steep glacier tongues are prone to snow and ice avalanches, which, relying on gravitational forces rather than radiant energy, contribute to mass loss and runoff. Despite their significance, such avalanches are often excluded from meltwater calculations. Moreover, in elevated glacial areas, such as edges, ice caps, and plateau glaciers, wind-driven snow removal results in considerable mass loss. In Antarctica, for instance, inland cold air flows toward coastal regions, generating strong katabatic winds that remove significant amounts of snow from glacier surfaces. This process is estimated to account for approximately 6% of the continent's annual glacier accumulation (Van Den Broeke and Bintanja 1995).

Evaporation and Sublimation from the Ice Surface (*EI*): Sublimation of snow cover and evaporation of snowmelt constitute critical components of glacier ablation. Both processes represent the latent heat absorbed by glaciers and are influenced by variables such as wind speed, near-surface temperature, snow (water) temperature, and substrate characteristics. The latent heat required for water evaporation is 7.5 times greater than that for ice melting, and the latent heat for ice sublimation is 8.5 times greater than that for ice melting. While runoff from melted ice and snow predominates in the ablation of continental glaciers in China, the significant energy expended for evaporation and sublimation is of notable importance, especially for continental glaciers situated in arid regions. Increases in temperature and wind speed facilitate the absorption of latent heat. Given that glacier meltwater remains at low temperatures (0–3 °C), its evaporation rate is lower than that of conventional water bodies. Snow sublimation in the Canadian Rockies has been modeled to represent 20–32% of total snowfall, with 17–19% attributed to sublimation losses from wind-transported snow (MacDonald et al. 2010). In the mid-latitude mountains of the Northern Hemisphere, snow sublimation exceeds 20% of the annual snowfall (Zhang et al. 2008), thus playing a direct role in influencing glacier mass balance. Refrozen snowmelt in winter (*WF*)—Refreezing prevents meltwater to be released as runoff. Previous studies have shown that substantial portions of meltwater refreeze in alpine regions, thereby increasing snowpack persistence and causing a delay of meltwater runoff from the snowpack. The refreezing has been estimated to account for up to 100% of the total melt, with a large spatial and temporal variation, predominantly related to elevation and air temperature (Ayala et al. 2017).

Refrozen Meltwater in the Superimposed Ice Zone (*FF*): Meltwater refreezing is a key process controlling the behavior of meltwater, specifically whether it becomes glacier runoff or is retained within the glacier. On Xiao Dongkemadi Glacier, Tanggula Mountains, central Tibetan Plateau, a portion of the meltwater refreezes at the snow/ice interface as superimposed ice. The formation of superimposed ice is determined by both the meltwater supply and the temperature conditions of the glacier. The snow-layer thickness on the glacier ice body is less than 2 m, even in the higher accumulation zone. About 60% of the meltwater generated in the accumulation zone from May to September is trapped at the snow/ice interface by refreezing and is not discharged from the glacier. Approximately 26% of the accumulated snow on the glacier surface is replaced by refreezing at the snow/ice interface in the accumulation zone. These facts indicate that superimposed ice formation is highly significant for water retention in glaciers under low-precipitation conditions (Fujita et al. 1996).

2. **Flow Routing**

The flow routing process of glacial meltwater, from its genesis to its egress within the glacial system, is relatively complex. Meltwater can reach the terminus of a glacier through three pathways: surface runoff, englacial conduits, and subglacial channels. These pathways are interlinked, forming an intricate glacial drainage system. The drainage systems and confluence processes vary significantly across different types and sizes of glaciers. For small continental and polar continental-type glaciers, during the ablation season, meltwater and liquid precipitation flow within a network of supraglacial channels, gradually converging toward the glacier terminus. Conversely, for medium to large-sized glaciers, particularly maritime glaciers, the majority of the meltwater enters the glacier's interior via surface crevasses and englacial conduits, with a minor proportion flowing directly out of the glacier along its margins and terminus. The water that penetrates the glacier then merges into one or several substantial streams, emerging from beneath the glacier terminus (Luo et al. 2018b; Doyle et al. 2018). The englacial and subglacial water bodies generally originate from four sources: (1) Surface water entry: Including meltwater and rainfall that enters the base of the ice bed through crevasses, moulins, or vertical shafts. (2) Basal melting of the glacier: Due to frictional heat or geothermal sources. (3) Melting of the walls of subglacial drainage channels: Caused by the turbulent heat exchange of water flow. (4) Influx from other water bodies: Such as tributary rivers and lakes entering the glacier.

Factors influencing glacial runoff confluence, beyond the conventional determinant of topographic gradient, are primarily associated with glacier type, length, surface moraine and ice lake characteristics, the spatial extent and thickness of snow accumulation, and the distribution features of crevasses within the glacier. For small continental glaciers, which typically lack the development of crevasses and subglacial channels, confluence principally occurs on the glacier surface, with confluence time largely dependent on surface slope and length. If the glacier surface is covered with supraglacial debris, then surface runoff is influenced by the characteristics of this cover, adding complexity to the confluence process and potentially delaying runoff. For medium to large continental glaciers, like the Tuyuksu-type valley glaciers, or maritime glaciers with subglacial channels, the pathways of confluence are varied and complex. Figure 3.14 illustrates the spatial structure and interconnectivity of the primary hydrological channels on the surface, within, and beneath most medium to large continental glaciers. During non-ablation or weak ablation periods, the accumulation zone contributes minimally to confluence due to scant meltwater and the impermeable action of firn and ice crystals. In contrast, the ablation zone, due to relatively stronger melting processes and the unblocked hydrological system on the surface, within, and beneath the ice, exhibits more pronounced confluence. In this phase, there are primarily three confluence pathways: (1) Meltwater at the base of saturated firn layers flowing directly to the glacier terminus via surface or englacial channels, or reaching the terminus via subglacial channels; (2) Meltwater from bare ice surfaces converging into subglacial channels via englacial conduits or directly reaching the glacier terminus; (3) Runoff from bare ice surfaces directly converging to the glacier terminus. During intense ablation periods, with severe glacier melting, even the accumulation zone contributes to confluence due to increased melting of the firn layer, with meltwater entering the glacier's interior through englacial and subglacial channel systems. The entire ice mass experiences strong melting, and the drainage system becomes more developed, leading to more complex and variable confluence pathways. The englacial and subglacial confluence processes are an interplay of water, heat, and glacial motion, where the morphology, size, and length of the englacial channel system dynamically change under their influence.

(1) **Ice Surface Flow Routing**

Influenced by micro-topography and the structure of the ice surface, meltwater rapidly traverses the ice, flowing toward lower elevations. When ice surface flow routing is high and accompanied by substantial volumes of water, the ice surface, through the combined forces of dynamic flushing and thermal erosion by currents, evolves into a fluvial ice feature that can be likened to a river on the ice surface. The depth of these ice surface rivers can vary from a few centimeters to several meters, depending on the duration of their formation, the quantity of water they channel, and other contributing factors. The riverbanks are typically steep, while the riverbeds are flat and smooth. On ice sheets, ice caps, and within the ablation zones of extensive

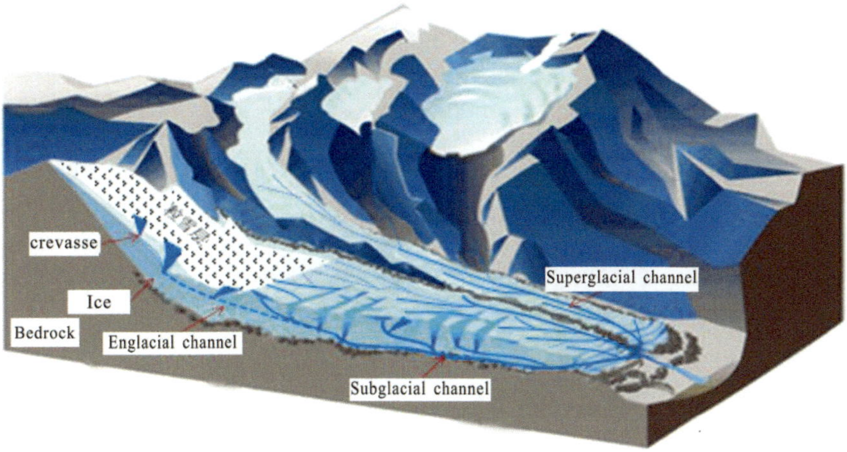

Fig. 3.14 Major drainage corridors and flow routing paths on the melted surface, inside, and beneath the ice

valley glaciers, these ice surface rivers may connect, forming a vast surface drainage network. This network, with its dendritic pattern, bears resemblance to the surface runoff networks found within drainage basins and exhibits distinctive characteristics: The drainage system is well-developed with immature main channels. Due to the intense surface ablation during summer months, the system boasts numerous and densely-packed ice surface rivers. However, the glacial movement and rapid surface melting inhibit these rivers from evolving into deep and broad main streams.

Locally, the ice surface rivers display a parallel distribution, often developing along the weaker sections of the ice, such as interfaces between ice of markedly different densities, the bases of semicircular arch structures, junctions between major and minor glaciers, and depressions on the surface. This parallel development, observable both transversely and longitudinally, is a defining characteristic of the distribution of these ice surface rivers. In contrast to natural valley drainage systems, the ice surface water system's density diminishes progressively from downstream to upstream. This inverse gradient is due to heightened temperatures and more extensive surface melting downstream, which, in turn, promotes a more elaborate drainage network. The positions of ice surface rivers are transient and subject to rapid change. The dynamism of glacier movement and the rapidity of surface melting lead to significant alterations in surface structure and micro-topography, thereby greatly influencing the development and routing of ice surface rivers (Figs. 3.15 and 3.16).

(2) **Englacial Flow Routing**

For most continental glaciers and small ice caps, the frequent mass exchange within the glacier limits the development of supraglacial hydrological systems. Only a minor portion of the meltwater reaches the glacier terminus via the glacier surface. The bulk of the meltwater enters the glacier interior or the subglacial environment through

Fig. 3.15 An ice surface river on Keqikaer Glacier in the Tianshan Mountains (Ding 2017)

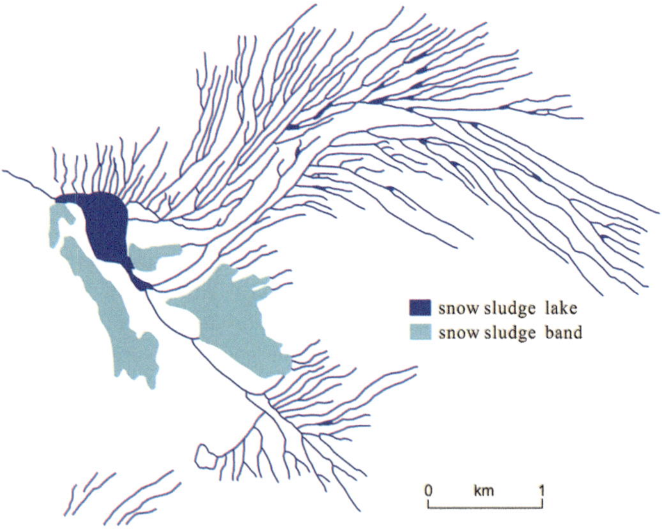

Fig. 3.16 An ice surface drainage system developed on an ice sheet in Greenland (Sugden and John 1976)

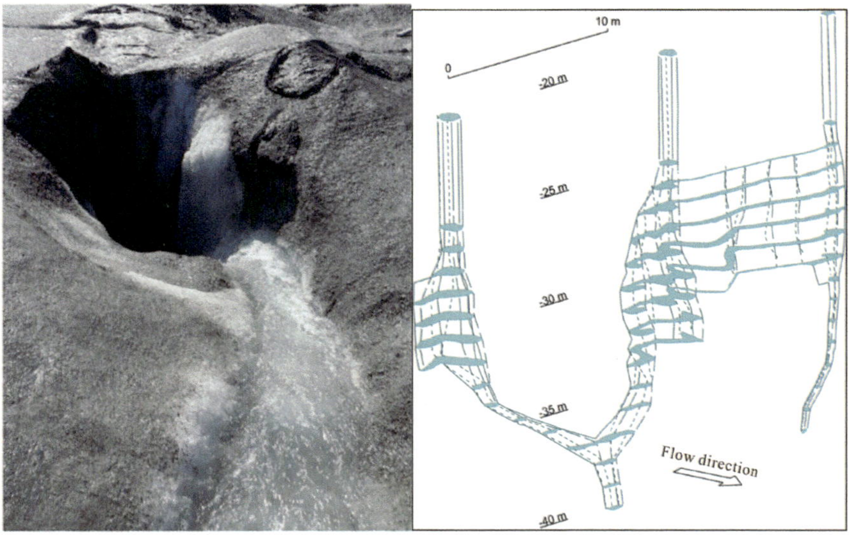

(a) (b)

Fig. 3.17 A glacial shaft (**a**) and the observed three-dimensional structure of an ice well in Storglaciären, Sweden (**b**) (Holmlund 1988)

crevasses or moulins and is transported to the outlet via the internal and basal drainage systems of the glacier. Crevasses tend to develop in the steeper upper-middle regions of the glacier where tensile stress induced by gravity exceeds the binding force of the ice, resulting in transverse fractures and the formation of crevasses. In the lower ablation zone of the glacier, crevasses can also form due to glacial motion, differential ablation, and the drainage of supraglacial lakes, although these crevasses are generally narrower than those found upstream. When crevasses occur on the cross-section of a supraglacial stream, the surface runoff is intercepted, leading to the formation of a moulin at the stream's entrance (Fig. 3.17a).

Moulins can persist for an extended period even after nearby crevasses close due to glacial movement. Moulins, similar in form to dolines found in karst topography, often develop in areas of the glacier where the structure is more fragile, such as at the junctions of crevasses and ice layers of different densities. By measuring the three-dimensional structure of moulins (Fig. 3.17b), it can be observed that moulins extend vertically downward from the surface and may present a step-like or spiral shape, gradually sloping toward the downstream direction of the glacier in the deeper sections. Lateral conduits may connect adjacent moulins. The water level within a moulin can change dramatically, influenced by the variability of water input and drainage capacity. When the input of meltwater increases sharply, the water may accumulate within the moulin because the expansion of internal channels cannot keep pace, causing the water level to rise. Conversely, when surface melting diminishes and meltwater input decreases, the water level may drop rapidly. If new crevasses

Fig. 3.18 A lateral drainage outlet inside Koxkar Glacier

form and create connections between two or more moulins, rapid changes in water levels may occur.

Apart from ice wells and connecting drainages, many horizontal drainages have developed inside glaciers (Fig. 3.18). With diameters ranging from several millimeters at the top of glaciers to several meters in the ablation zones, these drainages exhibit a dendritic or crisscross distribution. They may have originated from pipelines inside the firn in accumulation zones, preserved by glaciation, and later expanded and corroded by flowing water, or they may evolve from ice surface rivers, crevasses, and similar features.

(3) **Subglacial Flow Routing**

The development of subglacial drainage is influenced by meltwater, glacier basal temperature, and riverbed landforms. It can be broadly categorized into two types: the decentralized system (dendritic system) and the distributed system (radial system).

The distributed drainage system refers to a dendritic subglacial hydrological network where all tributaries eventually converge into several main channels that lead to the glacier's terminus. In such a system, conduits occupy only a small fraction of the glacial bed area. Due to the concentration of water flow, the distributed drainage system features high flow velocities and efficient water transport. The conduits comprising the distributed drainage system can be classified into several types: **R-channels** are characterized by their upper parts and sides being incised into the ice, while their bases are bedrock. Unlike surface streams, whose directions are controlled by slope, the extension of R-channels is primarily influenced by the subglacial hydraulic gradient. Hence, these channels may not necessarily develop downslope along the glacier but could also extend laterally or even uphill. **N-channels** cut into the glacier's bedrock from above, with their tops composed of glacial ice or basal till. N-channels form due to persistent erosion by meltwater under a significant hydraulic gradient, such as on V-shaped or rough glacier beds, which are conducive to the development of N-channels. **Tunnel valleys** form as N-channels continuously expand under favorable hydrological and geological conditions, creating broad U-shaped subglacial valleys. To meet the transport needs of glacial meltwater, N-channels tend to develop in the upper and middle parts of the glacial bed, while tunnel valleys are more likely to form in the lower parts and near the glacier terminus, where meltwater is more concentrated.

The distributed drainage system also refers to a planar hydrological network beneath the ice that occupies a large portion or the entire area of the glacial bed. This system includes several forms: **Water films** can form as thin layers between the glacier and the rock interface if the base of the glacier is at the pressure melting point and the bedrock is relatively impermeable. The thickness of water films is generally only a few millimeters. **Linked cavities** refer to a drainage system formed by the connection of cavities developed at the ice-rock interface (Fig. 3.19). The flow velocity in linked cavities is relatively slow due to the influence of narrow passages between cavities. If these passages narrow or close, water in the cavities may temporarily remain within the glacial system, acting as glacial storage. **Braided flow networks** arise when the glacial bed is composed of loose basal moraines, facilitating the formation of a network of N-channels that intersect and resemble braided streams found in wide river channels with low discharge. **Porewater flow**, also known as Darcian flow, is similar to porewater flow in soils. These distributed drainage systems may develop independently or in combination at the base of a glacier, depending on variables such as the supply of meltwater, glacier temperature, glacier structure, the composition of the glacial bed, and the topography of the base.

(4) **Glacial Water Storage Structure**

Glacial meltwater may become temporarily stored within the glacier as it migrates toward the glacier terminus due to suboptimal drainage conditions, leading to the accumulation of glacial meltwater. The water storage structures of a glacier include layers of snow and firn, ice crevasses, glacial lakes, englacial cavities, subglacial cavities, as well as the glacier's drainage network (refer to Fig. 3.14). The meltwater stored within the glacier is eventually discharged due to the influx of new meltwater and changes in the storage structures, with the timescale of this process ranging from several days to several years. This, in turn, significantly impacts the variability of meltwater runoff at the glacier terminus.

Englacial and Subglacial Water Storage: Ice crevasses and moulins are the primary structures through which glacial meltwater enters the englacial and subglacial environments. When drainage is impeded, some of the meltwater may be retained and stored within these crevasses or moulins. The storage capacity of moulins can be roughly estimated through measurements of their structure and fluctuations in water levels, but the volume of water that can be stored in these open glacial structures

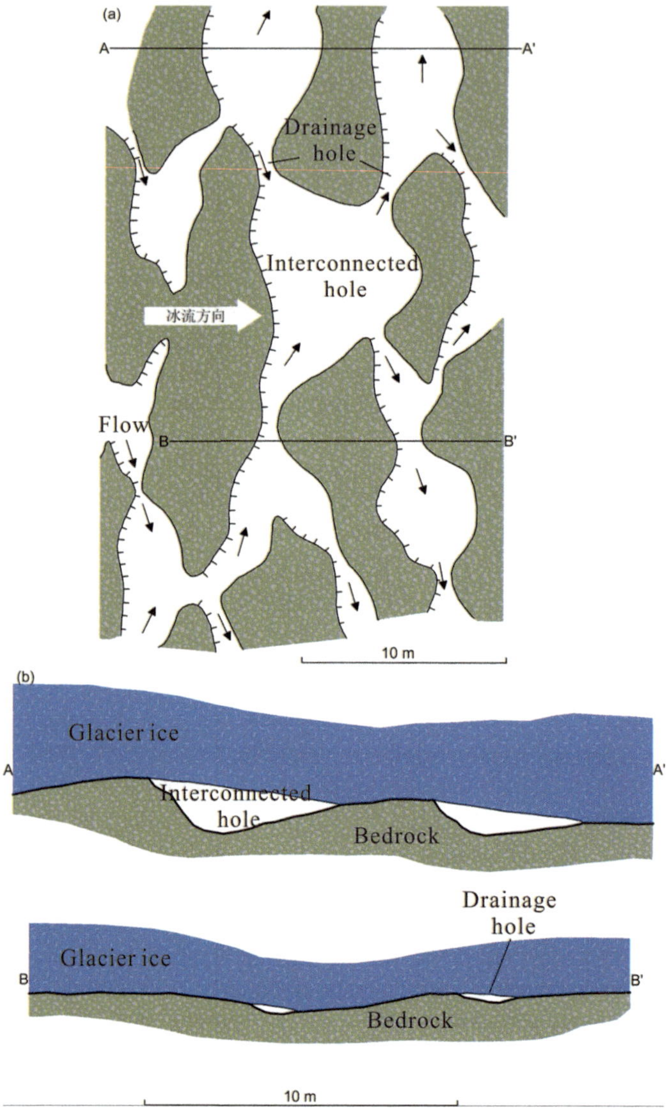

Fig. 3.19 Interconnected hole networks: **a** plan view; **b** sections A–A' and B–B' showing interconnected holes and drainage holes of a subglacial drainage system (Kamb 1987)

is quite limited. The lateral englacial drainage network also has a certain storage capacity, determined by the efficiency of the drainage conduits. However, due to the current lack of high-precision methods for detecting englacial drainage networks, the extent of their storage capacity remains uncertain. It is logical to assume that they only account for a small fraction of a glacier's total water storage.

Fig. 3.20 View of a fully stored ice surface lake on Keqikaer Glacier on July 5, 2007 (**a**) and the scene after its outburst on July 18, 2007 (**b**)

Glacial Lake Storage: Large water storage structures associated with modern glaciers or glacial activity can be collectively referred to as glacial lakes. Based on their location or structural characteristics, they are primarily divided into supraglacial lakes, ice-dammed lakes, and moraine-dammed lakes, among others (Fig. 3.20a). The development and evolution of supraglacial lakes vary significantly across different types of glaciers. On temperate glaciers, supraglacial lakes typically form during the initial stages of melting. As temperatures rise, the volume of water entering the lakes increases, leading to an expansion of the lake's capacity. Concurrently, as the temperature of the ice gradually reaches the pressure melting point, englacial drainage channels expand and eventually connect with the lake bottom or shore, leading to a drainage event known as a lake outburst.

In contrast, on other types of glaciers where the temperature remains below the pressure melting point, supraglacial lakes may persist for years until drainage channels are opened by dynamic or thermal processes, resulting in a lake outburst (Fig. 3.20b). In the ablation zones of glaciers with extensive supraglacial debris cover, the differential melting of ice often leads to the formation of numerous supraglacial lakes, with one side or the surrounding shores commonly characterized by steep or even overhanging ice cliffs. As these ice cliffs melt, the meltwater carries substantial amounts of rock debris and solutes into the lakes, making the water turbid and the bottom sediments thick. Some supraglacial lakes, due to the disappearance of ice cliffs and the accumulation of thick layers of bottom sediment, can become relatively independent storage structures, isolated from other sources of meltwater over extended periods. Besides supraglacial lakes, moraine-dammed and ice-dammed lakes formed at the terminus of glaciers can also intercept large amounts of glacial meltwater, thereby influencing glacial hydrological processes.

3.4 Impacts of Frozen Soil on Runoff Generation and Flow Routing

From a thermodynamic perspective, permafrost is the result of heat and mass exchange within the lithosphere-soil-atmosphere system. Consequently, geographical factors such as climate, geological structures, and topography are the primary influences on the formation of natural frozen soil (Ran et al. 2021; Zou et al. 2017). The presence of frozen soil alters the patterns of surface runoff and hydrological networks in cold regions, and the conditions for groundwater recharge, runoff, and discharge undergo fundamental changes due to frozen soil, resulting in unique hydrological and geological environments in these regions (Walvoord and Kurylyk 2016; Gao et al. 2021a, 2021b).

In cold regions and seasons, snowfall is the primary form of precipitation. During winter, snow accumulates on land or glaciers, while in spring and summer, rainwater and meltwater from snow and glaciers saturate the previously unfrozen upper layers of soil, with excess meltwater flowing into rivers, wetlands, and lakes as surface water. The stored surface water participates in the hydrological cycle through evaporation, infiltration, or lateral flow. Below the surface, groundwater reservoirs form during both freezing and thawing periods. During the freezing period, most of the water in the soil exists as subsurface ice, which can form seasonally or persist for decades or even thousands of years. In the thawing season, the melting of subsurface ice replenishes groundwater reservoirs, forming subsurface runoff that eventually flows into rivers, wetlands, lakes, and oceans.

3.4.1 *Seasonal Frozen Soil and Permafrost*

Frozen soil refers to soil or rock that remains at or below the freezing point of water (0 °C) and contains ice. In the field of permafrost studies, frozen soil is categorized by its duration: transient frozen soil (lasting from a few hours to half a month), seasonal frozen soil (lasting from half a month to within 2 years), and permafrost (lasting from two years to tens of thousands of years or more). Seasonal frozen soil forms when the mean annual ground temperature is above 0 °C, and the thawed ground loses heat and freezes during the cold season. Typically, as the temperature transitions from positive to negative values and progressively decreases during the cold season, seasonal frozen soil undergoes unidirectional freezing from the surface downward and thaws bidirectionally from the surface and deeper soil layers (Chou and Wang 2021). The thickness of seasonal frozen soil is determined by the maximum depth of seasonal freezing.

Within permafrost, the layer that freezes during the winter and thaws during the summer is known as the active layer. The thickness of the active layer is usually defined by the maximum depth of thawing that occurs over multiple years. The soil or rock that remains frozen year-round beneath the active layer is referred to as

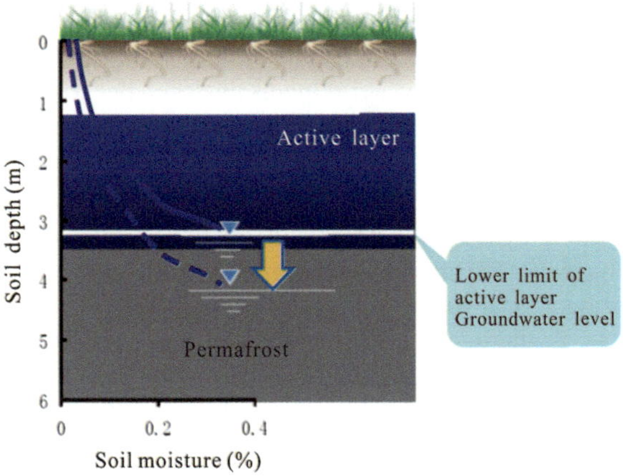

Fig. 3.21 Permafrost and active layer

permafrost (Fig. 3.21). The top boundary where perennial permafrost begins beneath the ground surface is commonly designated as the upper limit of permafrost; the position at the bottom of the permafrost layer where the ground temperature reaches 0 °C is known as the lower limit of permafrost; and the vertical distance between the upper and lower limits of permafrost is the thickness of the permafrost.

3.4.2 Thermal and Water Migration During the Freeze–Thaw Cycle

Frozen soil is a complex and heterogeneous system composed of solid, liquid, and gaseous phases. The thermal capacity and thermal conductivity of frozen soil are key thermodynamic parameters that influence its hydrothermal transfer characteristics. These parameters are primarily determined by the content, composition, structure, density, and distribution of materials such as minerals, liquid water, and ice within the frozen soil. They are also affected by convective processes of the liquid and gaseous phases in the soil. The water content, ice content, hydraulic conductivity, and soil water potential are critical hydrological parameters affecting the hydrothermal transfer characteristics of frozen soil, which are mainly related to the surface properties of frozen soil particles and the pore structure (Yang and Kane 2021).

During the soil freezing process, a phase change occurs from water to ice, significantly altering the soil's hydrological and thermodynamic parameters. Specifically, the thermal conductivity of ice is higher than that of water, and its specific heat is lower, which enhances heat conduction in the frozen soil layer. Meanwhile, the formation of ice reduces the liquid water content and porosity of the soil, leading to a decrease in capillary action and infiltration rate, ultimately reducing the hydraulic conductivity of the frozen soil. This difference in hydrothermal characteristics between frozen and thawed soil is primarily due to these changes. When soil freezes, some water remains unfrozen due to capillary action and the adsorption of soil particles; this water, known as unfrozen water, coexists with solid ice. The content, composition, and properties of unfrozen water in frozen soil are not static but are in dynamic equilibrium, varying with external conditions such as temperature.

Soil moisture can be categorized into several types: vapor water, strongly interactive water (adsorbed water), weakly interactive water (film water), capillary water, gravitational water, solid water (ice), crystalline water, and chemically bound water. Due to the interaction between soil water and the surrounding media in the soil–water system, the free energy of soil water is reduced, resulting in what is known as soil water potential. This potential represents the work required to transport a unit mass of water to different locations within the soil under isothermal conditions. It is generally accepted that the migration of liquid water during the freeze–thaw process in frozen soil is primarily driven by gradients in matrix potential, pressure potential, and gravitational potential within the soil water potential.

In saturated soils, the soil water potential is zero and in equilibrium. In unsaturated soils, the smaller the soil water content, the greater the absolute value of the soil water potential. During the freeze–thaw process in frozen soil, temperature and water content within the soil vary dynamically and non-uniformly due to external influences, disrupting the equilibrium state of soil water and creating gradients in soil water potential that drive moisture migration. During this migration, temperature plays a dominant role in regulating the content of unfrozen water and soil water potential. The freezing of water in soil occurs over a specific temperature range. As the temperature decreases, the water content in the soil gradually reduces, and the absolute value of the soil water potential increases. Initially, free water, including capillary and gravitational water, freezes, followed by the freezing of weakly bound water (film water), while strongly bound water (adsorbed water) freezes only at sufficiently low temperatures (Wu et al. 2023). As the freezing process progresses, the type of water in the soil transitions from gravitational and capillary water to film water, and the soil water potential correspondingly shifts from being dominated by gravitational and pressure potential to matrix potential. Ultimately, when the soil temperature drops below freezing, matrix potential becomes the primary driving force for the migration of unfrozen water.

In nature, the seasonal thawing layer in permafrost regions and the seasonal freezing layer in seasonal frost regions are both part of the near-surface soil layers. These layers exchange heat and moisture with the atmosphere through their annual freeze–thaw cycles.

3.4.3 *Freeze and Thaw Process of Frozen Soil*

The freezing and melting processes in the active layer of permafrost can be divided into four stages: the summer melting process, autumn freezing process, winter cooling process, and spring warming process. During the freeze–thaw cycle, the interface where soil water transitions between solid and liquid phases is referred to as the freezing front and the melting front, respectively. Based on ground temperature and moisture observations at the Barrow, Alaska station in 2006 (http://gtnpdatabase.org/activelayers/view/235), an analysis of moisture migration characteristics during each stage was conducted (Fig. 3.22).

Summer Melting Process: As the surface temperature rises during the summer, the melt front gradually moves downward, while gravitational water in the soil layer penetrates and migrates toward the melt front under the influence of gravity. As surface water continues to evaporate and the surface becomes dry, capillary water in the soil migrates to the surface. Driven by the temperature gradient, film water in the active layer migrates downward. In terms of heat transfer, both heat conduction and

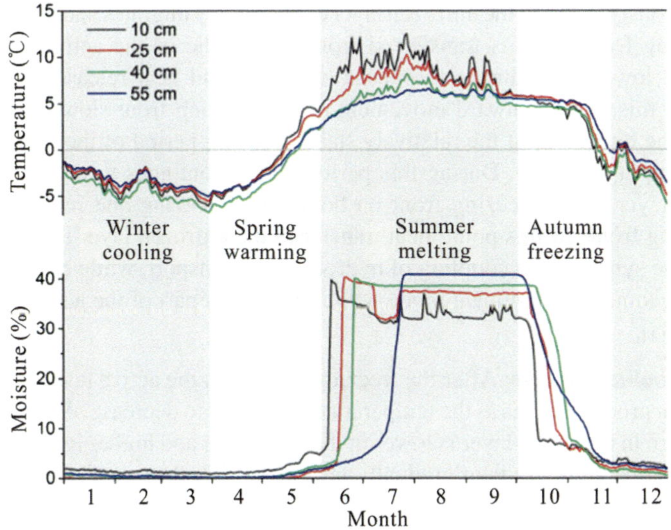

Fig. 3.22 Hydrothermal process in the active layer of permafrost (Ding 2020)

convective heat transfer are more active along the melt front. Below the melt front, conductive heat transfer is predominant.

Autumn Freezing Process: The autumn freezing process can be divided into two stages: the unidirectional freezing phase from the bottom of the active layer upwards, and the zero-curtain phase, where freezing occurs simultaneously from both the bottom and the surface. Unidirectional freezing begins when freezing initiates at the bottom of the active layer and continues upward until stable freezing forms at the surface. The "zero-curtain" phase starts with the formation of stable freezing at the surface and ends when the entire active layer is completely frozen.

During the unidirectional freezing stage, as the freezing front moves upward, water in the bottom of the active layer migrates or freezes from the unfrozen layer to the freezing front, driven by the temperature gradient and film water migration mechanisms. The overall trend of water migration is downward, while heat transfers from the unfrozen layer to the frozen layer. In the unfrozen layer, there is also a small amount of conductive heat transfer driven by temperature gradients and convective heat transfer driven by water vapor. As the freezing process progresses, convective heat transfer gradually increases and becomes the dominant mode of heat transfer.

Zero-Curtain Layer Stage: During the "zero-curtain" stage, bidirectional freezing occurs in the active layer, and the temperature profile in the soil exhibits higher temperatures in the middle and lower temperatures at both ends. Water migration primarily follows the thin film water migration mechanism. The freezing process during this stage can be divided into two periods: the rapid freezing period and the relatively stable freezing period. As a stable frozen layer forms on the surface, the freezing front in the upper part of the unfrozen layer quickly moves downward. Simultaneously, water in the unfrozen layer continuously migrates and freezes toward the freezing front. Heat is transferred from the middle of the active layer to the upper and lower sides during the phase transition and heat release of the water. Following this, the downward movement of the freezing front slows significantly, marking the beginning of the relatively stable freezing period of the "zero-curtain" stage (Song et al. 2020). During this period, water continues to migrate from the unfrozen layer to the freezing front on both sides, freezing and releasing heat at the freezing front. At this point, heat transfer in the unfrozen layer is fully realized through the synchronous coupling of hydrothermal transport, while conductive heat transfer becomes the dominant mode within the frozen part of the active layer (Luo et al. 2018a).

Winter Cooling Process: After the freezing process in the active layer is complete, the cooling process begins as the temperature continues to decrease. At this stage, the temperature in the active layer is lower in the upper part and higher in the lower part, with the temperature gradient gradually increasing. Conductive heat transfer is the dominant mode of heat transfer during this phase, accompanied by weak migration of unfrozen water.

Spring Warming Process: As spring temperatures rise, the active layer enters the spring warming process. During this time, evaporation of water from the soil's surface layer increases, leading to a decrease in water content and upward water migration from the interior of the active layer to the surface. However, due to the low temperatures and limited migration, heat transfer is still primarily dominated by conduction. In the late warming phase, a diurnal freeze–thaw cycle occurs near the surface, with the soil surface melting and water evaporating during the day, and refreezing at night, creating a freezing front. At this time, water in the active layer also tends to migrate toward the surface freezing front. If snow is present on the surface, the diurnal freeze–thaw process near the surface may be inhibited. Additionally, the water content of the soil surface will increase significantly due to the recharge from snowmelt, preventing further water migration from the active layer to the surface. The spring warming phase concludes when the surface no longer undergoes a daily freeze–thaw cycle, and the soil completely transitions to a thawed state.

After completing the above four processes, the active layer undergoes a complete freeze–thaw cycle. During the autumn freezing process and summer melting process, significant downward water migration occurs in the active layer. However, the migration of water during the winter cooling and spring warming processes is minimal. The amount of water that migrates to the freezing front during soil freeze–thaw cycles is related to the freezing rate: the slower the soil freezes, the greater the moisture increase at the freezing front. Near the base of the active layer, the freezing process is relatively slow due to the small range of temperature fluctuations and the gradual freezing rate. Consequently, after a freeze–thaw cycle, water in the active layer tends to accumulate near the bottom of the active layer, specifically near the upper limit of permafrost, leading to the gradual formation of an ice-rich zone in this area. This is the physical mechanism by which segregated ice forms near the upper permafrost through repeated freeze–thaw processes, and the ground ice formed through this mechanism is also known as pore ice (Cheng 1983). In Arctic Alaska, for example, the water content near the upper limit of permafrost increased by about 5% from 1963 to 1993 (Jorgenson et al. 2006).

The annual freeze–thaw process in seasonal frozen soils follows a distinct seasonal pattern and is generally divided into freeze–thaw and non-freeze–thaw periods, depending on whether soil freeze–thaw occurs. The freeze–thaw period can be further divided into four stages: unstable freezing, stable freezing, unstable thawing, and stable thawing (Xu et al. 2022). The period between the end of the stable thawing stage and the beginning of the unstable freezing stage is referred to as the freeze-free period. The timing of each stage can vary across different regions. During unstable thawing periods, temperatures fluctuate around 0 °C, and the thawing process does not consistently occur, with intra-day cycles of freezing and thawing sometimes taking place when temperatures drop below 0 °C. Similarly, during unstable freezing periods, the freezing process is not continuous, and thawing may occur when temperatures rise above 0 °C. In seasonal frozen soil, water migration occurs during both freezing and thawing. During the winter freezing period, the soil layer freezes from the surface downward, causing water to migrate and accumulate toward the freezing front, thereby increasing the total water content (including both ice and unfrozen water) within the freezing depth (Qin et al. 2021). When temperatures rise in summer, the soil layer thaws bidirectionally from the surface and the bottom of the freezing depth, causing water to migrate toward the interior of the frozen soil. The liquid water within the freezing depth becomes saturated or nearly saturated, while the shallow surface layer remains unsaturated due to evaporation.

3.4.4 Runoff Generation and Flow Routing in Frozen Soil

Runoff Generation

Due to the freeze–thaw action in frozen soil regions, runoff generation is intricately linked to the coupling effects of the thermal field and gravity field. The phase transitions of soil water between its liquid and solid states are central to this process, influencing how and when runoff occurs. Several key factors should be considered when evaluating runoff generation in these regions:

$$R = P - E + R_s + W(t) \tag{3.17}$$

where R is the runoff, P is the rainfall, E is the evaporation, Rs is the snow melt runoff, $W(t)$ represents the threshold temperature function of soil water capacity change during spring melting. It accounts for the influence of temperature on the soil's ability to retain or release water as it transitions from frozen to thawed states.

During the freeze–thaw cycle, complex hydrothermal coupling processes occur within the active layer and seasonally frozen soil, significantly impacting soil water runoff. The spatiotemporal variability of soil water phase changes in frozen soil regions adds complexity to the variations in thermal and hydrological parameters. During freezing, the phase change of water within the soil and rock alters the soil's thermal conductivity and heat capacity, while simultaneously reducing the effective porosity and field capacity of the soil and rock. This leads to changes in the soil water potential gradient and hydraulic conductivity, ultimately affecting the flux, direction, and velocity of unfrozen water. As a result, traditional theories and methods of pore water movement, which are primarily governed by gravity potential, are not applicable to soil water transport in frozen soil regions. The seasonal phase change of soil water triggers a transition from gravity-driven to matrix potential-dominated soil runoff processes, leading to temperature-influenced soil water characteristic curves and creating distinctive runoff and infiltration processes in frozen soil regions (Wang et al. 2017). The spatial variability in freeze–thaw cycles introduces heterogeneity in the unfrozen soil layer, resulting in temperature-influenced slope runoff patterns. Altitude is a key factor affecting temperature distribution, contributing to spatial variability in the unfrozen layers within the soil of mountainous regions, and shaping temperature-influenced slope runoff patterns (Gao et al. 2022). Consequently, in frozen soil regions, slope runoff, saturation overland flow, and Hortonian overland

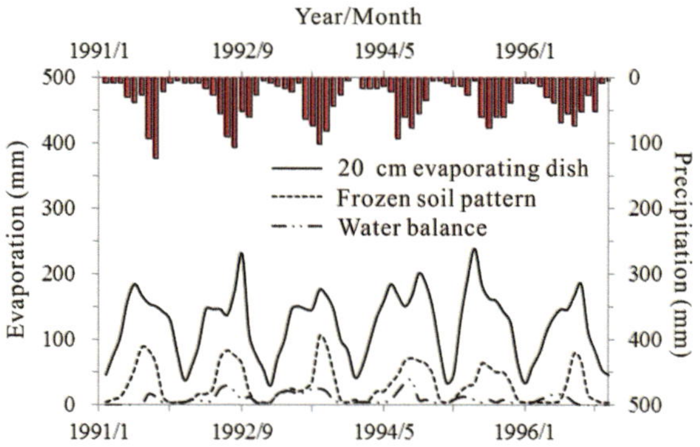

Fig. 3.23 Illustrates a comparison between simulated evaporation and observed evaporation at Maduo Station in Qinghai Province, considering scenarios both with and without the presence of frozen soils

flow coexist, forming a variable source area runoff pattern controlled by temperature and precipitation (Kidron 2021; Hu et al. 2020).

Both permafrost and seasonal frost can act as relatively impermeable layers, causing most of the snowmelt and rainfall to become direct runoff, thereby impeding the downward infiltration of water and the replenishment of deep groundwater (Mohammed et al. 2018). The hydrological cycle between surface water and groundwater is significantly influenced by the characteristics and distribution of permafrost. The thickness of the melting layer in permafrost and the variability of the frozen layer in seasonal frost are major factors affecting changes in groundwater levels (Brutsaert and Hiyama 2012). As permafrost thaws, the water content at its base increases, influencing the surface soil water content and water migration processes. The latent heat consumed by permafrost thawing reduces the amount of energy available for evaporation, thereby suppressing summertime evaporation (Fig. 3.23). In regions with seasonal frozen soil, water within the accumulated frost layer during the freezing period is often absorbed by the soil or converted into runoff during the thawing period, providing limited replenishment to the water below the frost layer. Additionally, the presence of the frost layer nearly eliminates the soil's evaporative capacity during the freezing period.

Compared to seasonal frozen regions, ground ice in permafrost areas acts as a unique "source" with a long-term regulatory effect on runoff (Ye et al. 2009). As the climate warms, this ground ice slowly melts and re-enters the hydrological cycle. Additionally, unique cryogenic geomorphological features in permafrost regions, such as frost mounds, ice wedges, and stone rings, influence local runoff processes. In permafrost areas, the interaction between surface water and groundwater is weaker, as water within the active layer typically infiltrates only to the bottom of this layer, making it difficult to replenish water below the permafrost. In contrast, in seasonal frost regions, the interaction between surface water and groundwater is primarily affected during the freezing period (Gao et al. 2022). Moreover, vegetation in permafrost areas tends to have shallower root systems and a weaker water retention capacity, which, along with differences in ecohydrological processes, also impacts runoff generation (Blume-Werry et al. 2019).

Flow Routing

The traditional single linear catchment theory is not suitable for flow routing processes in cold regions; instead, nonlinear theories and methods that account for the freeze–thaw cycle should be applied (Heinze 2021). The most critical factor in the cryosphere is frozen soil, as frozen layers block or inhibit the infiltration of soil water, thereby increasing surface runoff. In permafrost watersheds, the low permeability of the frozen soil layer makes it difficult for most of the melting snow and rainfall to infiltrate into the deep soil, leading to increased runoff and a shorter confluence time. From a spatial perspective, regions with high permafrost coverage typically exhibit hydrological characteristics such as a high runoff generation rate, a high direct runoff coefficient, a short response time of runoff to precipitation, and a short recession time (Yang and Kane 2021).

In permafrost regions, the peak of runoff typically occurs at the transition from spring to summer, when both precipitation and snowmelt contribute substantially to runoff. During this period, the impermeable action of the permafrost layer is pronounced, resulting in a low infiltration rate. As the season progresses into summer, the active layer gradually thaws, causing the permafrost table to descend and the barrier effect of the permafrost to diminish, leading to a decline in surface runoff from its peak. Once the active layer is fully thawed, infiltration rates increase, the basin's water storage capacity is enhanced, and evaporation intensifies. At this stage, the active layer can help attenuate flood peaks through runoff regulation. During winter, the permafrost layer obstructs the replenishment of runoff by groundwater, resulting in low winter runoff; in regions with 100% permafrost coverage, winter runoff may even approach zero. This pattern is vividly illustrated by the intra-annual distribution of runoff in sub-basins with varying permafrost coverage within the Lena River basin in Siberia (Fig. 3.24). At Site K, which has the highest permafrost coverage, peak runoff is rapidly reached during the spring flood period in May, and summer runoff remains high with significant seasonal variations. As permafrost coverage decreases, these characteristics gradually diminish, and at sites A, B, C, and D, where permafrost coverage is below 30%, the seasonal variation in runoff is already quite small.

Seasonal frozen soil also acts as an impermeable layer, with its hydrological effects diminishing as the seasonal frost melts, thereby influencing runoff concentration and baseflow recession. Seasonal frost directly impacts the spatiotemporal distribution of soil moisture content during freeze–thaw cycles. Water infiltrating during the unstable freezing period, water frozen at the front during the stable freezing period due to moisture migration from deep soil water and groundwater, water infiltrating from snowmelt and precipitation during the melting period, and increased infiltration from precipitation during the melting period—all contribute to changes in the water quantity within the permafrost layer. These water storage processes, coupled with the weak permeability of the frozen ground, result in higher runoff coefficients in seasonal frost regions during freeze–thaw periods compared to melting periods. The water storage regulation effect of seasonal frost delays the timing of groundwater replenishment. The water frozen in river channels and soil during the freezing period

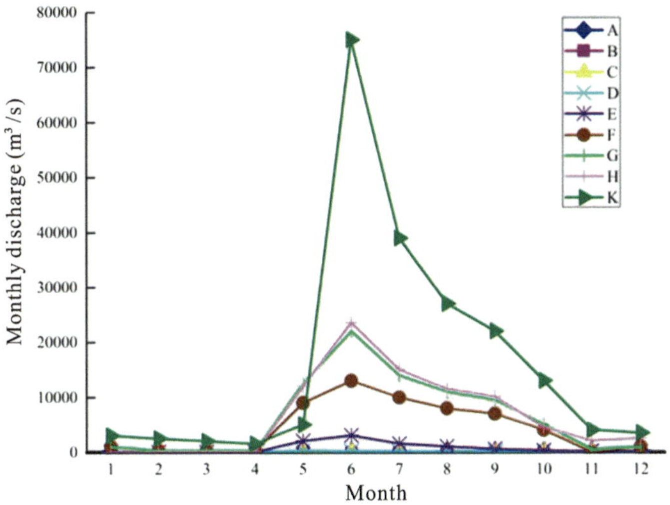

Fig. 3.24 Presents the average monthly discharge data for various observation stations within the Lena River Basin (Ye et al. 2009)

is released during the thawing period to replenish groundwater. Simultaneously, the water above the frozen layer can replenish groundwater in the form of gravity water after the thawing of seasonal frozen soil (Fig. 3.25).

3.5 Formation, Melting, and Migration of River-Ice, Lake-Ice, and Sea-Ice

River ice, lake ice, and sea ice are integral components of the cryosphere, with their processes not only reflecting climatic fluctuations but also exerting profound influences on the global hydrologic cycle (Yang et al. 2020a, 2020b). Unlike river and lake ice, which form on open inland water surfaces, sea ice develops over significantly larger areas. Its high albedo and the physical barrier it creates between the ocean and atmosphere have substantial impacts on radiative equilibrium and the exchange of heat, moisture, and gases. The cyclical freeze–thaw of large sea ice volumes modulates oceanic temperature and salinity, thereby influencing thermohaline circulation, which in turn affects atmospheric circulation and global climate patterns. Although river and lake ice are part of the terrestrial cryosphere and sea ice belongs to the marine cryosphere, they share many similarities in their hydrologic processes—such as ice formation, maturation, and ablation—due to their common transition from water to ice.

The development and variability of river, lake, and sea ice also have significant implications for the design, operation, and maintenance of hydraulic engineering infrastructure, such as slope revetments, dam structures, bridge piers, and riverside

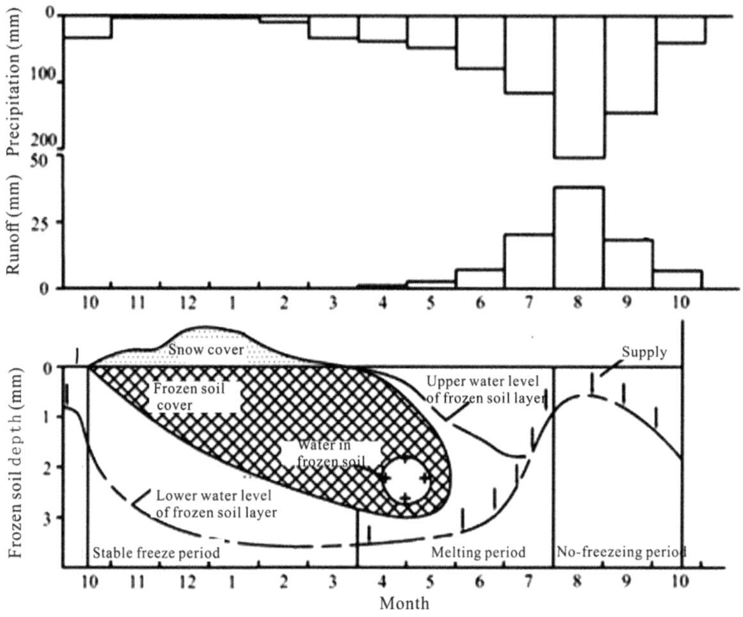

Fig. 3.25 Illustrates the dynamic changes in the hydrological characteristics of seasonal frozen soil throughout the freeze–thaw cycle (Ding 2020)

buildings. These ice-related processes can pose hazards that threaten the safety of human life and property. Understanding the unique characteristics of river, lake, and sea ice, and comprehending the principles governing their ice conditions, are essential for implementing effective disaster prevention and mitigation strategies.

3.5.1 Formation, Melting, and Migration of River-Ice and Lake-Ice

River and lake ice are solid formations resulting from the freezing of the surfaces of rivers, lakes, or reservoirs during cold seasons. River ice (Fig. 3.26a, b) typically forms under flowing water conditions, with its cyclical formation and decay involving complex thermodynamic and dynamic processes influenced by climatic and hydrological factors. In contrast, lake ice forms in relatively stable water conditions, and under similar temperature conditions, the duration of ice formation for lake ice is generally shorter than that of river ice. The life cycle of lake ice (Figs. 3.27 and 3.28) is primarily governed by thermal conditions. Additionally, wind forces and their shearing effects on the water surface generate surface currents and turbulent mixing, reducing stratification within the water body and significantly impacting regional energy and hydrological cycles. Furthermore, the winter freezing of plateau lakes, in addition to being influenced by climate, is also related to the mineralization level of the lake water.

River ice and lake ice develop within terrestrial water systems and are a common hydrological phenomenon in mid to high latitudes of the Northern Hemisphere and high-altitude regions, such as Northern Europe, the northern parts of the Eurasian continent, Northern North America, and the Tibetan Plateau, where river and lake surfaces freeze. The formation, ice conditions, and developmental processes of river and lake ice share many similarities, typically involving freezing during the autumn and winter seasons and thawing in the spring and summer, which results in distinct seasonal characteristics. The duration of river and lake ice presence is closely linked to the average monthly temperatures throughout the year but is also influenced by

Fig. 3.26 Hukou Icefall (**a**) in Jixian, Shanxi Province and River-ice (**b**) in the upper reaches of Shule River in Qilian Mountains, Qinghai Province

Fig. 3.27 Lake-ice in Lake Nam, depicting the seasonal ice formation on one of the highest and largest lakes in the Tibetan Plateau region

Fig. 3.28 Lake-Ice in Lake Superior, Minnesota, USA, showing the expansive ice cover typical of the winter months in this region

meteorological conditions such as precipitation, wind, and solar radiation (Zhang et al. 2024).

1. **Formation and Melting of River Ice and Lake Ice**

River and lake ice develop within terrestrial water systems and represent a hydrological phenomenon that is widespread in the mid to high latitudes of the Northern Hemisphere and in high-altitude regions such as Northern Europe, the northern parts of the Eurasian continent, Northern North America, and the Tibetan Plateau, covering the surfaces of rivers and lakes. The characteristics of river and lake ice in these freshwater systems exhibit a high degree of similarity in terms of formation scale, ice properties, and developmental processes. Typically, the ice forms during the autumn and winter seasons and melts in the following spring and summer, displaying pronounced seasonal characteristics. The persistence of river and lake ice is closely linked to mean monthly temperatures throughout the year, but it is also influenced by meteorological conditions such as precipitation, wind, and solar irradiance.

The Freezing and Melting of River Ice

When the water temperature drops to 0 °C—a point at which supercooling typically does not occur in rivers—minute ice needles begin to form on the river's surface. In areas near the banks, where the current is slower and the water is less turbulent, these ice crystals rise to the surface and coalesce into a continuous thin layer of ice, known as "shore ice." As the water body continues to lose heat, this layer grows thicker. As the ambient temperature decreases further, the density of frazil ice in the channel gradually increases. Under suitable hydrodynamic conditions, this can lead to the formation of a continuous ice cover, ice jams, or ice dams. With continued drops in temperature, freezing at the water–ice interface thickens the river ice. Additionally, the freezing of snowmelt on the ice surface can also contribute to the thickening of the ice layer.

When temperatures rise above 0 °C, river ice begins to melt. The thermal capacity of riverbanks, composed of gravel or rocks, is lower than that of the water body, causing the shores to warm up more quickly. As a result, shore ice melts and detaches from the banks first. In the deeper parts of the river, where the ice is thinner or where there are unfrozen open waters, the ice's resistance to the shearing forces of the river's hydrodynamics increases as temperatures continue to rise. Under the influence of water flow and wind, the ice mass fractures and breaks up, forming drift ice in the river. This drift ice gradually melts due to radiation on the ice surface and the thermal influence of water flowing beneath and along its sides.

The freeze–thaw cycle of river ice can be divided into three stages: the freezing period, the consolidation period, and the thawing period. The freezing period refers to the time from the onset of ice formation to when the river ice reaches its maximum extent for the year, typically spanning from the beginning of freezing to the lowest recorded temperatures. The onset of freezing is not characterized by a complete freeze-over of the river or lake but is instead determined by the critical formation of an initial ice configuration. The freezing process itself unfolds in three distinct stages: In early winter, the channels begin to freeze; in mid-winter, the drift ice continues to grow, and infiltrated water from the soil freezes, leading to the development of a flat and thick ice surface—a process aided by the infiltration of groundwater from upstream permafrost areas; in late winter, the ice mass continues to thicken and expand downstream (Ming-ko Woo 2012).

The consolidation period refers to the phase when the surface flow of a river or river segment becomes completely covered by a fixed ice layer, lasting until the onset of thawing. Even if the ice layer does not entirely cover the surface, the period is still considered consolidated if the area of open water is less than 20% of the total river segment area. Typically, the consolidation period is shorter than the freezing period.

The thawing period, also referred to as river opening, is the process by which the frozen river melts and breaks apart under thermal and hydrological influences until the ice completely disappears. River openings can be classified into two types: the 'gentle opening,' primarily driven by thermal effects, and the 'forceful opening,' dominated by hydrodynamic forces.

A gentle opening occurs as solar radiation intensifies and temperatures rise above 0 °C, warming the water and sequentially melting the ice near the banks and at the river's center. As the ice weakens, fragmentation increases, and melting accelerates. Characteristics of a gentle opening include stable water flow, an absence of extensive drift ice, and lower peak discharge; some ice chunks flow away with the water, while others collapse onto shallow banks and melt in place, prolonging the process.

Conversely, a forceful opening is driven by hydrodynamic forces. Before the opening, temperatures remain low, and the ice remains thick and solid. A sudden increase in upstream water flow causes a sharp rise in water levels, shearing and breaking the ice. This increase in water flow can be triggered by significant temperature rises upstream, causing substantial snowmelt or heavy precipitation, upstream river openings that release stored water, or increased discharge from upstream reservoirs. Forceful openings are common in rivers flowing from lower to higher latitudes and are characterized by rapid, unstable changes in water level, concentrated drift ice, and the potential for ice dams and floods (Fu et al. 2014).

Generally, river openings involve both thermal and hydrodynamic effects, resulting in what is known as 'gentle-forceful openings.' In this scenario, thermal effects partially melt and disintegrate the ice cover, followed by a gradual increase in upstream water flow that completes the opening process. This type of opening features characteristics intermediate between gentle and forceful openings. During the process, even if ice jams form in certain river sections, they typically do not cause severe ice flood disasters due to the weakened state of the ice.

In permafrost regions, the thawing of rivers presents unique challenges. As rivers begin to thaw, meltwater from ice and snow in shaded areas may refreeze during periods of freezing temperatures at night and thawing during the day. In some areas with particularly low summer temperatures, river ice may not completely melt before the onset of the next winter's freezing process. An example of this can be observed in the Moma River Valley, a tributary of the Indigirka River in Siberia, where river ice can persist from one winter to the next. Additionally, the extent and form of river ice within the same region may vary from year to year, influenced by the volume of water and its location during the freezing process. In these permafrost regions, the formation of river ice over consecutive winters is typically controlled by groundwater recharge, except in cases where rivers have multiple sources of recharge and high mineralization.

(1) **Lake-Ice Freezing and Melting Process**

Lake ice shares many similarities with river ice in terms of its freezing and thawing dynamics. Typically, lakes begin to freeze during the autumn and winter months, transitioning into thawing during the spring and summer of the following year. Similar to river ice, the freeze–thaw cycle of lake ice starts at the shorelines. This phenomenon occurs due to the differential heat capacities between the water body and the surrounding land, with the land losing heat more rapidly than the water, leading to the initial formation of ice along the lake's edges.

As temperatures decrease during the autumn and winter months, lake water that cools below the freezing point initiates the formation of ice crystals. Once a solid ice surface develops across the lake, a significant phenomenon occurs: the ice, with its considerably higher albedo of approximately 70%, reflects a large portion of the incoming solar radiation, in stark contrast to the water's low reflectivity of about 8%. This increased albedo substantially reduces the amount of solar radiation penetrating the lake, thereby intensifying ice formation due to the diminished heat flux and reduced solar shortwave radiation. In contrast, during the spring and summer seasons, rising temperatures trigger the melting of lake ice, following a distinct seasonal pattern (Sharma et al. 2020).

The thickness of the ice serves as a comprehensive metric for capturing the complexities of lake ice accretion and ablation. This parameter's variability arises not only from the growth and recession at the ice's basal layer but also from the dynamics occurring at the ice surface, including the formation and melting of snow ice. As temperatures rise and snow and ice begin to melt, the ice surface of the lake tends to thaw more rapidly than the underside.

Spatial and temporal heterogeneities in lake ice are particularly evident. In colder, high-latitude regions, lakes located at more southerly latitudes experience later freeze onset, while those further north exhibit delayed thaws (Huang et al. 2022). The duration of ice cover varies, ranging from approximately 7 months in subarctic zones to over 10 months, or even longer, in the Arctic. However, variations also occur within the same region. Warmer years can extend the period when the lake is ice-free, whereas in cooler summers, some Arctic lakes may retain their ice cover throughout the year. Additionally, smaller lakes in the same geographical area tend to have a shorter ice-free period compared to larger lakes. When considering the spatial distribution of ice thickness, subarctic lake ice is generally thinner, a result of milder and shorter winters compared to those in the Arctic. Moreover, within the same lake, ice thickness can vary; near the shore, the newly formed shore ice is limited in its ability to reach maximum thickness due to shallower water depths, resulting in thinner ice, particularly when a significant snowpack is present.

The encapsulation and reduction of lake ice depend on the net heat loss from the lake, with melting being a combination of thermal and dynamic processes. However, melting accelerates when warmer currents enter the lake, significantly weakening the integrity of the ice sheet. During the thawing phase, the morphological state of the lake ice changes due to internal melting and the infiltration and erosion of the ice structure by meltwater. Early in the melting process, the previously smooth ice surface becomes rugged, and pools of meltwater begin to form. Vertical fissures along the edges of the ice mass widen, increasing the surface roughness and albedo of the ice. The internal breakdown of the ice gradually undermines adjacent water bodies until the consolidated ice transforms into needle ice; ultimately, the ice disintegrates, losing its structural integrity.

(2) **Influence Factors of Formation and Melting of River Ice and Lake Ice**

The formation and melting of river and lake ice are primarily influenced by air temperature, water temperature, discharge, location, the shape of the river and lake, topography, and water conservation projects. Additionally, the salinity of lake water plays a significant role in the freezing and thawing processes of lakes.

The freezing and thawing of river and lake ice are primarily influenced by factors such as air temperature, water temperature, discharge, geographic location, the shape of the river or lake, topography, and the presence of water conservation projects. Additionally, the salinity of lake water plays a significant role in the freezing and thawing processes of lakes.

The cryogenic dynamics of riverine and lacustrine ice are largely governed by a combination of environmental parameters, including ambient air temperature, water temperature, hydrological discharge, locational factors, fluvial and lacustrine morphologies, regional topography, and the implementation of water conservation measures. Additionally, the salinity levels within the lacustrine environment play a crucial role in the freeze–thaw cycle of lake ice.

The accrual and recession of lake ice are deeply intertwined with the vertical transportation of energy and the regional meteorological forces at play. Insolation, particularly the influx of solar shortwave radiation, is instrumental in driving the interplay of ice accretion and ablation. The reflective property, or albedo, of lake ice is a critical determinant of the amount of solar radiation that penetrates the ice sheet, directly influencing the thermal balance between the ice and the underlying water. The presence of snow cover on lake ice modifies the albedo, with both snow depth and surface characteristics playing a role in adjusting this parameter. In snow-free ice conditions, the albedo is more sensitive to variations in ice thickness when the ice is relatively thin, while it is primarily influenced by surface properties when the ice is thick. Besides the role of shortwave radiation, the thermal flux from the underlying water body significantly affects the state of the lake ice. The rate of ice formation is closely linked to the conductive heat exchange at the ice-water interface, the temperature at the ice surface, and the existing ice thickness.

2. **The Migration of River Ice and Lake Ice**

The migration of river and lake ice, distinct from the expansive and often prolonged movement of snow and glaciers, occurs within confined systems such as river channels, lake basins, or expansive bodies like oceans. This confined nature means that upon melting, the resultant meltwater is quickly integrated into the surrounding water body. However, the process of ice migration, driven by currents and influenced by hydrothermal conditions, can lead to various dynamic changes within these systems, which significantly impact both local and large-scale hydrological environments.

(1) **The Migration of River Ice**

The mobility of river ice, particularly in its early stages, is primarily influenced by the river's current. As bank ice forms along the riverbanks, hydrodynamic forces may cause fragments to detach and drift into the river's main flow, where they become floating ice. The movement of this ice is generally faster towards the center of the river channel, where the current is stronger. Wind also contributes to the ice's mobility, affecting its speed and distribution across the water surface. Wind can cause ice to accumulate in specific areas, particularly if it pushes floating ice towards one side of the river or against obstacles such as riverbanks or islands.

As floating ice migrates downstream, it can encounter channel constrictions or other physical barriers, such as narrows in the river or bends where the flow slows down. These constrictions can cause ice to accumulate, as the river's breadth may not be sufficient to accommodate the volume of floating ice. When this happens, the ice masses can interlock, creating an obstruction that decelerates or even halts the motion of the ice, leading to surface freezing. This process is compounded by the continued flow of water underneath the ice, which brings more ice into the constricted area, increasing the mass of the ice jam.

As the ice cover expands upstream, it eventually reaches a point where the current is too fast for the ice to advance further. At this juncture, the leading edge of the ice sheet may break apart under the force of the current, producing ice chunks and frazil ice that are carried downstream. Frazil ice—consisting of small, needle-like ice crystals formed in supercooled water—can accumulate beneath the primary ice cover, contributing to the development of an ice jam. As more ice builds up beneath the initial freeze-up ice jam, the flow of water transitions from an open channel to sub-ice flow, increasing the wetted perimeter of the river's cross-section and decreasing the hydraulic radius.

The presence of ice sheets and jams significantly increases resistance to water flow, which further slows the current and promotes additional ice formation. This positive feedback loop can lead to rapid ice accumulation, transforming short-term ice build-ups into extensive ice jams. These jams can obstruct the river's flow, leading to the potential for ice floods, which pose significant risks to surrounding areas.

Given the dangers posed by ice jams, especially in rivers with gentle slopes, reduced velocities, and variable widths, it is crucial to closely monitor ice flux during the winter months. Proactive measures—such as ice-breaking operations or the installation of structures to control ice movement—should be employed early in the ice formation process to prevent the progression of adverse conditions and avert potential disasters. Effective management and monitoring of river ice conditions are essential for minimizing the risks associated with ice jams and floods.

As ice binding and jamming occur, the water flow diminishes, leading to a gradual rise in river levels due to partial blockages caused by ice floes. In cases of significant runoff or rapid deterioration of ice conditions, the water level can surge suddenly, potentially resulting in an ice flood (Fig. 3.29). Ice floods, primarily occurring in winter and spring when river discharge is low and flow velocities are reduced, are a direct consequence of ice binding and jamming on the river's surface. These floods are often triggered by the unexpected influx of Arctic air masses, which build upon the accumulation of dispersed river ice in earlier stages. In the initial phase of an ice flood, the rising water levels may cause irregular elevations in the ice sheet, leading to partial fractures. Water escaping through these fractures overflows and rapidly refreezes, thereby thickening the ice sheet or creating an ice dam, which exacerbates the flood. Ice floods increase the pressure on riverbanks, raising the risk of dike breaches. Additionally, ice jams and dams are inherently unstable structures formed by natural processes. As water levels rise, the basal pressure on these structures intensifies, and even minor disturbances—such as ice melt from warming temperatures or the erosive force of small water flows—can trigger their collapse.

As temperatures rise above the freezing point and the river enters the thawing phase, the ice surface begins to melt. Typically, the shoreline warms more quickly, causing the initial breakup and detachment of shore ice. As temperatures continue to increase, the ice progressively melts and, under the influence of currents and winds, breaks free, transitioning back into floating ice. The nature of this thaw can be classified as either a tranquil or violent breakup.

A tranquil breakup occurs when changes in flow rate are minimal, and the ice predominantly melts in place due to thermal effects, with little or no disruption from ice jams or dams. In contrast, a violent breakup arises when the flow rate increases rapidly before the ice is fully melted, leading to a mechanical breakup driven by hydraulic forces. The onset of a violent breakup is often associated with significant

Fig. 3.29 Ice flood and blasting in Heilongjiang in April 2015

temperature differences between upstream and downstream areas during the freezing period. In the spring, if upstream regions experience substantial snowmelt or an earlier thaw, resulting in increased discharge and a rapid rise in water levels, while downstream sections remain solidly frozen, the downstream ice becomes vulnerable to upstream forces. In such scenarios, particularly in narrow and meandering river sections, the accumulation of large ice fragments can easily form an ice dam, leading to elevated water levels and potentially triggering a flood.

(2) The Migration of Lake Ice

In smaller lakes or reservoirs, where water movement is slow or stagnant, a smooth and continuous ice sheet can readily form from the shore to the center during the winter cooling period. This formation occurs as the ice sheet and the lake shore freeze together, creating a bond that prevents movement. Conversely, during the thawing period, the lake banks, with their lower albedo, absorb more heat, leading to the initial melting of shore ice. As a result, the ice sheet gradually detaches from the shore, giving rise to floating ice. The varying melting rates across the ice surface may cause the lake ice to fragment into several independent floes. Due to the minimal water flow, these floes may drift under the influence of wind or vertical convection, though their overall displacement typically remains limited.

In larger lakes or reservoirs, the dynamics of ice formation and movement differ significantly due to the influence of periodic water surges driven by substantial tidal actions and fluctuations in water levels caused by inflows and outflows. At temperatures below 0 °C, tidal splashes initially freeze upon contact with the shores, gradually contributing to the formation of a suspended ice body that extends outward from the shore. This suspended ice body grows as tidal erosion continues and additional shore ice accumulates. Over time, this process may lead to the detachment of shore ice, which then drifts into the lake as floes. Some of these floating ice masses may be pushed back toward the shore by tidal forces, while others are carried along by the lake's water currents. If the lake discharges into river systems, the floating ice may be transported downstream into these rivers, or it may accumulate at the outflows, forming ice jams and dams. This phenomenon mirrors the ice jams observed in river systems and poses similar risks, including the potential for flooding and structural damage (Fig. 3.30).

The formation of floating ice is a frequent phenomenon in moraine lakes, ice-covered lakes, and glacier-dammed lakes associated with glacial activity. In addition to the natural process of freezing, the generation of floating ice is a major contributor to ice calving at glacier termini (see Fig. 3.31). Floating ice in glacial lakes typically consists of irregularly shaped blocks, with a significant portion of each block extending above the water's surface. As a result, the movement of these ice masses is influenced by both the lake's currents and the force of the wind.

In glacial lakes where water flow is minimal, floating ice often accumulates or drifts toward the windward shore. Conversely, in cases where water discharge is substantial, floating ice tends to migrate toward the dam or the lake's outflow, following the water current's direction. The typically large size of individual ice

Fig. 3.30 Frozen ice (grey) and calved ice (white) in the Longbasaba glacial lake of the Himalayas. (Ding 2020)

Fig. 3.31 Floating ice jam

masses in glacial lakes makes their movement toward an outlet prone to forming ice jams (as depicted in Fig. 3.31). While the mechanisms behind these jams are simpler than those found in rivers and non-glacial lakes, they can still trigger glacial lake outburst floods—catastrophic events with the potential for widespread destruction.

3.5.2 *Formation, Melting, and Migration of Sea Ice*

1. **Formation and Melting of Sea Ice**

Sea ice forms when seawater freezes, a process that is further influenced by the refreezing of precipitation on its surface and the accumulation of snow cover. Globally, sea ice spans approximately 25 million square kilometers. As a result of seawater freezing under cryogenic conditions, sea ice consists of a complex matrix of freshwater ice crystals, embedded salts, and air bubbles. Despite its saline origin, sea ice is hydrologically categorized as part of the freshwater ice resource, owing to the predominance of fresh ice crystals in its structure.

(1) **Formation of Sea Ice**

The growth and decay of sea ice are typically divided into three stages: the early ice period, the frozen period, and the final ice period (Gabison 1987). The early ice period spans from the first appearance of ice to the point where the ice becomes fully established, marking a phase of increasing ice formation. The frozen period begins once the ice is fully formed and lasts until the onset of thawing. During this period, ice conditions are at their most severe, with ice density exceeding 70%, a phase often referred to as the heavy ice period. The final ice period extends from the thawing day to the last day ice is present, during which the sea ice progressively melts as temperatures rise and sea temperatures increase.

Sea ice can be categorized using various criteria, including its form, freezing process, surface characteristics, size, and crystal structure. Among these, classification based on the formation process and existence form is particularly significant. Table 3.1 outlines the classification of sea ice according to these criteria.

(2) **Melting of Sea Ice**

The seasonal melting of sea ice follows a distinct cyclical pattern. Typically, first-year ice melts entirely during the melt season within a single year. Ice that survives this period transitions into second-year ice and may eventually become multi-year ice. Even when sea ice does not completely melt during the summer, its thickness is significantly reduced (Roach et al. 2022). Observational data show that multi-year ice, which can reach thicknesses of up to 4 m during winter, may thin to approximately 1 to 2 m by summer. The melting processes of sea ice are notably more complex than those of terrestrial ice, involving surface, bottom, lateral, and internal melting. Furthermore, the resulting meltwater affects both the physical and chemical properties of the ocean and the hydrological dynamics within the frozen zone.

Table 3.1 Classification of sea ice by existence form and freezing process

Classification principle	Sea ice type	Explanation/definition
Existential form	Fixed ice	Ice that remains stationary, not influenced by ocean currents or wind fields. It is typically connected to land masses, including coastal islands or shoals. Examples include ice reefs (attached to shores), shore ice (connected to shoals), and anchor ice (frozen to the seabed in shallow waters)
	Floating ice	Ice that is driven by ocean currents and winds. It can originate either from the freezing of seawater or from glaciers breaking off and drifting into the sea
Formation process	Nascent ice	Ice that forms when the sea temperature reaches the freezing point, or when snow falls on cold sea surfaces. Types include needle ice, grease ice, sticky ice, and spongy ice
	Nilas	Thin, elastic ice sheets less than 10 cm thick, formed from the continued growth of nascent ice. Nilas has a dull, dark surface and can bend or stack under wave action, potentially forming accumulated fatty ice
	Pancake ice	Also known as lotus leaf ice, pancake ice forms when ice pieces collide, causing their edges to curl upwards. It represents an intermediate stage of ice formation in moving water, typically ranging from 0.3 to 3 m in diameter and up to 10 cm thick. Pancake ice can rapidly cover large water areas
	One year ice	Sea ice that develops from nascent ice, with a thickness ranging from 0.3 to 3 m, which does not survive beyond one winter season
	Multi-year ice	Sea ice that has persisted for at least two summers without melting. Multi-year ice is thicker (up to 3–5 m) and contains more air bubbles but less salt than one-year ice, making it more challenging for icebreakers to navigate

Surface melting of sea ice is primarily driven by the absorption of solar radiation. When the intensity of solar radiation surpasses the thermal conductive flux within the sea ice, the excess heat raises the temperature of the ice surface, leading to melting. The resulting meltwater may either drain into the ocean or accumulate in melt ponds. These melt ponds, by lowering the albedo and increasing the absorption of solar energy, further amplify the melting process.

Underside melting occurs when warmer ocean waters come into contact with the ice bottom, where a significant portion of oceanic heat is concentrated. This process typically results in a more rapid rate of melting compared to surface melting. Recent research has demonstrated that in many areas of the Arctic Ocean, underside melting considerably exceeds surface melting, playing a critical role in the overall reduction of sea ice (Hartman et al. 2021).

Lateral melting of sea ice occurs during the summer when extensive sea ice breaks into smaller floes of varying sizes. The increased perimeter of these floes allows for greater exposure to warmer waters, leading to edge melting—a process known as lateral melting. This type of melting can result from direct heat transfer from the seawater, erosion caused by seawater infiltration, or physical fragmentation due to collisions between ice floes. Although the mechanisms differ, their individual contributions are often challenging to distinguish in observational data. Lateral melt rates can vary based on season, ice concentration, and geographic location. As the Arctic continues to warm and sea ice diminishes, the significance of lateral melting in the overall reduction of sea ice is expected to increase.

Internal melting within sea ice occurs when seawater infiltrates brine channels and air bubbles, transforming them into water-filled cavities. Since water absorbs solar radiation more effectively than ice, temperatures within these cavities rise, leading to the expansion of brine channels. Although this internal melting does not directly affect the ice's overall density or thickness, it significantly increases the ice's porosity. This increase in porosity weakens the structural integrity of the ice, making it more prone to fragmentation and thus accelerating the overall melting process.

The melting of snow on sea ice also plays a significant role in the overall melt dynamics. In spring, as snow begins to melt, the water collects in depressions on the rough ice surface, forming melt ponds. The most intense internal melting typically occurs at the bottom of these ponds. The melt pond water infiltrates the ice through brine channels, creating freshwater pools beneath the ice. This process, known as the freshwater flushing effect, accelerates the melting of sea ice by enhancing the flow of heat and fresh water through the ice structure, further promoting ice degradation.

Long-term fluctuations in the maximum and minimum extent of sea ice serve as key indicators of changes in sea ice mass balance. These multi-annual variations in sea ice mass not only reflect broader climatic shifts but are also influenced by various factors, including riverine inputs, wind patterns, meridional heat transport, and other dynamic processes. These interactions between climatic and environmental factors underscore the complex and interconnected nature of sea ice dynamics and their sensitivity to both global and regional changes.

2. **Migration of Sea Ice**

Analogous to the genesis of riverine and lacustrine ice, the formation of sea ice typically begins along coastlines, near the waterlines of ships, and in other locations where it can anchor itself before progressively expanding outward through freezing connections, leading to the establishment of fast ice. As tidal levels fluctuate, this fast ice can ascend and descend, extending from a few meters to hundreds of kilometers into the ocean. Due to tidal forces, sections of the ice along the outer margins of the fast ice may detach, subsequently being set adrift by various forces such as ocean currents, winds, and tides, thus forming pack ice.

The continental shelf regions at the peripheries of ocean basins are the principal zones of sea ice formation, where the movement of ocean currents and tidal actions are especially pronounced, making these areas hotspots for pack ice drift. During transit, pack ice may collide with fast ice, land, or slower-moving pack ice, leading to the accumulation and compression of ice, which in turn gives rise to features such as ice hummocks or pressure ridges. In bays and other maritime regions where land significantly influences ocean currents, pack ice tends to accumulate and drift sluggishly. For instance, the average drift velocity of sea ice in Bohai Bay ranges from 0.2 to 0.6 m/s, with some floes achieving speeds exceeding 1.0 m/s. The Arctic Ocean, encased by continents, displays extensive development of both shore-fast and pack ice. The surrounding landmasses impede ocean currents, resulting in the extensive accumulation of pack ice, which frequently occupies more than half of the sea surface area.

Beyond maritime migration, sea ice can be driven onto landmasses by potent winds, strong ocean currents, or sharp temperature shifts—a phenomenon known as an ice shove. When sea ice has reached a certain thickness, vigorous wave action can fracture it into sizable chunks, which then jostle and pile atop one another. The force of winds and ocean currents can propel these ice masses onto coastal areas in great numbers, often accompanied by a formidable sound.

In situations where the sea ice is thin or in its early stages of formation, a rapid temperature drop can lead to widespread freezing of seawater. However, due to the strong disturbance caused by waves, it is difficult for a large expanse of consolidated ice to form. Instead, the ice typically breaks into fragments, granules, or small pieces of debris. As this ice debris accumulates, the combined forces of wind, ocean currents, or the natural expansion of the ice can push the unconsolidated sea ice toward the coast. During an ice shove event, ice thickness can reach up to 12 m, potentially causing significant damage to coastal infrastructure. Ice shoves are sporadic occurrences in Arctic seas and may also manifest in larger lakes.

3.5.3 Ice Situation Characteristics of Sea Ice

Sea ice is commonly characterized by its extent, thickness, and density. Variations in these sea ice conditions profoundly influence atmospheric stratification, stability, and convection, as well as the broader thermohaline circulation of the oceans. The presence of sea ice alters the mechanisms of thermal and material exchange between the ocean and atmosphere, affecting not only the immediate marine ecosystem and atmospheric circulation but also potentially triggering regional or global climatic changes through intricate feedback loops. These feedback mechanisms, such as changes in albedo or the insulation of ocean heat, can amplify or mitigate climatic shifts, underscoring the critical role of sea ice in Earth's climate system.

(1) **Ice Situation of Arctic Sea Ice**

Arctic sea ice is composed primarily of multi-year ice and first-year ice, with its extent showing significant seasonal and interannual variability. The freezing phase for Arctic sea ice spans from October to March, while the thawing phase extends from April to September of the following year. Typically, sea ice formation begins in October, with a relatively rapid freezing rate during this period, continuing until December, when the ice reaches an average thickness of approximately 3 m. From

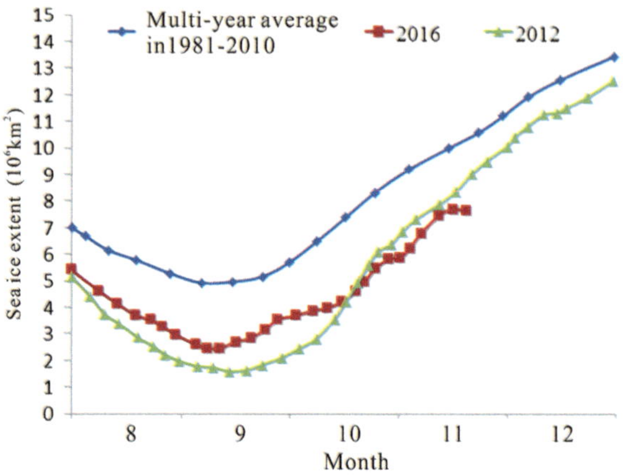

Fig. 3.32 Variation of Arctic Sea ice extent. Data sourced from the National Snow and Ice Data Center (NSIDC). (Download from http://nsidc.org/)

January to March, the growth rate slows, and the ice attains its peak thickness in March. Thawing begins in April, with the melting rate accelerating from May to August, before slowing down, resulting in the thinnest ice layer in September (Fig. 3.32).

An analysis of the interannual variations in Arctic sea ice extent reveals that prior to the 1970s, ice coverage was relatively stable. However, since the 1980s, there has been a noticeable downward trend, with the rate of decline becoming more pronounced in recent years; this decline varies across different regions. The most significant reduction in sea ice extent is observed in the East Siberian Sea, followed by the Chukchi Sea and the Beaufort Sea, while the Canadian Basin shows the least pronounced decline. Additionally, there has been a concurrent decrease in sea ice thickness. Seasonally, the extent of sea ice in the Northern Hemisphere peaks at approximately 15 to 16 million square kilometers in March and April and reaches its minimum, around 6 to 8 million square kilometers, in August and September.

(2) **Ice Situation of Antarctic Sea Ice**

Most of the Antarctic Sea ice is one-year ice, and the extent has regional differences in addition to significant inter-annual variations. In the past 30 years, the sea ice extent in the entire Antarctic region has been increasing at a rate of 1.3×10^4 km^2/yr, but it has not risen linearly. In the southern Hemisphere, sea ice extent is maximum in September (18–19 $\times$ 10^6 km^2) and minimum in March (2–3 $\times$ 10^6 km^2) (Fig. 3.33).

The majority of Antarctic sea ice is composed of one-year ice, and its extent exhibits both significant interannual variability and regional differences. Over the past 30 years, the overall extent of Antarctic sea ice has increased at a rate of approximately 1.3×10^4 km^2/yr. However, this growth has not followed a linear trajectory. In the Southern Hemisphere, the maximum extent of sea ice typically occurs in September, reaching between 18 to 19 $\times$ 10^6 km^2, while the minimum extent is observed in March, with a range of 2 to 3 million square kilometers (Fig. 3.33).

Regarding spatial variability, a decrease in sea ice extent has been observed in the Bellingshausen-Amundsen Seas, while the Ross Sea has experienced the most

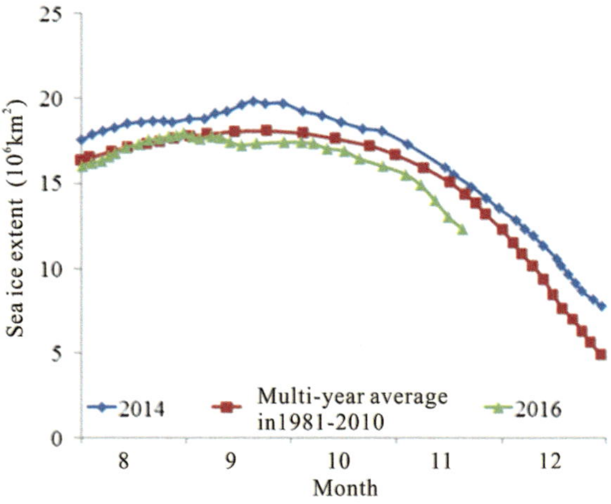

Fig. 3.33 Variation of Antarctic Sea ice extent over time, highlighting maximum and minimum extents in September and March, respectively. Data sourced from NSIDC

rapid increase, followed by the Weddell Sea. In contrast, the Indian and Pacific Oceans have exhibited a modest and gradual upward trend in sea ice extent. In terms of distribution, annual sea ice is predominantly concentrated in the Weddell Sea region near the Antarctic Peninsula and the southern reaches of the Ross Sea. The sea ice distribution in the Weddell Sea area near the Antarctic Peninsula is more concentrated in the east and less so in the west, whereas the Ross Sea displays an inverse distribution pattern, with higher concentrations in the west and lower in the east.

In addition to seasonal sea ice, permanent ice persists on the Abbott, Cook, Shackleton, West, and Fimbul ice shelves, while seasonal sea ice dominates the remaining regions surrounding the Antarctic continent. When analyzing the spatial distribution of sea ice density, it becomes apparent that the Weddell Sea near the Antarctic Peninsula exhibits higher density in the east and lower density in the west, whereas the Ross Sea shows the opposite pattern, with higher density in the west and lower in the east.

During the annual melt period, which spans from September to February, sea ice density initially decreases around 60°S in the eastern Weddell Sea, near the Ross Ice Shelf in the Ross Sea, and along the periphery of the Antarctic continent. The Weddell Sea functions as a critical positive feedback nucleus for Antarctic sea ice dynamics, exerting a primary influence on changes in Antarctic sea ice conditions. In contrast, the Ross Sea, located within the Pacific sector of Antarctica, acts as a negative feedback hub, mitigating positive feedback alterations in sea ice and exerting control over the sea ice in the Pacific Ocean sector of the Ross Sea. This influence extends beyond the Antarctic region, impacting the Arctic Pacific and, consequently, playing a pivotal role in the dynamics of Atlantic Arctic sea ice, which in turn affects the sea ice in the Antarctic Weddell Sea.

References

Alean J (1985) Ice avalanches: some empirical information about their formation and reach. J Glaciol 31:324–333. https://doi.org/10.3189/S0022143000006663

Auer AH (1974) The rain versus snow threshold temperature. Weatherwise 27:67

Ayala A, Pellicciotti F, Peleg N, Burlando P (2017) Melt and surface sublimation across a glacier in a dry environment: distributed energy balance modelling of Juncal Norte Glacier, Chile. J Glaciol 63 (241):803–822. https://doi.org/10.1017/jog.2017.46

Benn DI, Åström JA (2018) Calving glaciers and ice shelves. Adv Phys X 3:1513819. https://doi.org/10.1080/23746149.2018.1513819

Black T, Kurtz D (2023) Maritime glacier retreat and terminus area change in Kenai Fjords National Park, Alaska, between 1984 and 2021. J Glaciol 69:251–265. https://doi.org/10.1017/jog.2022.55

Blume-Werry G, Milbau A, Teuber LM, Johansson M, Dorrepaal E (2019) Dwelling in the deep-strongly increased root growth and rooting depth enhance plant interactions with thawing permafrost soil. New Phytol 223(3):1328–1339. https://doi.org/10.1111/nph.15903

Brutsaert W, Hiyama T (2012) The determination of permafrost thawing trends from long-term streamflow measurements with an application in eastern Siberia. J Geophys Res Atmos 117(22):1–10. https://doi.org/10.1029/2012JD018344

Cheng G (1983) The mechanism of repeated-segregation for the formation of thick layered ground ice. Cold Reg Sci Technol 8:57–66. https://doi.org/10.1016/0165-232X(83)90017-4

Chou Y, Wang L (2021) Seasonal freezing-thawing process and hydrothermal characteristics of soil on the Loess Plateau, China. J Mt Sci 18:3082–3098. https://doi.org/10.1007/s11629-020-6599-9

Dang K, Wu D (1991) Influence of snowfall distribution on different stands at Huoditang forest region in Qinling mountain. J Northwest For Univ 02:1–8

DeWalle DR, Rango A (2008) Principles of snow hydrology. Cambridge University Press

Di Mauro B, Garzonio R, Baccolo G, Franzetti A, Pittino F, Leoni B, Remias D, Colombo R, Rossini M (2020) Glacier algae foster ice-albedo feedback in the European Alps. Sci Rep 10:4739. https://doi.org/10.1038/s41598-020-61762-0

Ding Y (ed.). (2017) Introduction to Hydrology in Cold Regions [M]. Beijing: Science Press.

Ding Y (ed.). (2020) Cryospheric Hydrology [M]. Beijing: Science Press.

Ding Y, Zhang S, Chen R, Han T, Han H, Wu J, Li X, Zhao Q, Shangguan D, Yang Y, Liu J, Wang S, Qin J, Chang Y (2020) Hydrological basis and discipline system of cryohydrology: from a perspective of cryospheric science. Front Earth Sci 8(December):1–12. https://doi.org/10.3389/feart.2020.574707

Doyle SH, Hubbard B, Christoffersen P, Young TJ, Hofstede C, Bougamont M, Box JE, Hubbard A (2018) Physical conditions of fast glacier flow: 1. Measurements from boreholes drilled to the bed of store glacier, West Greenland. JGR Earth Surf 123:324–348. https://doi.org/10.1002/2017JF004529

Duan K, Yao T, Wang N, Shi P, Meng Y (2022) Changes in equilibrium-line altitude and implications for glacier evolution in the Asian high mountains in the 21st century. Sci China Earth Sci 65(7):1308–1316. https://doi.org/10.1007/s11430-021-9923-6

Edward A, Holmlund P (1988) Studies of the drainage and the response to climatic change of Mikkaglaciären and Storglaciären. Number 220 in Meddelanden från Naturgeografiska Institutionen vid Stockholms Universitet. A. Stockholm: Univ. OCLC: 46136236.

Feiccabrino J, Lundberg A (2008) Precipitation phase discrimination in Sweden. In: 65th Eastern snow conference, Fairlee (Lake Morey), Vermont, USA, pp 239–254

Francou B, Ribstein P, Saravia R, Tiriau E (1995) Monthly balance and water discharge of an inter-tropical glacier: Zongo Glacier, Cordillera Real, Bolivia, 16° S. J Glaciol 41:61–67. https://doi.org/10.3189/S0022143000017767

Fu C, Popescu I, Wang C, Mynett AE, Zhang F (2014) Challenges in modelling river flow and ice regime on the Ningxia-Inner Mongolia reach of the Yellow River, China. Hydrol Earth Syst Sci 18(3):1225–1237. https://doi.org/10.5194/hess-18-1225-2014

Fujita K, Seko K, Ageta Y, Jianchen P, Tandong Y (1996) Superimposed ice in glacier mass balance on the Tibetan Plateau. J Glaciol 142(142):454-460. https://doi.org/10.3189/S0022143000003440

Gabison R (1987) A thermodynamic model of the formation, growth, and decay of first-year sea ice. J Glaciol 33:105–119. https://doi.org/10.3189/S0022143000005414

Gao H, Ding Y, Zhao Q, Hrachowitz M, Savenije HHG (2017) The importance of aspect for modelling the hydrological response in a glacier catchment in Central Asia. Hydrol Process 31(16):2842–2859. https://doi.org/10.1002/hyp.11224

Gao H, Feng Z, Zhang T, Wang Y, He X, Li H, Pan X, Ren Z, Chen X, Zhang W, Duan Z (2021a) Assessing glacier retreat and its impact on water resources in a headwater of Yangtze River based on CMIP6 projections. Sci Total Environ 765:142774. https://doi.org/10.1016/j.scitotenv.2020.142774

Gao H, Wang J, Yang Y, Pan X, Ding Y, Duan Z (2021b) Permafrost hydrology of the Qinghai-Tibet Plateau: a review of processes and modeling. Front Earth Sci 8(January):1–13. https://doi.org/10.3389/feart.2020.576838

Gao H, Han C, Chen R, Feng Z, Wang K, Fenicia F, Savenije H (2022) Frozen soil hydrological modeling for a mountainous catchment northeast of the Qinghai-Tibet Plateau. Hydrol Earth Syst Sci 26(15):4187–4208. https://doi.org/10.5194/hess-26-4187-2022

Gao P, Wei W, Liu M (2012) Characteristic analysis on temperature change in snow layer in western Tianshan mountain in China. 31(4):1074–1080

Groot Zwaaftink CD, Mott R, Lehning M (2013) Seasonal simulation of drifting snow sublimation in Alpine terrain. Water Resour Res 49:1581–1590. https://doi.org/10.1002/wrcr.20137

Han C et al (2010a) A discuss of the separating solid and liquid precipitations. J Glaciol Geocryol

Han C, Chen RS, Liu J, Yang Y, Qing W (2010b) A discuss of separating solid and liquid precipitation. J Glaciol Geocryol 32:249–256

Han H, Wang J, Wei J, Liu S (2010c) Backwasting rate on debris-covered Koxkar glacier, Tuomuer mountain, China. J Glaciol 56:287–296. https://doi.org/10.3189/002214310791968430

Han P, Long D, Zhao F, Slater LJ (2023) Response of two glaciers in different climate settings of the Tibetan Plateau to climate change through year 2100 using a hybrid modeling approach. Water Resour Res 59:e2022WR033618. https://doi.org/10.1029/2022WR033618

Hartman JD, Sangiorgi F, Barcena MA, Tateo F, Giglio F, Albertazzi S, Trincardi F, Bijl PK, Langone L, Asioli A (2021) Sea-ice, primary productivity, and ocean temperatures at the Antarctic marginal zone during late Pleistocene. Quatern Sci Rev 266:107069. https://doi.org/10.1016/j.quascirev.2021.107069

Heilig A, Mitterer C, Schmid L, Wever N, Schweizer J (2015) Seasonal and diurnal cycles of liquid water in snow—measurements and modeling. J Geophys Res 120:2139–2154. https://doi.org/10.1002/2015JF003593

Heinze T (2021) A multi-phase heat transfer model for water infiltration into frozen soil. Water Resour Res 57:e2021WR030067. https://doi.org/10.1029/2021WR030067

Hock R (2005) Glacier melt: a review of processes and their modelling. Prog Phys Geogr 29(3):362–391. https://doi.org/10.1191/0309133305pp453ra

Holmlund P (1988) Studies of the drainage and the response to climatic change of Mikkaglaciären and Storglaciären. Number 220 in Meddelanden från Naturgeografiska Institutionen vid Stockholms Universitet. A. Stockholm: Univ. OCLC: 46136236.

Hu L, Bao W, Shi P, Wang J, Lu M (2020) Simulation of overland flow considering the influence of topographic depressions. Sci Rep 10:6128. https://doi.org/10.1038/s41598-020-63001-y

Huang L, Timmermann A, Lee S-S, Rodgers KB, Yamaguchi R, Chung E-S (2022) Emerging unprecedented lake ice loss in climate change projections. Nat Commun 13:5798. https://doi.org/10.1038/s41467-022-33495-3

Huss M, Hock R (2018) Global-scale hydrological response to future glacier mass loss. Nat Clim Chang 8(2):135–140. https://doi.org/10.1038/s41558-017-0049-x
Huss M, Jouvet G, Farinotti D, Bauder A (2010) Future high-mountain hydrology: a new parameterization of glacier retreat. Hydrol Earth Syst Sci 14(5):815–829. https://doi.org/10.5194/hess-14-815-2010
Jorgenson MT, Shur YL, Pullman ER (2006) Abrupt increase in permafrost degradation in Arctic Alaska. Geophys Res Lett 33:2005GL024960. https://doi.org/10.1029/2005GL024960
Jouberton A, Shaw TE, Miles E, McCarthy M, Fugger S, Ren S, Dehecq A, Yang W, Pellicciotti F (2022) Warming-induced monsoon precipitation phase change intensifies glacier mass loss in the southeastern Tibetan Plateau. Proc Natl Acad Sci USA 119:e2109796119. https://doi.org/10.1073/pnas.2109796119
Kamb B (1987) Glacier surge mechanism based on linked cavity configuration of the basal water conduit system. J Geophys Res 92:9083–9100. https://doi.org/10.1029/JB092iB09p09083
Kidron GJ (2021) Comparing overland flow processes between semiarid and humid regions: does saturation overland flow take place in semiarid regions? J Hydrol 593:125624. https://doi.org/10.1016/j.jhydrol.2020.125624
Lehning M, Thomas G, Schirmer M (2011) Mountain snow distribution governed by an altitudinal gradient and terrain roughness. Geophys Res Lett 38(19). https://doi.org/10.1029/2011GL048927
Li H, Wang J, He X (2012) Influence of blowing snow on snow mass and energy exchanges in the Qilian Mountain. J Glaciol Geocryol 34(05):1084–1090
Liu H, Cai T, Man X, Chai R, Lang Y (2012) Effects of major forest types of Xiaoxing'an mountain on the process of snowfall, snow cover and snow melting. J Beijing For Univ 34(02):20–25
Lu H, Wei W, Liu M, Gao P, Han Q (2011) Densification and accumulation rate of snow in the stable snow cover period in the Tianshan mountains. J Glaciol Geocryol 33(02):374–380
Luo DL, Jin HJ, He RX, Wang XF, Muskett RR, Marchenko SS, Romanovsky VE (2018a) Characteristics of water-heat exchanges and inconsistent surface temperature changes at an elevational permafrost site on the Qinghai-Tibet Plateau. JGR Atmos 123. https://doi.org/10.1029/2018JD028298
Luo Y, Wang X, Piao S, Sun L, Ciais P, Zhang Y, Ma C, Gan R, He C (2018b) Contrasting streamflow regimes induced by melting glaciers across the Tien Shan—Pamir—North Karakoram. Sci Rep 8. https://doi.org/10.1038/s41598-018-34829-2
MacDonald MK, Pomeroy JW, Pietroniro A (2010) On the importance of sublimation to an alpine snow mass balance in the Canadian Rocky Mountains. Hydrol Earth Syst Sci 14:1401–1415. https://doi.org/10.5194/hess-14-1401-2010
Mayo L (1984) Glacier mass balance and runoff research in the U. S. A. Geografiska Annaler. Series A, Physical Geography, 66(3):215–227. https://doi.org/10.2307/520695
Meehl GA, Arblaster JM, Collins WD (2008) Effects of black carbon aerosols on the Indian Monsoon. J Clim 21:2869–2882. https://doi.org/10.1175/2007JCLI1777.1
Mohammed AA, Kurylyk BL, Cey EE, Hayashi M (2018) Snowmelt infiltration and macropore flow in frozen soils: overview, knowledge gaps, and a conceptual framework. Vadose Zone J 17. https://doi.org/10.2136/vzj2018.04.0084
Naegeli K, Huss M (2017) Sensitivity of mountain glacier mass balance to changes in bare-ice albedo. Ann Glaciol 58:119–129. https://doi.org/10.1017/aog.2017.25
Phillips E, Finlayson A, Bradwell T, Everest J, Jones L (2014) Structural evolution triggers a dynamic reduction in active glacier length during rapid retreat: evidence from Falljökull, SE Iceland. JGR Earth Surf 119:2194–2208. https://doi.org/10.1002/2014JF003165
Pomeroy JW, Jones HG (1996) Wind-blown snow: sublimation, transport and changes to polar snow. In: Wolff EW, Bales RC (eds) Chemical exchange between the atmosphere and polar snow, NATO ASI Series I, Global environmental change, vol 43. Springer, Berlin, pp 453–489
Pomeroy JW, Marsh P, Gray DM (1997) Application of a distributed blowing snow model to the Arctic. Hydrol Process 11:1451–1464. https://doi.org/10.1002/(SICI)1099-1085(199709)11:11%3c1451:AID-HYP449%3e3.0.CO;2-Q

Qi W, Feng L, Liu J, Yang H (2020) Snow as an important natural reservoir for runoff and soil moisture in Northeast China. JGR Atmos 125:e2020JD033086. https://doi.org/10.1029/2020JD033086

Qin Y, Bai Y, Chen G, Liang Y, Li X, Wen B, Lu X, Li X (2021) The effects of soil freeze–thaw processes on water and salt migrations in the western Songnen Plain, China. Sci Rep 11:3888. https://doi.org/10.1038/s41598-021-83294-x

Ran Y, Li X, Cheng G et al (2021) Mapping the permafrost stability on the Tibetan Plateau for 2005–2015. Sci China Earth Sci 64:62–79. https://doi.org/10.1007/s11430-020-9685-3

Roach LA, Eisenman I, Wagner TJW, Blanchard-Wrigglesworth E, Bitz CM (2022) Asymmetry in the seasonal cycle of Antarctic Sea ice driven by insolation. Nat Geosci 15:277–281. https://doi.org/10.1038/s41561-022-00913-6

Rounce DR, Hock R, McNabb RW, Millan R, Sommer C, Braun MH, Malz P, Maussion F, Mouginot J, Seehaus TC, Shean DE (2021) Distributed global debris thickness estimates reveal debris significantly impacts glacier mass balance. Geophys Res Lett 48(8) https://doi.org/10.1029/2020GL091311

Sakai A, Nakawo M, Fujita K (1998) Melt rate of ice cliffs on the Lirung Glacier, Nepal Himalayas, 1996. Bull Glaciol Resour 16:57–66

Shakoor A, Ejaz N (2019) Flow analysis at the snow-covered high-altitude catchment via distributed energy balance modeling. Sci Rep 9:4783. https://doi.org/10.1038/s41598-019-39446-1

Shanley JB, Chalmers A (1999) The effect of frozen soil on snowmelt runoff at Sleepers River, Vermont. Hydrol Process 13(12–13):1843–1857. https://doi.org/10.1002/(SICI)1099-1085(199909)13:12/13%3c1843::AID-HYP879%3e3.0.CO;2-G

Sharma S, Meyer MF, Culpepper J, Yang X, Hampton S, Berger SA, Brousil MR, Fradkin SC, Higgins SN, Jankowski KJ, Kirillin G, Smits AP, Whitaker EC, Yousef F, Zhang S (2020) Integrating perspectives to understand lake ice dynamics in a changing world. JGR Biogeosci 125:e2020JG005799. https://doi.org/10.1029/2020JG005799

Skiles SM, Flanner M, Cook JM, Dumont M, Painter TH (2018) Radiative forcing by light-absorbing particles in snow. Nat Clim Change 8:964–971. https://doi.org/10.1038/s41558-018-0296-5

Song L, Wang L, Li X, Zhou J, Luo D, Jin H, Qi J, Zeng T, Yin Y (2020) Improving permafrost physics in a distributed cryosphere-hydrology model and its evaluations at the Upper Yellow River Basin. J Geophys Res Atmos 125(18):1–22. https://doi.org/10.1029/2020JD032916

Sturm M, Holmgren IJ, Nig IMK, Morris K (1997) The thermal conductivity of seasonal snow. 43(143):26–41

Sturm M, Holmgren J, Mcfadden JP, Liston GE, Chapin FS, Racine CH, Holmgren G (2001) Snow-shrub interactions in arctic tundra: a hypothesis with climatic implications. J Clim 14:336–344

Sugden D, John B (1976) Glaciers and landscape: A geomorphological approach. Edward Arnold.

Sun L, Zhao L, Li R, Yao J, Xiao Y, Liu G (2010) The infulence of snow cover on the surface albedo and ground temperature. J Mt Sci 28(03):266-273.

Van Den Broeke M, Bintanja R (1995) The interaction of katabatic winds and the formation of blue-ice areas in East Antarctica. J Glaciol 41:395–407. https://doi.org/10.3189/S0022143000016269

Walvoord MA, Kurylyk BL (2016) Hydrologic impacts of thawing permafrost—a review. Vadose Zo J 15(6):1–20. https://doi.org/10.2136/vzj2016.01.0010

Wang G, Mao Y, He B, Wu Q, Shen Y (2012) Changes in snow covers during 1961–2011 and its effects on frozen ground in Altay Region, Xinjiang. J Glaciol Geocryol

Wang L, Zhou J, Qi J, Sun L, Yang K, Tian L, Lin Y, Liu W, Shrestha M, Xue Y, Koike T, Ma Y, Li X, Chen Y, Chen D, Piao S, Lu H (2017) Development of a land surface model with coupled snow and frozen soil physics. Water Resour Res 53(6):5085-5103. https://doi.org/10.1002/2017WR020451

Warren CR, Sugden DE (1993) The Patagonian Icefields: a glaciological review. Arct Alp Res 25:316. https://doi.org/10.2307/1551915

Woo M (2012) Permafrost hydrology. Springer Heidelberg Dordrecht London New York, Berlin, pp 73–116

Wu K, Liu S, Zhu Y, Liu Q, Jiang Z (2020) Dynamics of glacier surface velocity and ice thickness for maritime glaciers in the southeastern Tibetan Plateau. J Hydrol 590:125527. https://doi.org/10.1016/j.jhydrol.2020.125527

Wu T, Li H, Lyu H (2023) Effect of freeze–thaw process on heat transfer and water migration between soil water and groundwater. J Hydrol 617:128987. https://doi.org/10.1016/j.jhydrol.2022.128987

Xu S, Fu Q, Li T, Meng F, Liu D, Hou R, Li M, Li Q (2022) Spatiotemporal characteristics of the soil freeze-thaw state and its variation under different land use types—a case study in Northeast China. Agric For Meteorol 312:108737. https://doi.org/10.1016/j.agrformet.2021.108737

Yang D, Kane D (2021) Arctic Hydrology, Permafrost and Ecosystems. In: Yang, D., Kane, D.L. (eds) Arctic Hydrology, Permafrost and Ecosystems. Springer, Cham. https://doi.org/10.1007/978-3-030-50930-9_31

Yang D, Zhao Y, Armstrong R, Brodzik MJ, Robinson D (2020a) Yukon river discharge response to seasonal snow cover change

Yang X, Pavelsky TM, Allen GH (2020b) The past and future of global river ice. Nature 577(7788):69–73. https://doi.org/10.1038/s41586-019-1848-1

Ye B, Yang D, Zhang Z, Kane D (2009) Variation of hydrological regime with permafrost coverage over Lena Basin in Siberia. J Geophys Res-Atmos, 114(D7):1291-1298. https://doi.org/10.1029/2008JD010537

Ye B, Yang D, Zhang Z, Kane DL (2009) Variation of hydrological regime with permafrost coverage over Lena Basin in Siberia. J Geophys Res Atmos 114(7):1–12. https://doi.org/10.1029/2008JD010537

Zhang T, Frauenfeld O, Serreze M, Etringer A, Oelke C, McCreight J, Barry R, Gilichinsky D, Yang D, Ye H, Ling F, Chudinova S (2005) Spatial and temporal variability in active layer thickness over the Russian Arctic drainage basin. J Geophys Res Atmos 110. https://doi.org/10.1029/2004JD005642

Zhang W, Zhou J, Wolfgang K, Cheng G, Ye B, He X, Li H (2013) Monitoring and modeling the influence of snow cover and organic soil on the active layer of permafrost on the Tibetan Plateau. J Glaciol Geocryol 35(03):528–540

Zhang Y, Ishikawa M, Ohata T, Oyunbaatar D (2008) Sublimation from thin snow cover at the edge of the Eurasian cryosphere in Mongolia. Hydrol Process 22:3564–3575. https://doi.org/10.1002/hyp.6960

Zhang Y, Qiu Y, Li Y, Leppäranta M, Jia G, Jiang Z, Liang W (2024) Spatial-temporal variation of river ice coverage in the Yenisei river from 2002 to 2021. J Hydrol 637(June). https://doi.org/10.1016/j.jhydrol.2024.131440

Zou D, Zhao L, Sheng Y, Chen J, Hu G, Wu T, Wu J, Xie C, Wu X, Pang Q, Wang W, Du E, Li W, Liu G, Li J, Qin Y, Qiao Y, Wang Z, Shi J, Cheng G (2017) A new map of permafrost distribution on the Tibetan Plateau. Cryosphere 11(6):2527–2542. https://doi.org/10.5194/tc-11-2527-2017

Chapter 4
Cryospheric Meltwater Runoff

Shiqiang Zhang, Hongkai Gao, and Qiudong Zhao

The rivers in the cryospheric area receive different proportions of meltwater runoff, which is closely linked to changes in air temperature. As a result, the characteristics and inter-annual variability of meltwater runoff are distinct from those of rainfall runoff processes. Climate change affects the replenishment of rivers with meltwater in diverse ways. This chapter summarizes the various types of river recharge in the cryospheric area, analyzes the response characteristics of meltwater runoff to climate change on seasonal and interannual scales, and projects the runoff of different cryospheric rivers at the global and typical watershed scales under future changes.

4.1 Types of River Recharge in Cryospheric Regions

Cryospheric regions, including the high mountains of North America, the Arctic, and the headwaters of rivers on the Qinghai-Tibet Plateau, are widely distributed with mountain glaciers, snow, permafrost, and seasonal frozen soil. Rivers in these regions are characterized by multiple water sources. Based on river discharge hydrographs

S. Zhang (✉)
Northwest University, Xi'an, China
e-mail: zhangsq@nwu.edu.cn

H. Gao
School of Geographic Sciences, East China Normal University, Shanghai, China
e-mail: hkgao@geo.ecnu.edu.cn

Q. Zhao
Northwest Institute of Eco-Environment and Resources, Chinese Academy of Sciences, Lanzhou, China
e-mail: dsslab@163.com

Y. Ding et al., *Cryohydrology*, https://doi.org/10.1007/978-981-96-7360-5_4

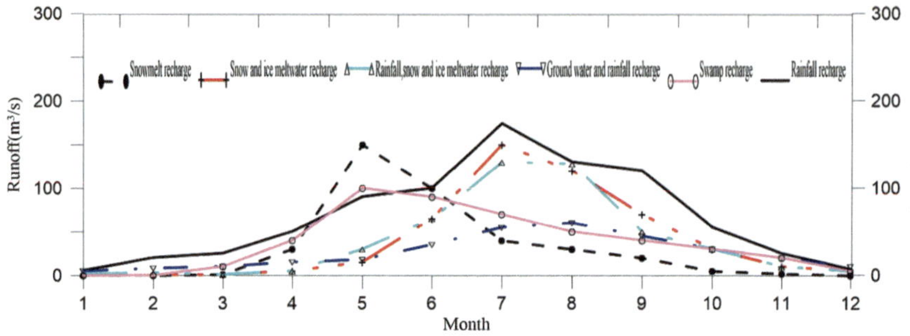

Fig. 4.1 Hydrographs of rivers with different cryospheric recharge types

and hydrological characteristics, rivers in cryospheric regions can be classified into different types according to their recharge sources (Fig. 4.1):

(a) **Snowmelt-replenished rivers**: These rivers are primarily fed by spring snowmelt. Their high-water period is driven by snowmelt, with the flood season occurring relatively early, typically peaking between May and June. For instance, in the rivers of the Tacheng district in Xinjiang Uygur Autonomous Region, runoff from April to June is predominantly supplied by snowmelt, accounting for 60–70% of the annual runoff, with the flood peak occurring in May. In the northern Altay region, the flood peak typically appears in June.

(b) **Snow and ice meltwater-replenished rivers**: These rivers are mainly replenished by snowmelt and glacier meltwater. They are characterized by high glacier coverage within the basin, intense summer glacier melting, and significant contributions of meltwater to river flow. Snowmelt and ice meltwater can constitute 30–80% of the total runoff, with minimal interannual variation. Hydrologically, the flood period occurs in midsummer, with a slow rise in water levels, and the flood recession process is almost symmetrical. Late summer and early autumn often bring sudden, high-peak, short-duration floods. These rivers are primarily found on the southern slopes of China's Tianshan Mountains, the western Qilian Mountains, the Pamirs, Karakoram Mountains, Kunlun Mountains, and other regions.

(c) **Rainfall and snow-ice meltwater mixed-recharge rivers**: These rivers are recharged by a combination of snow-ice meltwater and rainfall runoff. This combination drives river flooding, with runoff from June to August typically constituting 40–60% of the annual total, and snow-ice meltwater accounting for 20–30% of the annual runoff. Spring floods are relatively minor, and interannual runoff variation is minimal. The river hydrograph varies with basin size; smaller rivers exhibit steeper flood peaks and recessions, while larger basins show more gradual changes. These rivers are mainly found in the Ili Valley on the western part of the northern slope of the Tianshan Mountains and in the middle and eastern parts of the Qilian Mountains in China.

(d) **Groundwater-fed and rainfall-replenished rivers**: These rivers are primarily sustained by groundwater and precipitation. Their defining feature is a water source mainly derived from groundwater in areas with discontinuous permafrost. River runoff remains relatively stable and continuous through winter. During heavy rains, a thin peak flow is observed, as seen near Great Bear Lake in Western Canada. In permafrost regions, groundwater-fed runoff typically accounts for about 15% of the annual precipitation.

(e) **Swamp-replenished rivers**: These rivers are mainly fed by swamps. In early spring, before the soil has thawed, surface runoff rapidly fills depressions, causing floods; however, these floods are generally less intense than snowmelt floods. In summer, significant water is intercepted by depressions and organic soils in swamps, reducing surface runoff. These rivers are typically found in the tundra regions of Canada.

4.2 Characteristics and Changes in Cryospheric Meltwater Runoff

Cryospheric meltwater runoff is closely linked to temperature variations, which exhibit diurnal cycles and seasonal patterns. Additionally, cryospheric meltwater runoff demonstrates different change patterns under climate change.

4.2.1 Diurnal Cycle of Cryospheric Runoff

Whether from a continental or marine glacier, the diurnal runoff of meltwater is closely linked to the cycle of air temperature, exhibiting a daily "peak-valley" pattern, with the peak typically occurring during the day. For example, during the ablation period of Glacier No.1 in the headwaters of the Urumqi River, the lowest water level appears around 10:00 AM (Beijing Time), while the highest water level is observed between 17:00 and 18:00 PM (Beijing Time), with runoff peaks and valleys lagging behind the changes in air temperature (Fig. 4.2) (Gao et al. 2013). The length of the lag time varies depending on the type, size, and nature of glaciers in different basins. The diurnal characteristics of runoff also differ across various months of the ablation period. As the climate warms, the diurnal patterns of glacier meltwater runoff during different months of the ablation period are expected to change accordingly.

Similar to the diurnal cycle of glacier meltwater, the diurnal cycle of snowmelt also fluctuates with air temperature, with snowmelt typically peaking in the afternoon. The diurnal cycle of snowmelt exhibits several key characteristics: first, snowmelt on sunny days primarily occurs during daylight hours, particularly in the afternoon. On cloudy days or during the later stages of snowmelt, the ablation process may extend throughout the day. For example, hourly observations of snow melting, refreezing, and runoff at the Weissfluhjoch field in the Swiss Alps from May 18th to May 19th,

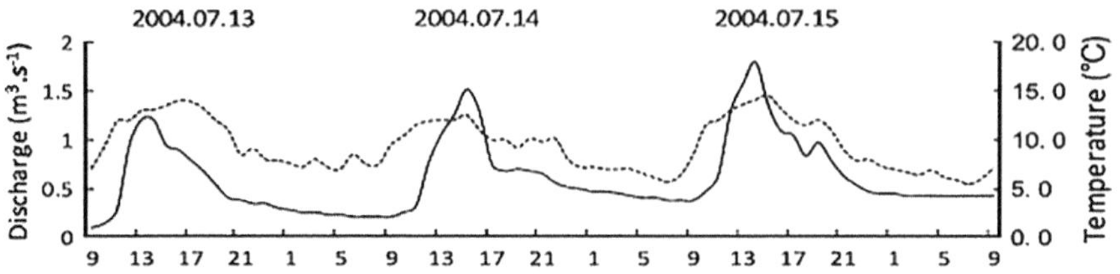

Fig. 4.2 Diurnal variation in discharge (solid line) and air temperature (dashed line) on selected clear-weather days (Gao et al. 2013)

1985, demonstrated a clear diurnal cycle, with peak melting between 3:00 and 5:00 PM. Due to delayed confluence, the maximum snowmelt runoff was observed around 2:00 AM. As the snowmelt process progresses, the amplitude of diurnal variation in snowmelt runoff gradually decreases.

Second, when the hourly average temperature is below 0 °C, the snowmelt rate is low; however, when the hourly average temperature rises above 0°C, the snowmelt rate increases significantly. Third, the snowmelt rate shows a gradually increasing trend. In the early stages of ablation, the large cold storage of snow requires external heat to first warm the snow. The ablation process typically begins at the snow surface. As the snow temperature rises, the entire snow layer gradually reaches the critical melting point, and even a small external energy input can cause rapid snowmelt. Finally, the amount and duration of precipitation significantly impact the snowmelt process. As the climate warms, the diurnal snowmelt process will also change in response to air temperature variations.

4.2.2 Seasonal Distribution of Cryospheric Meltwater Runoff

The seasonal distribution of cryospheric meltwater runoff closely follows the pattern of air temperature, with ablation and runoff occurring primarily during periods when air temperatures exceed 0 °C. Long-term changes in the seasonal distribution of cryospheric runoff are evident in the earlier onset of the ablation period, increased ablation in the early stages, and a delayed end of ablation. For glaciers, surface snow melts first, followed by the melting of the ice surface as temperatures rise. Ablation is most intense during the summer and gradually weakens as temperatures decrease, typically concluding in the autumn. Snowmelt runoff primarily occurs in the spring as temperatures rise and snow cover melts in both low-altitude and alpine areas. In permafrost regions, changes are mainly observed in the slowing of the winter recession process. These changes are discussed in more detail below.

Glacier meltwater exhibits a distinct seasonal distribution pattern. In Asia, glacier meltwater runoff typically begins in May and concludes in October, with variations depending on the type of glacier. The seasonal runoff of continental glaciers is highly variable and unevenly distributed, with a short ablation period from May to September and meltwater concentrated between July and September. Baseflow is minimal, and streams may even dry up during winter. For instance, the Laohugou Glacier in the Qilian Mountains contributed 85.8% of its annual runoff between June and August from 1959 to 1961. In contrast, marine glacier meltwater runoff shows less variation throughout the year and is more evenly distributed. The ablation period is longer, extending from April to October, with a peak ablation period from May to September. For example, during the intense melting period of the Hailuogou Glacier on Gongga Mountain, glacier meltwater accounts for about 80% of the annual runoff, with the maximum occurring in July or August. Due to global warming, the onset of glacier meltwater runoff has advanced, peak runoff has increased, and the end time of runoff has been delayed.

A typical snowmelt runoff event was observed at the Kuwei Hydrological Station in the Kayites River Basin, Xinjiang. From April 16 to the end of June, a significant positive correlation was observed between runoff and air temperature. However, starting in early July, the relationship between runoff and temperature became negligible. Before June 15, runoff changes consistently lagged behind air temperature changes, with the lag time gradually increasing, indicating that melting was primarily occurring in high-altitude areas. From mid-June to the end of June, despite the continued significant positive correlation between runoff and temperature, temperature changes began to lag behind runoff, suggesting that the influence of snowmelt on runoff was weakening and had largely disappeared by this time.

The period of snowmelt varies across different regions. For example, in the Binggou watershed of the Qilian Mountains, China, snowmelt begins at the end of March and is nearly complete by the end of May, with the primary contribution to runoff occurring in April and May. In the Altai Mountains, the snowmelt runoff process starts in mid-April and continues until mid-to-late June, with maximum daily runoff varying significantly from year to year due to differences in snow cover. In the European Alps, snowmelt runoff typically begins in May, and the length of the melting period depends on the amount of winter snow accumulation and the rate of spring temperature rise. In North America, the snowmelt runoff period is generally short, lasting about 2 weeks.

Frozen ground hinders vertical water flow, often resulting in saturated soil conditions in areas of continuous permafrost, while subsurface flow through perennially unfrozen zones in discontinuous permafrost. The freeze–thaw cycle in the active layer significantly affects the direction and velocity of soil water movement, as well as soil storage capacity and hydraulic conductivity. The aquiclude effect of frozen soil reduces surface infiltration, increases soil moisture, and enhances surface runoff generation. It is well documented that permafrost increases peak flow, reduces groundwater recharge, and amplifies seasonal variability in streamflow. The impacts of permafrost on runoff are strongly seasonal and closely tied to soil temperature. Research has shown that the runoff coefficient is larger in spring and autumn and smaller in summer. An increase in winter flow has been observed in many cold regions.

Low runoff during the early thawing season (LRET) was observed in a typical catchment in the upper reach of the Heihe River basin, where approximately 64% of the area is covered by permafrost and 36% by seasonally frozen ground (Fig. 4.3). For instance, from June 5 to 9, 2013, a 45.7 mm rainfall event occurred with air temperatures ranging from 3.0 to 11.9 °C, yet only 0.68 mm of total runoff was generated. This LRET phenomenon occurs annually in the catchment (Gao et al. 2022). During LRET, the topsoil is dry at the start of the thawing season. As thawing progresses, the soil thaws from both the surface and downward, gradually increasing soil moisture. Although the soil above the frozen layer may become saturated with ponding, no percolation or runoff generation occurs due to the underlying frozen soil (Fig. 4.3), indicating a disconnection between the soil and groundwater systems during this period. The LRET phenomenon has also been reported in other regions of the world (Gao et al. 2022).

Discontinuous baseflow recession (DBR) was observed in the frozen basin. In the Hulu Basin, the baseflow exhibited a clear linear recession with a coefficient of 80 days, which decreased to 60 days at the end of the recession period (Fig. 4.4) (Gao et al. 2022), indicating a sudden release of groundwater.

4.2.3 *Impact of Climate Change on Cryospheric Meltwater Runoff*

In the context of climate change, cryospheric runoff is profoundly affected. The responses of glacier, snowmelt, and permafrost meltwater runoff to climate change are outlined below.

(1) **Glacier meltwater runoff**: As temperatures rise during the melting season, the glacier meltwater runoff coefficient for individual glaciers increases significantly. Although the glacier's area may not change substantially, the volume of glacier runoff increases. With continuous warming and minimal changes in precipitation, glacier runoff will initially increase, reach a peak, and then begin to decrease as the glacier area shrinks. Simulations of glacier dynamics suggest that the peak value and timing of glacier meltwater runoff depend on the rate of warming and the glacier's size. Glaciers that are more sensitive to climate change will exhibit higher peak runoff values and earlier peak timing.

As the climate warms, the intensity of glacier melting increases, and the area of bare ice within glaciers expands, leading to a decrease in albedo, which further accelerates glacier melting. Rising temperatures also elevate the glacier equilibrium line, reducing the area and thickness of the glacier's firn and snow cover, exacerbating glacier shrinkage. The extent of changes in glacier accumulation areas caused by the same rise in equilibrium line height can vary across regions and glacier sizes. Smaller glaciers experience relatively greater changes in their accumulation areas compared to larger glaciers, resulting in different responses of meltwater runoff to climate change across various climatic regions and glacier sizes.

In the same basin, glaciers of varying sizes are typically present. The trend of glacier meltwater runoff changes in the basin is influenced not only by the size and

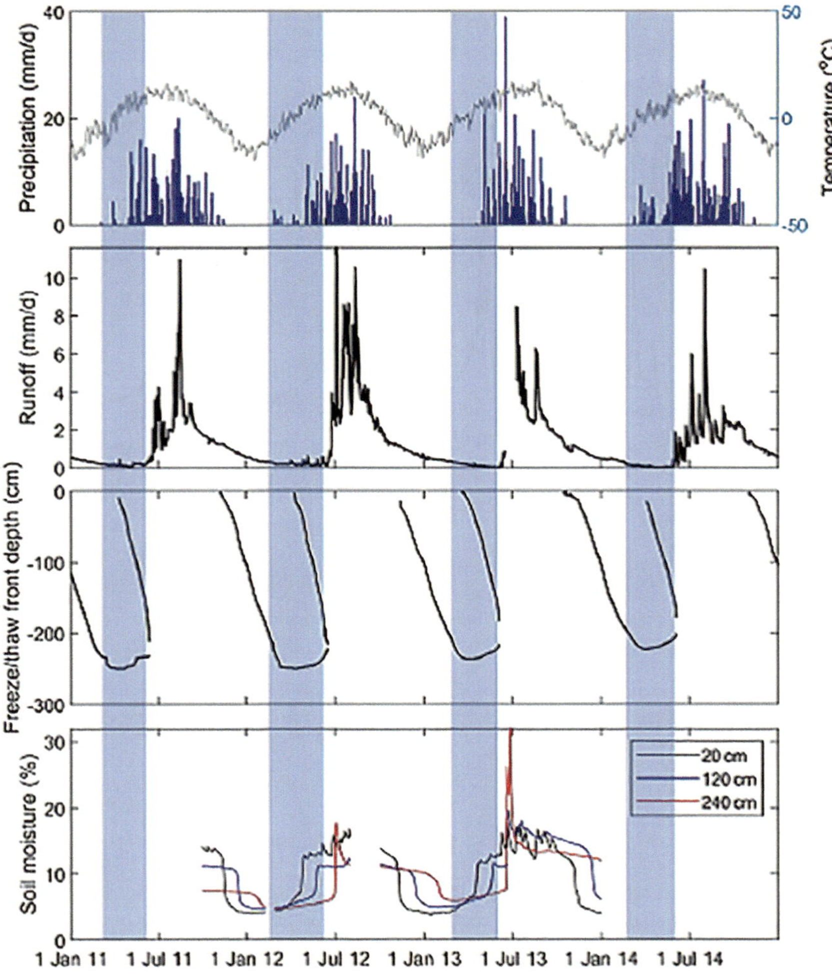

Fig. 4.3 Observed daily precipitation, air temperature, runoff depth, freeze/thaw front depth, and soil moisture at 20, 120, and 240 cm depths in the Hulu catchment (Gao et al. 2022)

coverage of the glaciers but also by differences in glacier mass balance and types. As air temperatures in the basin rise, the rate of glacier melting increases, leading to a continued reduction in glacier area and a general decline in the total amount of glacier meltwater runoff. Consequently, at a certain tipping point, glacier meltwater runoff in the basin will begin to decrease, and the glaciers' capacity to regulate river runoff will gradually diminish until it is eventually lost.

For example, the runoff changes of Glacier No. 1 (43° 05′ N, 86° 49′ E), located at the source of the Urumqi River, which has the longest observation record in China, reflect the response of glacier meltwater runoff in the basin to climate change. The control basin area of the observation station is 3.34 km^2, with glacier

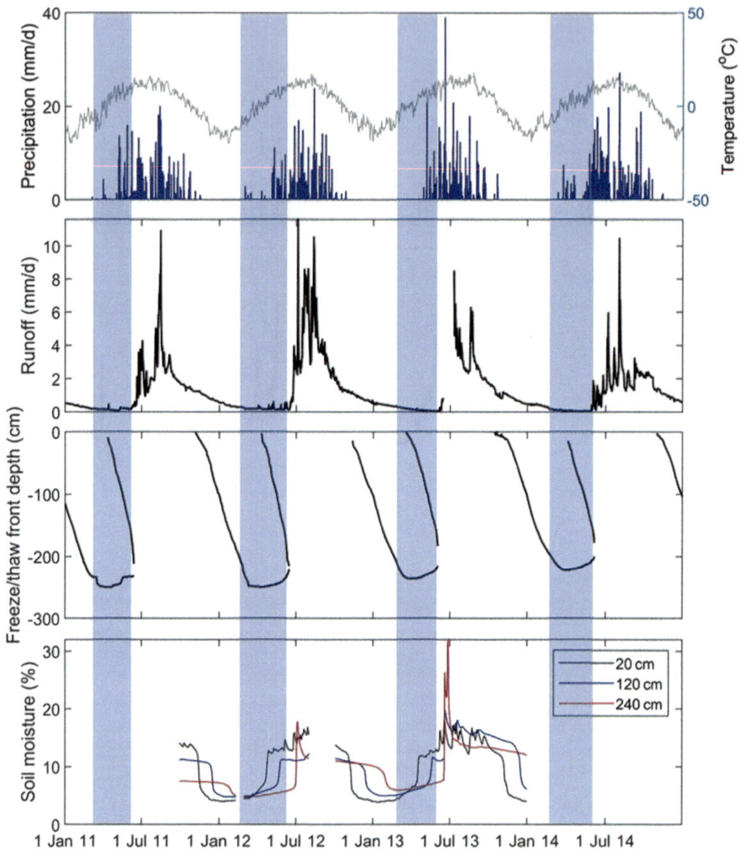

Fig. 4.4 Groundwater recession from 2011 to 2014 on a logarithmic scale, showing a linear recession parameter *Ks* of 80 days in the early recession periods and 60 days at the end of the recession periods (Gao et al. 2022)

coverage decreasing from 55.6% in 1980 to 50.0% in 2006. Measured runoff data (Fig. 4.5) indicate that the average annual runoff depth of glacier runoff increased from 583.1 mm during 1980–1995 to 839.7 mm during 1996–2006, an increase of 256.6 mm, accounting for 37.3% of the average annual runoff depth over the observation period. During the same period, precipitation increased by 84.8 mm (18.1%), contributing approximately 12.3% to the increase in glacier meltwater runoff depth. The remaining 171.6 mm (approximately 25%) is attributed to the loss from the negative glacier mass balance, which was −349.2 mm during the same period. Considering the 53% glacier coverage, this contribution to runoff depth is equivalent to 185 mm (26.9%). The consistency between runoff changes and the basin's water balance suggests that the intense warming has led to significant glacier melting, with the increase in precipitation also playing a reinforcing role.

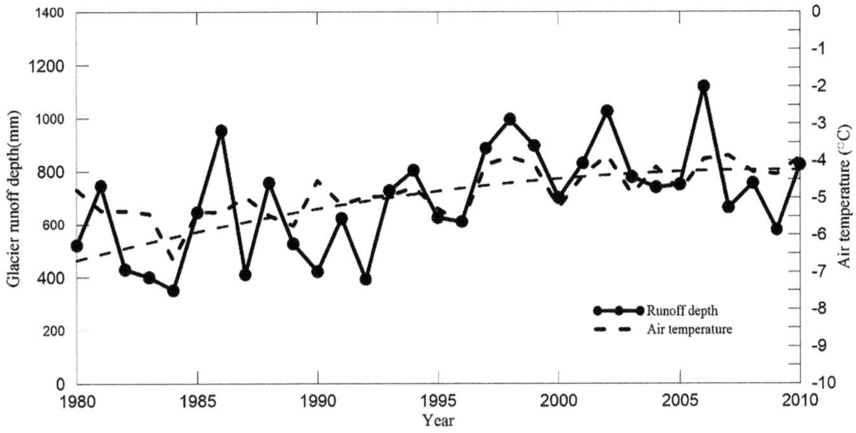

Fig. 4.5 Annual runoff depth changes of Glacier No. 1, the source of the Urumqi River. The dotted line represents a second-degree polynomial regression, based on data from the "Urumqi No. 1 Glacier Annual Report"

From the perspective of glacier meltwater runoff in China, temperatures in western China have generally shown an upward trend over the past 50 years, particularly since 1980. As temperatures continue to rise, glaciers are shrinking, and the volume of glacier meltwater has significantly increased. According to the modified monthly degree-day model (Zhang et al. 2012), changes in glacier meltwater across various river systems in western China indicate that, in the Tarim River system, glacier meltwater increased by 59.34×10^8 m^3 from 2000 to 2016 compared to the 1960s. The Qinghai-Tibet Plateau's inflow river system experienced the largest relative change, with glacier meltwater increasing by 162.1%. In the outflow river system, the Ganges River system, which has the largest glacier area, also saw the most substantial change, with an average increase of about 2.87×10^8 m^3/a.

The average annual glacier meltwater volume in western China was 629.56×10^8 m^3 from the 1960s to 2006, increasing from 517.65×10^8 m^3 in the 1960s to 601.57×10^8 m^3 in the 1970s and 1980s, and further rising to 695.48×10^8 m^3 in the 1990s. After 2000, glacier meltwater volume reached its highest level in 46 years. During this period, the average annual meltwater volume was 794.67×10^8 m^3, 26.2% higher than the multi-year average. Compared to the 1960s, the amount of meltwater from 2000 to 2006 increased by 54%.

From a spatial distribution perspective, the average annual runoff depth of glacier meltwater in China increased in most basins from 2001 to 2006 compared to the 1960s. The largest increase occurred in southeastern Tibet, reaching up to 2000 mm. Another significant increase was observed in the western section of the Himalayas, with a rise of 1300 mm, followed by the Xinjiang region, which saw an increase of 100–500 mm, and the Hexi area, with an increase of 200–400 mm. The maximum relative increase occurred in southern, southeastern, and central Tibet, with a relative change of up to 160%. Xinjiang experienced a relative increase of 20–30%, while the Hexi area showed an increase of 30–60%.

(2) **Snowmelt runoff**: Changes in snowmelt runoff are closely linked to variations in snow cover. Snowfall trends vary widely on a global scale, and due to climate warming, the snow cover period in the Northern Hemisphere during spring and summer has significantly shortened, leading to an earlier snowmelt date and increased early snowmelt runoff. The end of the snowmelt season has been delayed, resulting in a decrease in late snowmelt runoff but an overall increase in total snowmelt runoff. This shift has altered the seasonal distribution of runoff in basins. For rivers primarily fed by snowmelt, the seasonal distribution of runoff varies significantly, with distinct spatial patterns across different regions.

For example, in North America, the start of snowmelt has advanced by 15–20 days from 1948 to 2002, with the peak snowmelt runoff in many rivers occurring 10–30 days earlier (Fig. 4.6). In China, the Kelan River, dominated by snowmelt, has seen an increase in snow cover and a significant advancement in the snowmelt runoff concentration period. The peak runoff month has shifted from June to May, with maximum monthly runoff increasing by 15%. The proportion of runoff during the snowmelt season (April to June) has risen from 60% to nearly 70% of the total annual runoff, with snowmelt floods occurring earlier and peak flood discharge increasing. The maximum observed peak discharge has risen from 200 m^3/s in the 1970s–1980s to 350 m^3/s since the 1990s, increasing flood hazards.

In China, the western high-altitude regions have shown an overall increasing trend in snowmelt runoff from 1960 to 2012/2014. Significant increases in snowmelt runoff have been observed on the southern slopes of the Tianshan Mountains, the western section of the Qilian Mountains, the source of the Yangtze River, and the Changbai Mountains. However, most parts of Northeast China have experienced a significant decrease in snowmelt runoff.

(3) **Permafrost meltwater runoff**: In the context of global warming, permafrost degradation causes the thickening of the active layer, altering soil water storage capacity and impacting the runoff coefficient and recession processes of basins. Changes in the maximum soil freezing depth of seasonal frozen soil also influence runoff. However, directly observing the long-term impact of permafrost on runoff is challenging. The impact is often inferred from changes in river runoff across different regions with varying permafrost coverage.

For example, in Arctic basins, permafrost degradation has led to increased permafrost temperatures, thickening of the active layer, and expansion of the melting zone. This has resulted in a significant increase in the annual runoff of major Arctic rivers, though the changes vary widely across different basins. In the Lena and Yenisei Rivers in central Siberia, winter runoff has increased significantly, whereas this phenomenon is less pronounced in the Mackenzie and Yukon Rivers in Canada.

The Qmax/Qmin ratio (maximum monthly runoff divided by minimum monthly runoff) is used as an indicator to quantify the impact of permafrost on hydrography recession and is widely applied in other regions. Many studies have found that under

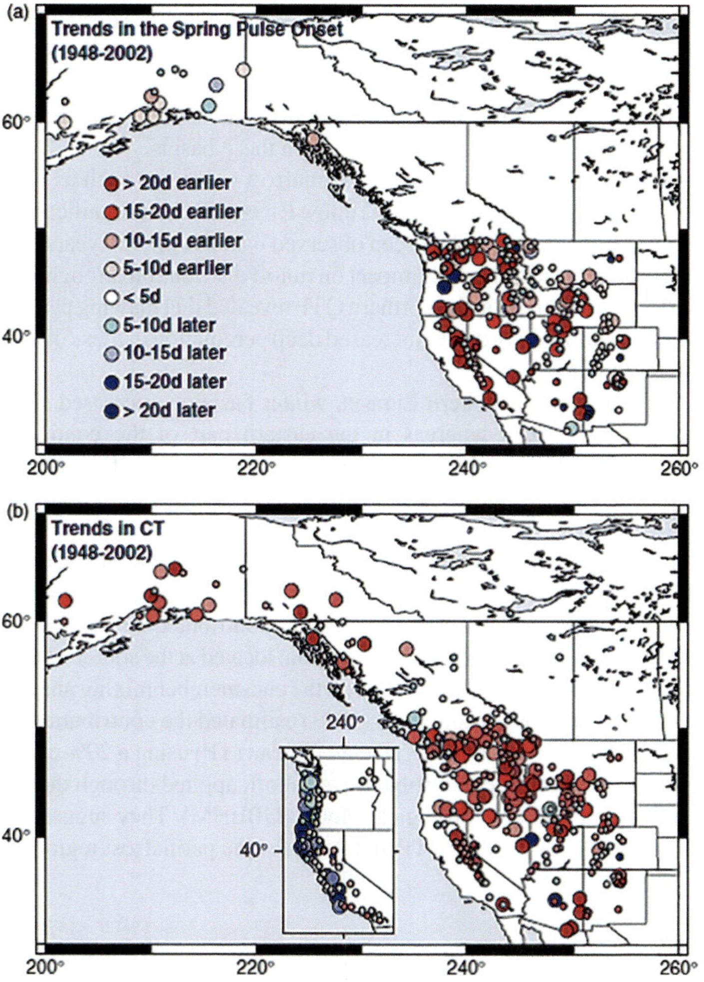

Fig. 4.6 Trends in **a** spring pulse onset and **b** date of center of mass of annual flow (CT) for snowmelt-dominated and (inset) non-snowmelt-dominated gauges across western North America. Shading indicates the magnitude of the trend, expressed as the change in timing (days) over the 1948–2000 period. Larger symbols indicate statistically significant trends at the 90% confidence level. Note that spring pulse onset dates could not be calculated for Canadian gauges (Stewart 2009)

the influence of climate change on the Qinghai-Tibet Plateau (QTP), the Qmax/Qmin ratio has decreased. This decrease is attributed to a reduction in Qmax during the thawing summer season, caused by increased infiltration and percolation, while Qmin in winter has increased due to enhanced groundwater recession.

Figure 4.7 illustrates changes in the recession coefficient and negative accumulated temperature at the control hydrological stations of Zamusi, Yingluoxia, Changmabao, and Tangnaihai, located in the source areas of the Shule River, Heihe

River, Shiyang River, and Yellow River, respectively. From 1950 to 2010, rivers with high permafrost coverage, such as the Shule River (73%) and Heihe River (58%), experienced an increase in the winter recession coefficient, indicating a significant slowdown in the recession process. This mitigation is consistent with the continuous decrease in negative accumulated temperature in these basins.

In contrast, in watersheds with lower permafrost coverage, such as the Shiyang River (33%) and the upper reaches of the Yellow River (43%), no significant change in the winter recession coefficient has been observed over the past 50 years, suggesting that permafrost degradation has less impact on runoff distribution throughout the year. However, a long-term study on the northern QTP revealed that thawing permafrost has reduced winter runoff, likely due to increased deep percolation in these discontinuous permafrost regions.

In the Arctic region of western Eurasia, winter baseflow increased significantly across basins of all sizes, whereas in the eastern part of the continent, runoff showed no obvious changes. In the Arctic region of eastern North America, winter runoff decreased, while in the western part of the continent, winter runoff increased (Fig. 4.8).

Some studies have attempted to quantitatively evaluate the contribution of meltwater of frozen ground to river runoff. Li et al. (2020) reported that supra-permafrost water, precipitation, and glacier/snow meltwater contributed 51%, 26%, and 23%, respectively, to river water at the Tuotuohe station, located at the source of the Yangtze River, from June 2016 to May 2018, based on the end-member mixing analysis model of stable isotope $\delta^{18}O$ (δD). Wang et al. (2023) estimated the contribution of ground ice storage loss to river runoff on the Tibetan Plateau (TP) using a 20% change simulation in ground ice meltwater contribution to runoff, applied through the distributed Geomorphology-based Ecohydrological Model (GBEHM). They suggested that this contribution is 2.2% for the entire TP and 9.6% for the permafrost region within the TP.

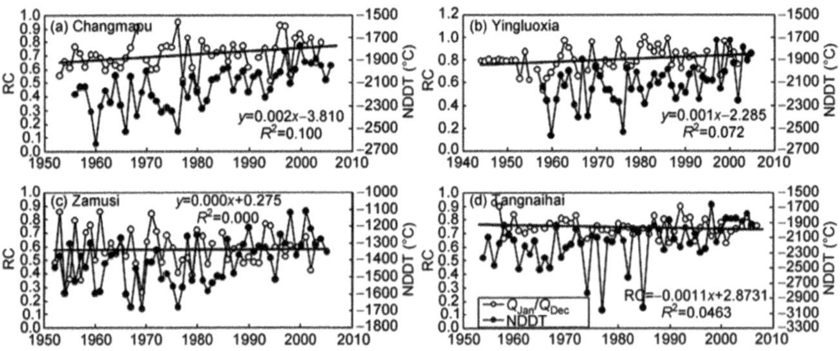

Fig. 4.7 Changes in the recession coefficient (RC) and annual negative degree-day temperature (NDDT) in the source areas of the Shule River (**a**), Heihe River (**b**), Shiyang River (**c**), and Yellow River (**d**) from 1950 to 2007 (Niu et al. 2011)

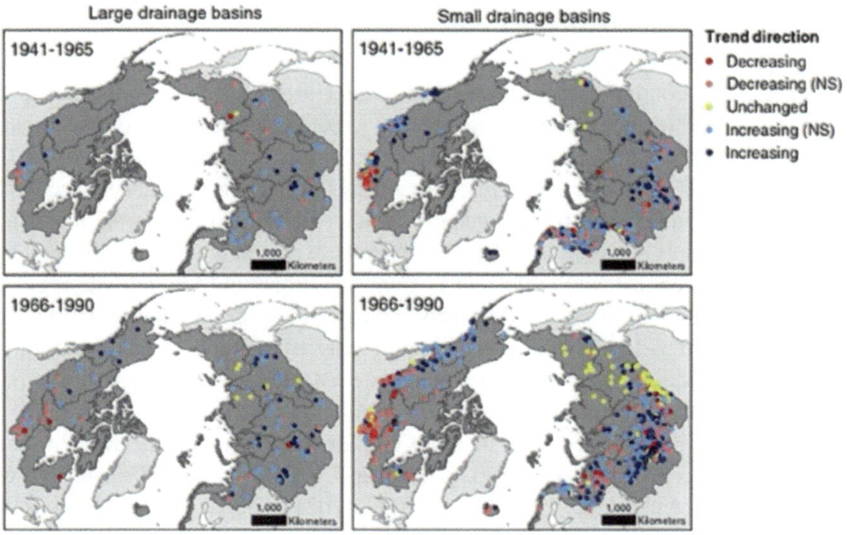

Fig. 4.8 Spatial distribution of short-term trends (25 years) at stations monitoring large (>50,000 km^2) and small (10–50,000 km^2) river basins during the mid- and late-twentieth century (1941–1965 and 1966–1990, respectively) (Rennermalm et al. 2010)

Additionally, in the context of climate warming, ice conditions and the timing of river and lake thawing in high-altitude and mid- to high-latitude regions are also changing. The onset of the freezing period for rivers is being delayed, the duration of the freezing period is shortening, and the thawing period is also becoming shorter, leading to an earlier thawing date.

4.3 Projection of Cryospheric Meltwater

Building on the successful simulation of changes in cryospheric meltwater runoff in typical river basins over the past few decades, several studies have projected future changes in meltwater and river runoff across different types of cryospheric meltwater-replenished rivers. Some studies have also provided partial projections of changes in cryospheric meltwater runoff on a global scale. However, the trends projected by different models show certain discrepancies. Therefore, this section briefly introduces the projected future changes in glacier meltwater runoff, snowmelt runoff, and permafrost meltwater, followed by an analysis of the uncertainties in these estimations.

4.3.1 Projection of Glacier Meltwater Runoff

As the climate continues to warm, global glacier area and volume are expected to decrease in the future. According to three greenhouse gas emission scenarios (RCP2.6, RCP4.5, and RCP8.5), global glacier volume is projected to decline by 25% $\pm$ 6%, 33% $\pm$ 8%, and 48% $\pm$ 9%, respectively, by the end of the twenty-first century (IPCC 2021). Under the RCP4.5 scenario, glacier volume in the northern Arctic, Antarctic, and Sub-Antarctic regions of Canada is expected to decrease by 20% by the end of the century, while glaciers in Central Europe and low-latitude regions could see a reduction of up to 90%. This indicates that, on a global scale, glacial meltwater runoff is likely to decrease significantly by the end of the twenty-first century, greatly diminishing its ability to regulate river runoff. This poses substantial challenges for water resources management in inland arid regions.

The IPCC AR6 assessed with medium confidence that under the RCP2.6 and RCP8.5 scenarios from 2015 to 2100, glaciers are expected to lose 18% and 36% of their early 21st-century mass, respectively (IPCC 2021). Global glacier mass loss by 2100 is projected to be 18 $\pm$ 13% with 0.9–2.3 °C of global warming and 36 $\pm$ 20% with 3.2–5.4 °C of global warming. The projections of regional glacier loss rates are unevenly distributed and vary significantly between scenarios. In most regions, "peak water" has either already been reached or is expected to be reached before mid-century, with an earlier "peak water" under RCP2.6 compared to RCP8.5. The expected decrease in glacier runoff by the end of the twenty-first century will be more pronounced during droughts and dry seasons, potentially leading to water shortages for over 200 million people in the high mountains of Asia.

At the basin scale, the trend of glacier meltwater runoff varies significantly across different climatic regions and basins, closely linked to the size distribution of glaciers and future climate change scenarios. Basins dominated by larger glaciers tend to experience a slower reduction in glacier area, and meltwater runoff may continue to increase until the mid-twenty-first century. Conversely, basins dominated by smaller glaciers are more sensitive to climate warming, leading to a rapid decrease in glacier area and potentially reaching a tipping point.

For example, projections for the Yarkant River Basin in China, where the average glacier size is 1.94 km^2, suggest that the depth of glacier meltwater runoff will continue to increase until 2050. Due to the slower shrinkage of glacier area, glacier meltwater runoff, including rainfall runoff from glacier retreat areas, is expected to continue rising until 2050. The average annual glacier meltwater runoff from 2011 to 2050 is projected to increase by 13–35% compared to the average from 1961 to 2006.

In contrast, in the Beida River Basin in the Qilian Mountains, located in the northern temperate zone, where the average glacier area is 0.45 km^2, projections indicate no clear trend in glacier meltwater. Due to the rapid decrease in glacier area, runoff from glacier areas continues to decline. The annual glacier runoff, including rainfall runoff in glacier retreat areas, is expected to reach a tipping point, shifting from increases to decreases between 2011 and 2030.

4.3.2 Snowmelt Runoff Projection

Projections from various models indicate a continued decline in snow water equivalent (SWE) and snow cover extent (SCE), though there is considerable variation among models. The SCE in the Northern Hemisphere is expected to decrease by approximately 20% relative to the 1995–2014 mean by around 2060 and then stabilize under the RCP2.6 scenario. In contrast, the RCP8.5 scenario projects snow cover losses of up to 60% by 2100 (IPCC 2021). Regionally, the loss of SWE is likely to result in more frequent snow droughts. For instance, the frequency of consecutive snow droughts is projected to increase to 80–100% of years at 4 °C warming in western Canada and to 42% of years under the RCP8.5 scenario in the western USA by 2100.

At the basin scale, the impact of projected changes in snowmelt runoff on river runoff varies significantly. For example, in the Rhine River Basin, air temperature is projected to increase by 1.0–2.4 °C by the middle of the twenty-first century, shifting the dominant runoff source from snowmelt and rainfall to primarily rainfall. This shift will likely result in increased winter runoff, decreased summer runoff, higher peak flows, and more frequent peak flow events. Low runoff in summer will become more frequent and prolonged, significantly impacting local social and economic systems.

In the eastern mountainous regions of Turkey, where river runoff is mainly derived from snowmelt, projected temperature increases of about 6 °C by the end of the twenty-first century will lead to reduced snowfall in winter and an earlier snowmelt period. The peak runoff in the region's four watersheds is expected to occur an average of four weeks earlier, with a significant increase in winter flow. As a result of earlier snowmelt, runoff will be significantly reduced from April to May. These changes in runoff will have a considerable impact on the safety of local reservoirs and water resources management.

There is medium confidence that under a 4 °C warming scenario, 40% of the current irrigated demand in sub-basins that rely primarily on snowmelt runoff would need to be supplemented by other water sources. Basins where alternative sources are unavailable will likely face agricultural water scarcity. Additionally, the projected changes in snow and glacier melt runoff are expected to reduce water inputs to hydropower, leading to a decline in hydroelectricity production in mountain basins, such as those in India, Switzerland, and the USA (IPCC AR6 WGI, 2021).

4.3.3 Permafrost Meltwater Projection

Permafrost is expected to undergo increasing thaw and degradation worldwide during the twenty-first century. CMIP6 models project that the annual mean frozen volume in the top 2 m of soil could decrease by 10–40% for every degree of global temperature increase. The southern boundary of the permafrost is projected to shift northward by 1°–3.5° (relative to 1986–2005) with a 1.5 °C temperature rise (IPCC 2021).

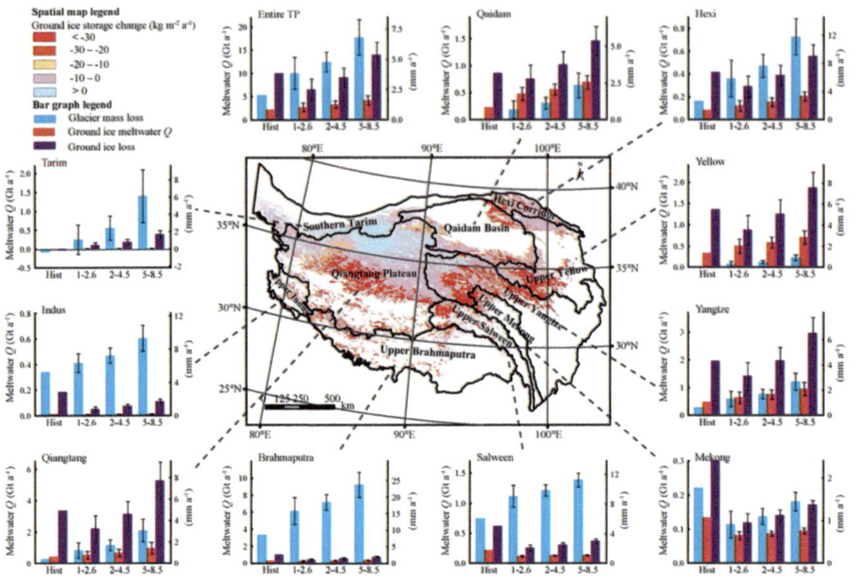

Fig. 4.9 Ground ice loss and meltwater runoff across the Tibetan Plateau (TP) during historical and future periods. The central map shows the spatial distribution of the rate of change in ground ice storage from 1980 to 2019. Areas where permafrost was absent throughout the historical period are masked in white. The bar graphs depict the amount of glacier mass loss (blue), ground ice loss (purple), and ground ice meltwater runoff from thawing permafrost (Qpf) (red) for the entire TP and 10 sub-regions during the historical period (1980–2019, Hist) and the future period (2020–2100) under three emission scenarios (SSP1-2.6, SSP2-4.5, and SSP5-8.5). Negative values indicate increased ground ice and/or glacier mass. Q values are presented as both annual runoff volume (Gt a^{-1}) and annual average runoff depth over the entire TP or each sub-region (mm a^{-1}). Error bars represent the uncertainty range from different CMIP model projections. *Note* 1 Gt of water or ice = 10^9 m^3 of water (Wang et al. 2023)

Future changes in permafrost and ground ice, along with their contributions to river runoff on the Tibetan Plateau (TP) from 2020 to 2100, have been projected using the ensemble mean of climate projections from 38 CMIP6 models under the SSP1-2.6, SSP2-4.5, and SSP5-8.5 scenarios. In a warmer future climate, the area underlain by permafrost is expected to decrease by 23.3% ± 4.5% (SSP1-2.6), 26.3% ± 3.3% (SSP2-4.5), or 30.5% ± 3.7% (SSP5-8.5) during 2020–2100 (Fig. 4.9). Correspondingly, ground ice storage is projected to decline by 28.8% ± 10.1%, 40.1% ± 9.5%, or 61.6% ± 12.7% under these scenarios, respectively (Wang et al. 2023).

By tracing meltwater flow paths, the contribution of ground ice melt in permafrost to total river runoff across the TP is estimated to be 0.6% ± 0.1% (SSP1-2.6), 0.7% ± 0.1% (SSP2-4.5), and 0.9% ± 0.2% (SSP5-8.5). This contribution is more pronounced in specific regions, including the upper Yellow River and the Qaidam Basin (Fig. 4.9), and at certain elevation ranges, such as around 4600 m in the upper

Yangtze, 4800 m in the upper Mekong, and 4000 m in the Hexi Corridor (Wang et al. 2023).

At the local scale, about 17.3% (SSP1-2.6), 22.0% (SSP2-4.5), or 30.0% (SSP5-8.5) of the permafrost region shows a >5% contribution to river runoff, indicating a higher reliance on permafrost meltwater in future climate scenarios compared to the historical period (Wang et al. 2023).

4.3.4 Uncertainty of Projections

The uncertainties in projecting future changes in cryospheric runoff primarily arise from three factors: the uncertainty of climate scenarios, the uncertainty of downscaling methods, and the uncertainty of hydrological models and parameters.

(1) **Uncertainty of Climate Scenarios**: Current climate change projections are primarily based on outputs from CMIP6 GCM models, which are inherently incomplete and imperfect in simulating real-world climate conditions. Significant errors in precipitation, which is crucial for hydrological simulations, contribute to this uncertainty. Additionally, the greenhouse gas emission scenarios that underpin these models introduce further variability, as their assumptions directly influence model outputs. Even under the same emission scenario, temperature and precipitation projections can vary substantially between different GCMs. As a result, the uncertainty of climate scenario data remains the most significant source of uncertainty in cryospheric meltwater runoff projections.

(2) **Uncertainty of Downscaling Methods**: Climate variables predicted by global climate models (GCMs) are generated on a global scale, creating a spatial resolution mismatch when applied to small-scale hydrological models. To address this issue, dynamic or statistical downscaling methods are employed. However, despite the development of various downscaling techniques, statistical downscaling from different regional climate models and land surface parameterization schemes for dynamic downscaling can produce significant discrepancies in regional climate projections. These differences further increase the uncertainty in projected runoff changes.

(3) **Uncertainty of Hydrological Models and Parameters**: The uncertainty in hydrological models primarily arises from the model structure, input data, and model parameters. Errors stemming from the model's structure contribute to the uncertainty of the projected results. The hydrological processes in cryospheric regions are complex, and the long-term impacts of permafrost meltwater are difficult to observe directly, leading to significant uncertainty in simulating water-heat transfer processes and frequent phase changes in the soil. Model parameters are typically calibrated and validated using historical observational meteorological data and simulated runoff or other datasets. However, the accuracy of these parameters depends on several factors: the quality of the calibration

data, methods of parameter identification and optimization, spatial distribution of parameters, and the inherent uncertainty of the model's physical parameters. Unreasonable or inaccurate model parameters further increase the uncertainty in simulation and projection.

Questions

1. Based on the characteristics of water replenishment, how can rivers in the cryosphere be classified?
2. What are the characteristics of cryospheric melt runoff in response to climate change over the past few decades?
3. What are the main sources of uncertainty in estimating future changes in cryospheric meltwater runoff?

References

Gao M, Han T, Ye B, Jiao K (2013) Characteristics of melt water discharge in the Glacier No. 1 basin, headwater of Urumqi River. J Hydrol 489:180–188

Gao H, Han C, Chen R, Feng Z, Wang K, Fenicia F, Savenije H (2022) Frozen soil hydrological modeling for a mountainous catchment northeast of the Qinghai-Tibet Plateau. Hydrol Earth Syst Sci 26(15):4187–4208

IPCC (2021) Climate change 2021: the physical science basis. Contribution of working group I to the sixth assessment report of the Intergovernmental Panel on Climate Change. In: Masson-Delmotte V, Zhai P, Pirani A, Connors SL, Péan C, Berger S, Caud N, Chen Y, Goldfarb L, Gomis MI, Huang M, Leitzell K, Lonnoy E, Matthews JBR, Maycock TK, Waterfield T, Yelekci O, Yu R, Zhou B (eds) Cambridge University Press, Cambridge, United Kingdom and New York, NY, USA, 2391 pp. https://doi.org/10.1017/9781009157896

Li Z, Li Z, Feng Q, Zhang B, Gui J, Xue J, Gao W (2020) Runoff dominated by supra-permafrost water in the source region of the Yangtze river using environmental isotopes. J Hydrol 582:124506

Niu L, Ye BS, Li J, Sheng Y (2011) Effect of permafrost degradation on hydrological processes in typical basins with various permafrost coverage in Western China. Sci China Earth Sci 54:615–624

Rennermalm AK, Wood EF, Troy TJ (2010) Observed changes in pan-arctic cold-season minimum monthly river discharge. Clim Dyn 35(6):923–939

Stewart IT (2009) Changes in snowpack and snowmelt runoff for key mountain regions. Hydrol Process Int J 23(1):78–94

Wang T, Yang D, Yang Y, Zheng G, Jin H, Li X, Yao T, Cheng G (2023) Unsustainable water supply from thawing permafrost on the Tibetan Plateau in a changing climate. Sci Bull S2095-9273(23):00291–8

Zhang S, Ye B, Liu S, Zhang X, Hagemann S (2012) A modified monthly degree-day model for evaluating glacier runoff changes in China. Part I: model development. Hydrol Process 26(11):1686–1696

Extended Reading

Chen R, Zhang S, Zhao Q et al Impact of cryosphere change on clod region runoff in Western China (in Chinese). Science Press, Beijing

Ding Y, Zhang S, Chen R (2017) Introduction on cold region hydrology (in Chinese). Science Press, Beijing

Chapter 5
Hydrochemistry for Snowpack, Glaciers and Permafrost

Xiangying Li, Jinkui Wu, and Tianding Han

Climate warming has led to the melting of snowpack, the retreat of glaciers, and the degradation of permafrost. The resulting meltwater carries a large amount of chemical species downstream, impacting streamflow water quality and the primary productivity of aquatic ecosystems, thereby affecting elemental cycles (Fig. 5.1). Additionally, glacier retreat and permafrost degradation can release or sequester greenhouse gases, generating feedback effects on the climate system. Generally, the chemical species in meltwater comprise both inorganic and organic components. While major ions predominantly dictate meltwater chemistry, the presence of minor elements imparts unique properties and functions to water bodies. Despite this, research on organic species remains limited, with current investigations primarily focusing on organic carbon and nitrogen within glacial and permafrost catchments.

X. Li (✉)
Shaanxi Key Laboratory of Earth Surface System and Environmental Carrying Capacity, Northwest University, Xi'an, China
e-mail: shaanxilxy@163.com

J. Wu · T. Han
Northwest Institute of Eco-Environment and Resources, Chinese Academy of Sciences, Lanzhou, China
e-mail: jkwu@lzb.ac.cn

T. Han
e-mail: tdhan@lzb.ac.cn

Y. Ding et al., *Cryohydrology*, https://doi.org/10.1007/978-981-96-7360-5_5

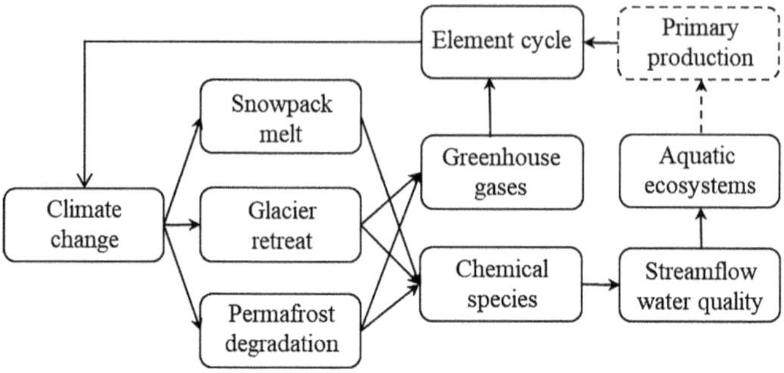

Fig. 5.1 Relationships between snowpack, glaciers, permafrost and streamflow water quality, aquatic ecosystem, and carbon cycles

5.1 Snowpack Hydrochemistry

The biogeochemical cycling of chemicals is influenced by snowfall and snowpack dynamics. Snowpacks act as reservoirs, storing chemicals throughout the winter and reflecting the chemical nature of both wet and dry deposition, encompassing naturally occurring substances and pollutants. Snowpack chemistry is also impacted by interactions with plant canopies and the activity of organisms within the snow. During melting or rain-on-snow events, fractionation processes redistribute and release these chemicals from the snowpack. In polluted environments, acidic snowmelt runoff can lead to toxic levels of dissolved heavy metals in streams, causing fish kills in early spring. Additionally, stable isotope fractionation occurs in snow during winter and can be used to study the phase changes that are important in snow hydrology.

5.1.1 Snowpack Chemistry

Snowpacks accumulate and lose chemicals through various pathways. Major pathways for chemical gains include snowfall, rainfall, and dry deposition. These processes create chemically distinct layers within snowpacks, which often persist with identifiable characteristics until they come into contact with liquid water. Chemical losses occur only when liquid water carrying chemicals drains from the snowpacks, as liquid water within snowpacks primarily serves to redistribute the chemicals. Snowpacks in windy areas also accumulate or lose chemicals due to drifting or scouring by the wind. In contrast, forest snowpacks are generally protected from the wind but can receive large amounts of organic debris from the overhead canopy.

The blowing-snow particles can become enriched through several processes such as the sublimation of pure water and the scavenging of gases and aerosols. The deposition of blowing snow significantly impacts species concentrations in snowpacks. If these particles are completely vaporized, the gases and aerosols can be re-released into the atmosphere. Snow crystals at the snowpack surface can also sublimate without wind transport. Other pathways for chemical gains or losses include vapor transport and microbiological processing. Additionally, the movement of gases within snowpacks can influence snowpack chemistry.

The internal processing of snowpack chemical loads can either increase or decrease the amount of dissolved solids. In forest snowpacks, chemicals are often lost from meltwater due to assimilation by organisms such as lichens, algae, bacteria, and fungi associated with organic debris from the canopy. These chemical losses may be more significant in forest snowpacks, which contain a greater abundance of microflora and fauna. Overall, snowpack chemistry reflects variations in wet-deposition chemistry due to human pollution (e.g., coal combustion) and natural factors (e.g., aeolian dust). Notably, the most effective way to assess the impact of snowpacks on chemical cycling is to measure the chemistry of the water leaving the snowpack.

5.1.2 Snowpack Meltwater Chemistry

Snowpack meltwater represent a chemical flux density delivered to the soil surface. The chemical flux density (F) for a chemical species is computed as:

$$F = M \times c \tag{5.1}$$

where M is the meltwater expressed as a liquid volume per unit ground area ($m^3\ m^{-2} = m$), and c is the concentration of chemical species in meltwater.

The chemical flux density of chemical ions in meltwater can be compared with the ion fluxes in wet and dry deposition. However, ion concentrations in meltwater do not merely reflect snowpack chemistry. This discrepancy arises from the fractionation and preferential elution processes occurring within the snowpacks.

1. **Melt fractionation process**

Impurities in snowpack generally reside on the surface of ice grains. This allows early liquid water passing through a snowpack layer to flush out a greater amount of substances than water passing through the same layer later. This process, known as melt fractionation, refers to the separation of constituents based on their solubility and availability. During the formation of ice crystals, most impurities are excluded and migrate to the surface. Substances scavenged from the atmosphere also accumulate on ice crystal surfaces. Once in the snowpack, the metamorphosis of ice grains causes further migration of impurities to the crystal surface. As a result, early flushes of meltwater from a snowpack tend to wash out more impurities compared to late-season meltwater, which is relatively depleted of impurities.

The melt fractionation of snowpack chemicals into early meltwater has been studied extensively. For example, concentrations of species such as SO_4^{2-} and NO_3^-

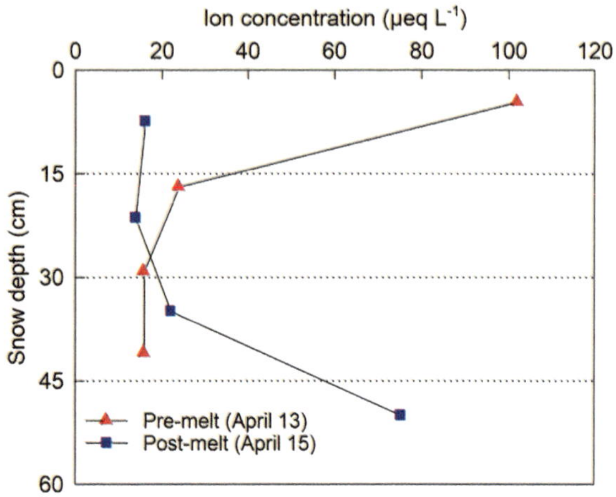

Fig. 5.2 Change in total ion concentrations (Ca^{2+}, Na^+, Cl^-, NO_3^-, and SO_4^{2-} in snowpack at 240 sites) with snow depth in British Columbia, Canada (modified from Hudson and Golding 1998)

in early meltwater were several times higher than those in the initial snowpack, while late meltwater contained only a small fraction of these concentrations. Additionally, concentrations of some ions shifted from higher levels in the surface layers to higher levels in the deeper layers due to the fractionation process (Fig. 5.2).

In polluted environments, the initial flush of acid snowmelt can lead to stream acidification and elevated concentrations of heavy metals, which are toxic to aquatic animals. The winter climate and the frequency of melt-freeze cycles help control the magnitude of acidic snowmelt release. In regions with deep snowpacks and infrequent winter thaws, the effects of fractionation isolate in spring. The highest concentrations occur during longer-duration melt and lower-melt-rate events. Conversely, in regions with shallow snowpacks and frequent winter thaws, fractionation reduces acidity due to frequent liquid–water release, thereby lessening the severity of snowmelt.

The mechanism of meltwater delivery from snowpacks interacts with fractionation effects. During a daily melt cycle, the initial meltwater typically contains high concentrations of species stored at the base of snowpacks. Subsequent meltwater comes from current melting of surface snow, which has lower species concentrations. Thus, the meltwater released each day forces the stored water from the snowpack base to flow in a process similar to translatory flow often seen in watershed soils. This results in meltwater concentrations varying from high to low throughout the cycle.

2. **Preferential elution process**

In addition to fractionation effects, species are not flushed from snowpacks with the same efficiency in early meltwater. The tendency for some ions to be washed out before others is known as preferential elution. According to Brimblecombe et al. (1985), the elution sequence from high to low mobility in snowpacks is:

$$SO_4^{2-} > NO_3^- > NH_4^+ > K^+ > Ca^{2+} > Mg^{2+} > Na^+ > Cl^-$$

The reasons for preferential elution are not entirely clear, but it may be related to the ability of ions to incorporate into the ice crystal lattice. Preferential elution results in fresh snow, which initially contains a wide range of anion concentrations, gradually becoming depleted of SO_4^{2-} and dominated by Cl^-. Additionally, meltwater from snowpacks is relatively enriched in SO_4^{2-} compared to the within-pack snow.

The effects of fractionation and preferential elution lead to a more acidic release of snowmelt early in the season. Not only are more impurities released from snowpacks early on, but more acidic anions and H^+ are also released sooner. Consequently, in regions with deep acidic snowpacks, these processes create an early acidic pulse in meltwater, significantly influencing soil–water and streamflow chemistry.

5.1.3 Solute Release and Streamflow Water Quality

Streamflow during snowmelt consists of a mixture of event water from snowmelt or rain-on-snow and pre-event water from subsurface soil and groundwater. Pre-event water, which has been in contact with organic and inorganic soil and rock material, may contain high levels of dissolved solids and considerable acid neutralizing capacity (ANC) due to interactions with easily weathered minerals and prolonged subsurface residence time. The acidity of streamflow during snowmelt depends on the fractional contributions and chemistry of the meltwater and the buffering capacity of the pre-event water. If the meltwater contribution is significant and the buffering capacity of the pre-event water is insufficient, the resulting streamflow can become highly acidic, leading to toxic conditions for aquatic organisms.

For example, one stream, Wildcat Run, had high ANC pre-event water due to groundwater contact with easily weathered limestone (Fig. 5.3). In contrast, the adjacent basin, McGinnis Run, had pre-event water with low ANC because the groundwater was exposed to weather-resistant sandstone. During a large rain-on-snow event, both watersheds experienced significant increases in streamflow and drops in pH. However, the pH in Wildcat Run never fell below 5.6, whereas pH in McGinnis Run dropped to nearly 4.4 at peak flow (Fig. 5.3). The low pH in McGinnis Run was accompanied by high dissolved-aluminum concentrations. Notably, McGinnis Run lacks resident trout, while Wildcat Run supports a healthy, reproducing trout population. Acidic meltwater can significantly impact stream chemistry, but its effects will vary depending on the biogeochemical nature of the watershed.

5.2 Glacial Hydrochemistry

Glacial meltwater chemistry is closely related to the configuration and dynamic nature of the subglacial hydrological system. As meltwater sources change and transport pathways evolve, chemical weathering reactions—and thus meltwater chemistry—are primarily controlled by hydrological flow paths. Chemical weathering can

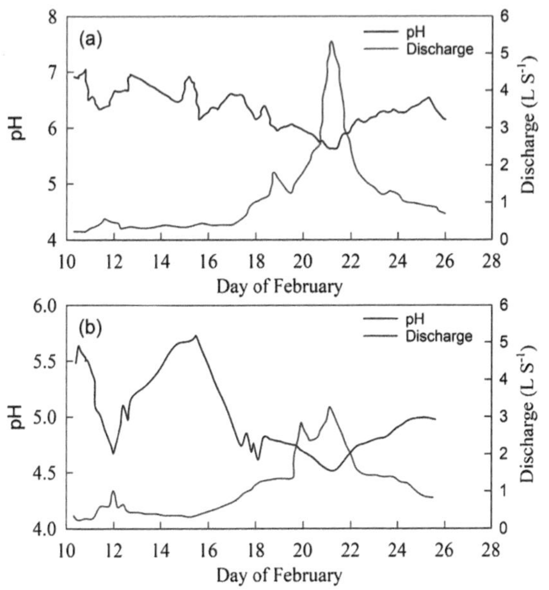

Fig. 5.3 Comparison of discharge and pH in the Wildcat Run (**a**) and McGinnis Run (**b**) basins during a rain-on-snow event in southwestern Pennsylvania (modified from Sharpe et al. 1984). Note that the Wildcat Run with well-buffered pre-event water showed a much smaller pH decline during the event than the McGinnis Run with poorly buffered pre-event water

influence climate change by regulating atmospheric CO_2 concentrations over geological timescales. As glacier melt increase, chemical species stored in glaciers and resulting from chemical weathering are released downstream, significantly impacting streamflow water quality and elemental biogeochemical cycles.

5.2.1 Major Ions and Minor Elements

The chemical composition of glacial meltwater is influenced by bedrock type, the geochemical characteristics of the meltwater, and the evolution of the drainage system. In glacial environments, chemical species in meltwater primarily consists of major ions and minor elements derived from chemical weathering processes, as well as aerosol and marine inputs. Major cations from chemical weathering can be used to evaluate the chemical denudation rates in a basin, while some elements can influence downstream water quality. Glacial catchments exhibit different temporal variations in streamflow chemistry compared to catchments under different hydrological and climatological conditions, as bulk streamflow is decoupled from precipitation events.

Many studies have focused on the major ions in glacial meltwater, including Na^+, K^+, Ca^{2+}, Mg^{2+}, HCO_3^-, SO_4^{2-}, Cl^-, and NO_3^-. Generally, the concentration orders for anions and cations are generally $HCO_3^- > SO_4^{2-} > Cl^-/NO_3^-$ and $Ca^{2+} > Mg^{2+} > Na^+/K^+$ respectively (Table 5.1). However, in polar glacial basins, the order of cation concentrations is slightly different, typically $Ca^{2+} > K^+/Na^+ > Mg^{2+}$. The mean concentration of most ions in meltwater is lower than the global means for river water, with some exceptions. For instance, the mean concentration of Ca^{2+} from Qiyi Glacier in the Qilian Mountains of central Asia and K^+ from Bench Glacier in Alaska are higher than the global river averages (Table 5.1). Meltwater chemistry is primarily characterized by HCO_3^--Ca^{2+}, HCO_3^--(Ca^{2+} + Mg^{2+}) and (HCO_3^- + SO_4^{2-})-(Ca^{2+} + Mg^{2+}), similar to global surface water. This indicates that meltwater chemistry is largely controlled by Ca^{2+}, Mg^{2+}, HCO_3^-, and SO_4^{2-}. Compared to major ions, fewer studies have examined dissolved elements in meltwater. These elements typically include minor elements such as Fe, Al, Sr, B, Ba, Mn, Li, Zn, Cu, Ti, Ni, Mo, Cs, Cr, Co, and Cd (Table 5.2), as well as rare elements like As, La, Ce, Nd, Yb, Pu, and Lr.

Concentrations of some chemical species in meltwaters exhibit distinct temporal variations. On a diurnal scale, from 13:00 to 03:00 the next day, major ion concentrations are relatively low, while from 03:00 to 08:00, they are relatively high. This pattern is linked to the diurnal melt process, with higher ion concentrations occurring when meltwater discharge is lower and lower concentrations when discharge is higher (Li et al. 2022). On a seasonal scale, the concentrations of major ions and certain minor elements (e.g. Li, Sr, and Ba) are lower during high flow seasons compared to low flow seasons (Fig. 5.4a–d). These chemical species display an inverse relationship with meltwater discharge, underscoring the impact of hydrological processes on hydrochemical dynamics. Conversely, the concentrations of other minor elements (e.g., Al, Bi, Co, Zn, Sn, and Th) exhibit random variations, showing no significant relationship between elemental concentration and meltwater discharge (Fig. 5.4e, f). This indicates that physicochemical processes also play a crucial role in determining hydrochemical characteristics (Li et al. 2019).

The seasonal variation of solute concentration in meltwaters is closely related to the evolution of the glacier drainage system. Meltwater typically consists of two flow types: fast flow and delayed flow. Fast flow transports water rapidly through a channelized drainage system in ice-walled conduits and basal channels, whereas delayed

Table 5.1 Major ion concentrations ($\mu g\ L^{-1}$) in meltwater draining typical glaciers

Ions	Dongkemadi Glacier, Asia	Qiyi Glacier, Asia	Haut Glacier d'Arolla, Switzerland	Bench Glacier, Alaska	Leverett Glacier, Greenland
Cl^-	113	1278	88	72	175
NO_3^-	207	557	400	N/A	56
SO_4^{2-}	777	25,020	5200	12,576	1539
HCO_3^-	38,315	110,525	17,000	26,047	7088
K^+	339	568	370	2318	1061
Na^+	1011	2328	370	575	944
Mg^{2+}	1149	8173	400	432	211
Ca^{2+}	9993	31,024	6600	11,000	1419

N/A denotes no available data

Data from Li et al. (2013, 2016), Mitchell and Brown (2007), Anderson et al. (2000), Hindshaw et al. (2014)

Table 5.2 Comparison of species concentrations (μg L^{-1}) in glacier-dominated rivers draining Urumqi Glacier No.1 (UG1), Dongkemadi Glacier (DG), Qiyi Glacier (QG) and Haut Glacier Arolla (HGA) with those in glacier-fed rivers at the head of Yangtze (HYA), Brahmaputra (HBR) and Indrawati/Dudh Koshi (HID) as well as the guideline of the World Health Organization (WHO; Li et al. 2019)

	Glacier-dominated rivers at glacial basins				Glacier-fed rivers at alpine basins			WHO
	UG1	DG	QG	HGA	HYA	HBR	HID	
pH	7.5–8.9	7.0–9.3	7.4–7.7	7.3–8.7	8.4–9.3	N/A	6.6–8.8	6.5–8.5
B	0.82–6.47	3.10–55.6	N/A	N/A	0.04–0.23	15.9–926	N/A	500
Al	10.0–42.4	1.14–41.0	3.29–73.1	35	N/A	147–4524	N/A	200
Ti	0.06–0.30	0.10–0.96	0.11–2.35	7.3	0.002–0.003	0.5–68.4	N/A	N/A
Cr	0.03–0.75	0.02–0.2	0.06–0.45	0.9	N/A	2.2–8.3	0.32–13.7	50
Mn	0.03–10.5	0.10–3.52	0.51–4.05	9.4	N/A	84.2–287	0.75–125	400
Fe	12.5–106	21.5–113	0.42–97.8	390	N/A	129–7791	N/A	2000
Co	0.02–0.26	0.01–0.45	0.01–0.04	0.1	N/A	N/A	0.00–3.94	N/A
Ni	0.10–3.13	0.06–1.87	0.08–0.22	0.6	N/A	N/A	0.10–14.2	20
Cu	0.10–2.47	0.07–2.47	0.29–0.76	0.7	N/A	N/A	0.00–8.33	2000
Zn	0.25–1.71	0.07–16.7	0.38–4.99	N/A	N/A	N/A	0.00–62.6	3000
As	0.18–0.29	0.22–1.41	N/A	0.8	0.01–0.04	8.5–195	0.00–15.6	10
Mo	0.09–1.75	0.04–0.58	0.21–0.29	N/A	N/A	N/A	N/A	70
Cd	0.00–0.04	0.00–0.03	0.00–0.01	0.2	N/A	N/A	0.00–0.05	3
Sb	0.01–0.05	0.03–0.85	0.01–0.02	N/A	N/A	N/A	N/A	N/A
Ba	1.19–7.43	0.88–19.2	16.8–56.4	0.7	0.04–0.07	8.5–45.5	0.35–76.9	700
Pb	0.03–0.16	0.00–2.14	0.16–1.39	0.2	0.01	N/A	0.00–7.83	10
Th	0.00–0.03	N/A	N/A	N/A	N/A	N/A	N/A	N/A
U	0.04–5.05	0.04–0.65	N/A	0.7	N/A	N/A	0.06–21.9	15
Hg	N/A	N/A	N/A	0.3	0.000–0.003	N/A	N/A	1

N/A denotes no available data

flow transports water slowly through a distributed drainage system at the ice-bedrock interface (Tranter et al. 1993). Fast flow limits the potential for solute acquisition, whereas delayed flow enhances it. During early flow season, supraglacial meltwater is routed through the ice-marginal and glacier interior drainage systems. The longer residence time promotes prolonged and intimate contact of meltwater with sediments and deposits, leading to higher solute concentrations. During peak flow season, meltwater runs off the glacier surface and enters subglacial environments. The duration of meltwater-sediment interaction decreases, resulting in lower solute concentrations. During late flow season, the supraglacial meltwater store is largely depleted, and meltwater storage within the contracting englacial and subglacial conduits is renewed. This results in higher solute concentrations due to the longer residence time of meltwater in subglacial environments (Li et al. 2019).

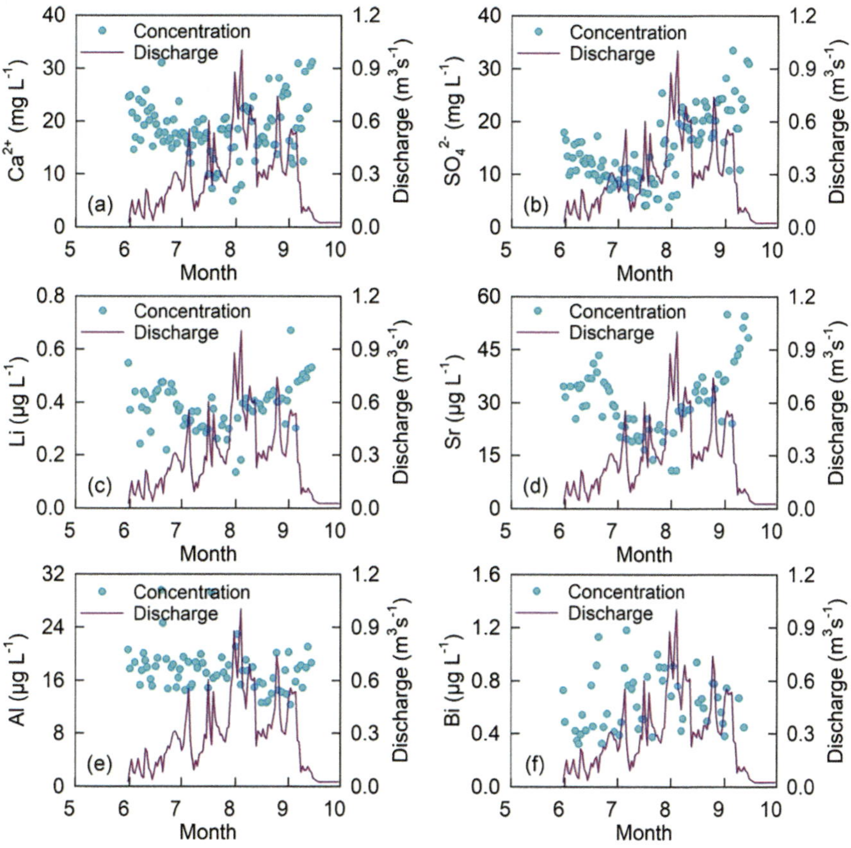

Fig. 5.4 Seasonal variation of concentrations of major ions and minor elements in meltwater draining Glacier No.1 in Asia during 2013 (modified from Li et al. 2019). Note that SO_4^{2-} and Ca^{2+} represent major ions, and Li, Sr, Al and Bi represent minor elements

5.2.2 Bioavailable Elements and Species

With the accelerated retreat of glaciers, the interaction between meltwater and rock are intensifying in glacial environments. As meltwater increases, glaciers may release large quantities of nutrients, significantly impacting aquatic ecosystems and elemental biogeochemical cycles. This section focuses on the seasonality and export of nutrients such as iron, silicon, phosphorus, and organic carbon released from glaciers.

1. **Iron**

Iron (Fe) limitation for primary producers is prevalent in large sectors of the world's oceans, most notably the Southern Ocean, the Pacific Northwest, and some regions of the North Atlantic, all areas proximal to significant glacial activity. These Fe-limited oceans are believed to impact the global climate by regulating atmospheric CO_2 concentrations through their influence on the strength of the biological pump. Traditionally, oceanic Fe inputs have been attributed to aeolian dust, riverine discharge, benthic recycling, and sea ice. Notably, glaciers and ice sheets (including icebergs) are also recognized as significant contributors of Fe to the oceans.

Glacially derived Fe exists in three forms: truly dissolved Fe (DFe; pore size <0.02 μm), colloidal/nanoparticulate Fe (CNFe; 0.02 to 0.45 μm), and sediment-bound nanoparticulate Fe (PFe; >0.45 μm). Studies in Greenland and Antarctica have shown that sediment-bound Fe is the most abundant form, while truly dissolved Fe is typically the least abundant. Quantifying the transport of glacial Fe is critical because it limits primary productivity in aquatic ecosystems. Several studies have investigated Fe export from glaciers and ice sheets, highlighting the importance of understanding the biogeochemical processes of Fe on the Earth's surface.

The Greenland ice sheet (GrIS) is a significant source of Fe in the North Atlantic. The annual export of DFe from the ice sheet is 0.3 Tg yr^{-1}, which is equivalent to that of dFe from dust aerosols to the North Atlantic (Hawkings et al. 2014). Annual releases of total Fe from Greenland and the Antarctic ice sheets range from 409 to 2540 Gg yr^{-1} and 57 to 170 Gg yr^{-1}, respectively, with PFe being the largest component and DFe is the smallest (Fig. 5.5). Release of Fe from ice sheets is primarily controlled by sediment in meltwaters, comparable to Fe export from dust aerosols to the waters around Greenland and Antarctica. For glaciers worldwide, the combined release of DFe and CNFe release is 185 Gg yr^{-1}, far exceeding those from ice sheets (19.5 to 52.1 Gg yr^{-1}; Li et al. 2019). Overall, Fe release from glaciers and ice sheets is expected to increase as the climate warms. However, research on PFe remains limited. It is crucial to understand the temporal and spatial variations of glacial Fe and its transport to fully comprehend how aquatic ecosystems respond to glacier melt.

The stable isotope of Fe ($\delta^{56}Fe$) is a crucial tool for studying the biogeochemical cycle of Fe. Globally, $\delta^{56}Fe$ values in terrestrial rivers vary from −1.2 to 2.8‰, depending mainly on the chemical characteristics of river water and the pore size of the filtered membrane used before chemical analysis. However, $\delta^{56}Fe$ values of DFe in glacial meltwater-fed rivers are similar to the crustal $\delta^{56}Fe$ values found in Copper River tributaries, which are also fed by glacial meltwaters. In some Arctic glacial catchments, $\delta^{56}Fe$ values in river water range from −0.11 to 0.09‰ and do not change significantly with variations in DFe concentration. Future studies that better define $\delta^{56}Fe$ values in glacial meltwaters will enhance our understanding of the global distribution and impact of glacially-derived Fe on downstream ecosystems.

2. **Silicon**

Silicon (Si) plays a crucial role in global biogeochemical cycles, serving as a vital nutrient for many marine organisms, particularly diatoms. These organisms use Si to build their cell frustules and contribute to up to half of oceanic carbon fixation (Cermeño et al. 2015). Diatoms are especially significant in the biological pump and the carbon cycle when silica input is high. Additionally, the chemical weathering of silicate minerals in rocks can sequester atmospheric CO_2 on geological timescales.

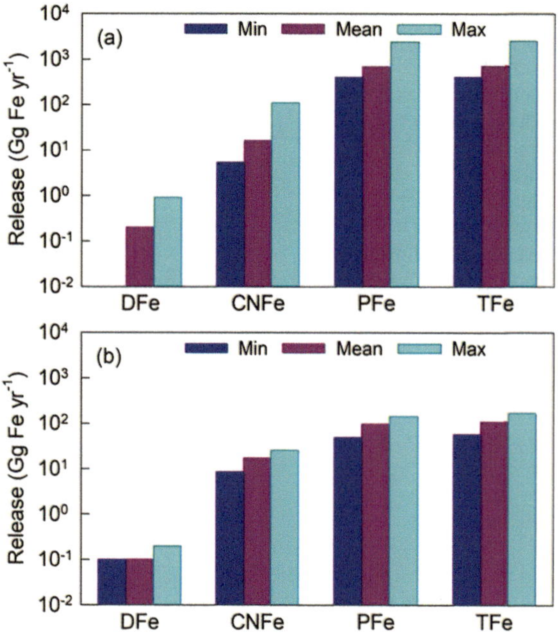

Fig. 5.5 Release of dissolved (DFe), colloidal/ nanoparticulate (CNFe) and particulate (PFe) iron, as well as total iron (TFe) from the Greenland ice sheet (**a**) and Antarctic ice sheet (**b**; modified from Hawkings et al. 2014). Note that Min and Max denote minimum and maximum values, respectively

Therefore, understanding the components of the Si cycle, constructing silica budgets, and evaluating past and future changes are of significant importance.

In the oceans, most dissolved silica (DSi) originates from riverine discharge, groundwater, aeolian dust, hydrothermal activity, and seafloor weathering. During glacial cycles, glaciers and ice sheets, which can cover up to 30% of the land surface area, discharge vast quantities of freshwater and sediment into coastal ocean. These glaciers and ice sheets are recognized as significant contributors, delivering substantial fluxes of bio-essential Si to downstream ecosystems. However, their role in the global Si cycle remains not fully established. Recent evidence suggests that silica dissolution rates in glacial environments may be higher than previously believed, underscoring the importance of glacially derived Si in oceanic Si budgets.

Previous studies have shown pronounced seasonal variations in DSi concentrations in meltwaters. For instance, in the Leverett Glacier basin of Greenland, higher DSi concentrations are observed during the early melt season. This is due to a less efficient subglacial drainage system beneath the GrIS and a weak dilution process by supraglacial meltwater (Fig. 5.6a). As ablation progresses, DSi concentrations gradually decrease, correlating with the increase in meltwaters and its strong dilution effect. In comparison, seasonal variation in ASi concentrations is less pronounced, displaying a fluctuating trend (Fig. 5.6b). ASi is the most important component of Si in meltwaters. Although the mean dry weight of ASi from Leverett Glacier is 0.91%, its mean concentration is as high as 392 μM (Fig. 5.6b). While DSi concentrations are lower, the erosion rate is higher than that of ASi and comparable to the erosion rate of DSi in some Arctic rivers. In contrast, the mean concentration of ASi from

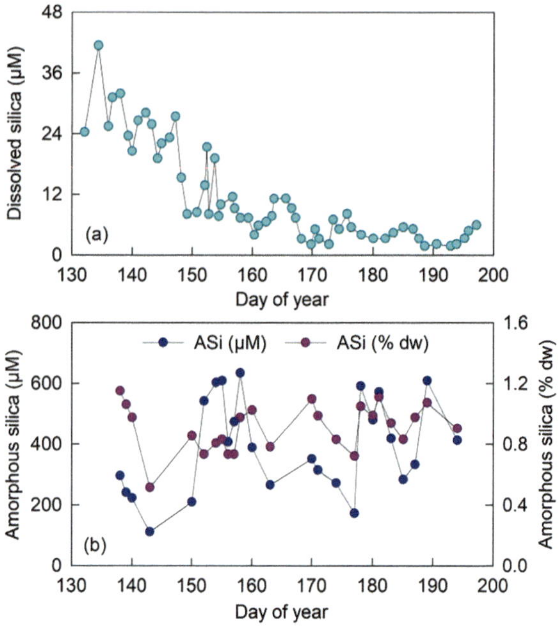

Fig. 5.6 Time series of concentrations of dissolved silica (DSi; **a**) and amorphous silica (ASi; **b**) in meltwaters as well as the percentage dry weight of sediment (% dw; **b**) from proglacial river of the Leverett Glacier in the Greenland (modified from Hawkings et al. 2017)

icebergs is lower than that in meltwaters, with ASi concentration from dirty ice at the bottom of icebergs being higher than from clean ice.

Ice sheets play a crucial role in global silicon (Si) cycles, particularly in the export of Si. Previous studies have shown that the release of amorphous silica (ASi) from the GrIS is 0.16 Tmol yr^{-1}, significantly higher than from icebergs (0.03 Tmol yr^{-1}; Hawkings et al. 2017). The Si export from the GrIS accounts for 37% of the Si flux entering the Arctic Ocean, comparable to inputs from dust deposition, groundwater, and hydrothermal activity. The release of Si from the Antarctica ice sheet (AnIS) is 0.08 Tmol yr^{-1}, about half that of the GrIS. Combined, the total export of Si from both ice sheets amounts to 0.3 Tmol yr^{-1}, approximately 3% of the global Si budget. Notably, current studies lack data on Si export from glaciers.

3. **Phosphorus**

Phosphorus (P) is a bioessential macronutrient that significantly influences ecosystem productivity and biogeochemical cycles in both marine and terrestrial environments. It is considered the ultimate limiting nutrient on geological timescales, partly because it cannot be fixed from the atmosphere and may currently limit productivity in certain regions. However, many aspects of the global P cycle remain poorly understood, particularly the terrestrial fluxes of P to the oceans, the principal speciation of exported P, and the associated bioavailability of these fluxes.

Studies on the transport of P to the oceans have primarily focused on river runoff, atmospheric deposition, and submarine groundwater discharge. Among these, riverine runoff is the dominant source of P flux. Additionally, glaciers and ice sheets have been recognized as significant contributors of nutrients to downstream ecosystems. A few studies have highlighted their role in the global P cycle (Hawkings et al. 2016).

Previous studies have shown that two distinct environments likely dominate P export from the GrIS. The supraglacial environment is the principal source of snow/ice melt, which are relatively dilute and contain low concentrations of rock-derived P. Habitats associated with supraglacial debris may recycle and export P (Stibal et al. 2009). This environment plays a critical role in ice sheet hydrology, as most meltwater is generated at the surface and routed to the bed. The second source is the subglacial environment, characterized by high rock-water ratios and microbially mediated chemical weathering of bedrock. Meltwater drainage through the subglacial system is thought to be an important source of P (Hawkings et al. 2015).

In Greenland, the temporal variation of dissolved P species in meltwaters shows significant differences. Elevated concentrations of soluble reactive P (SRP) at Leverett Glacier are closely associated with the flushing of stored waters from the subglacial system due to supraglacial lake drainage events and the subsequent drainage of subglacial waters (Fig. 5.7a). For dissolved organic P (DOP), there is little temporal variation, with values similar to those of supraglacial meltwaters, indicating a supraglacial source. However, at Kiattuut Sermiat Glacier, the temporal variation of SRP and DOP concentrations is not as pronounced (Fig. 5.7b). Bulk meltwater DOP concentrations are similar to supraglacial DOP concentrations, indicating a supraglacial source, as observed at Leverett Glacier. Notably, DOP concentrations are lower than SRP concentrations at both glaciers. For particulate phosphorus (PP), its mean content at Leverett Glacier is similar to that measured in shield rocks (Hodson et al. 2004). In contrast, suspended sediments at Kiattuut Sermiat Glacier contain more total P than those at Leverett Glacier. This higher P content is likely due to the catchment geology, which, although similar to Leverett Glacier, includes basaltic intrusions with higher P concentrations than shield rocks.

The total P flux from Leverett Glacier is estimated as 1986 Mg yr^{-1} (Hawkings et al. 2016). This includes 18.3 Mg yr^{-1} from dissolved P, with SRP being the dominant form, and 34.4 Mg yr^{-1} from labile PP, which is twice the SRP flux. At Kiattuut Sermiat Glacier, the total P flux is one to two orders of magnitude lower than that at Leverett Glacier. Additionally, the dissolved P fluxes are less than 10% of those at Leverett Glacier, primarily due to lower discharge. Unlike Leverett Glacier, the labile PP flux at Kiattuut Sermiat Glacier is lower than the dissolved P flux, with only a small fraction of the total P export being labile; most is exported as more refractory P in sediments. Notably, PP dominates the total P flux from both glaciers.

Overall, glaciers transport large quantities of P to surrounding coastal regions. High rates of physical erosion, combined with abundant meltwater at the glacier bed, drive the biogeochemical weathering of P minerals. Dissolved P yields from the GrIS are at least equal to those of some of the world's largest rivers. Labile fluxes from GrIS meltwater may constitute as much as 15% of the total Arctic riverine flux (Hawkings et al. 2016), which has previously been considered the primary source of nutrients for the surrounding ocean, and are significantly larger than the flux from atmosphere. The GrIS's total P flux of ~400 Gg yr^{-1} rivals that of the largest river

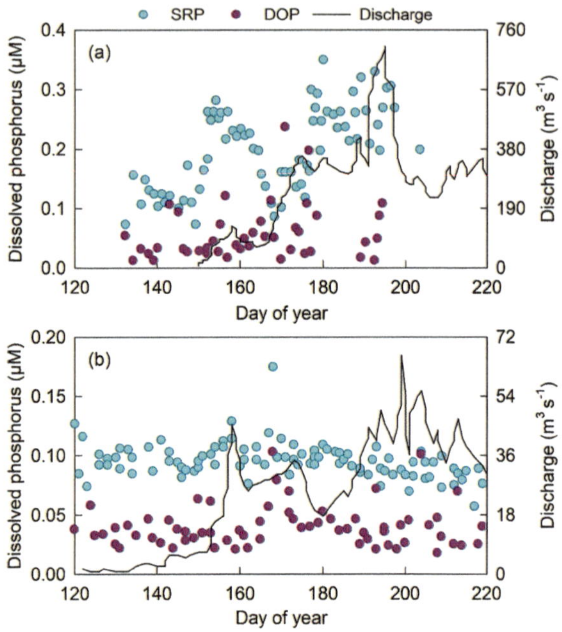

Fig. 5.7 Time series of dissolved phosphorus species from proglacial rivers of Leverett Glacier (**a**) and Kiattuut Sermiat Glacier (**b**) in the Greenland (modified from Hawkings et al. 2016). Note that SRP and DOP denote soluble reactive phosphorus and dissolved organic phosphorus, respectively

systems in the world (Hawkings et al. 2016). The significance of the GrIS P fluxes depends on the deposition and burial of sediments in the fjord and coastal regions.

4. **Organic carbon**

Glaciers have been recognized as a biome with significant implications for the global carbon cycle. These ecosystems accumulate organic carbon (OC) through both primary production and the deposition of carbonaceous material. OC is released via meltwater at the glacier surface, flowing to the glacier terminus and into proglacial streams and fjords. Additionally, ice-locked OC can enter marine and lacustrine environments through iceberg calving. Microbial activity also breaks down OC, generating CO_2 that reenters the atmosphere. The balance between these processes determines the export of OC. OC not consumed by microbes on-site is transported into downstream aquatic ecosystems by meltwater flushing (Hood et al. 2015).

The total storage of dissolved organic carbon (DOC) is estimated at 4.48 petagrams of carbon (Pg C), with 4.19 Pg stored in the AnIS, 0.22 Pg in the GrIS, and 0.07 Pg in other glaciers (Fig. 5.8a). In contrast, the storage of particulate organic carbon (POC) is 1.39 Pg C, with 1.27 Pg in the AnIS, 0.07 Pg in the GrIS, and 0.06 Pg in other glaciers. Combined, the global OC storage in glaciers and ice sheets amounts to approximately 6 Pg C. This is equivalent to about 6% of the total OC pool in North American Arctic soils (Ping et al. 2008). Although the global store of glacier OC is small compared to the total OC in permafrost, glacial OC, particularly as DOC, is rapidly transferred to proglacial aquatic ecosystems, bypassing potential biotic removal in soils. Thus, glacier ice represents a significant, yet largely unrecognized,

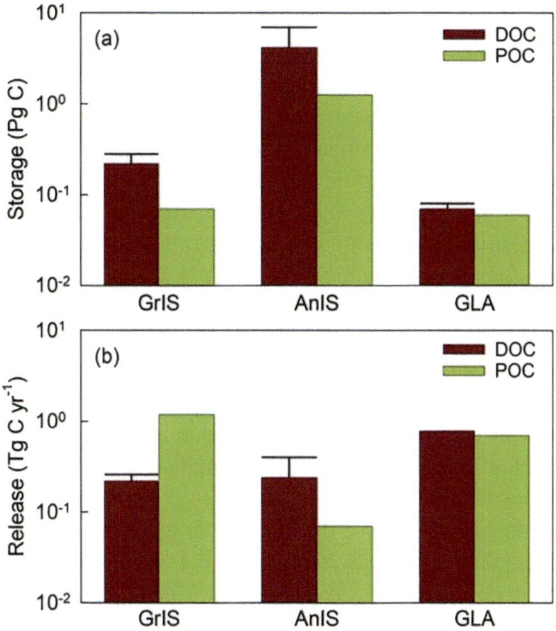

Fig. 5.8 Storage and release of organic carbon from glaciers (GLA), Greenland ice sheet (GrIS) and Antarctic ice sheet (AnIS; modified from Hood et al. 2015 and Li et al. 2018). DOC and POC indicate dissolved organic carbon and particulate organic carbon, respectively

hydrological reservoir of OC that deserves consideration in studies addressing global riverine carbon dynamics and land-to-ocean OC transport (Schuur et al. 2008).

Glaciers release 1.04 TgC per year of DOC into proglacial ecosystems, with the largest contribution (0.58 TgC yr^{-1}) coming from glaciers, compared to the GrIS (0.22 TgC yr^{-1}) and the AnIS (0.24 TgC yr^{-1}). The global distribution of DOC export differs from the distribution of DOC storage. POC export is closely tied to suspended sediment yields. Glaciers and ice sheets release 1.97 TgC yr^{-1} of POC to downstream ecosystems, with the largest contribution (1.19 TgC yr^{-1}) from the GrIS, compared to glaciers (0.70 TgC yr^{-1}) and the AnIS (0.08 TgC yr^{-1}). Globally, annual release of glacial OC is ~3 TgC yr^{-1}. In the context of land-to-ocean OC export, annual glacial OC export is approximately 0.6% of the global riverine flux of DOC to the coastal ocean and about 1% of the global POC flux (Beusen et al. 2005; Dai et al. 2012).

5.2.3 Solute Release and Streamflow Water Quality

Glaciers export large amounts of harmful elements to downstream due to water–rock interactions and the release of ice-locked storage. This has raised concerns about drinking water and health, particularly for people living in high mountain regions.

Studies have compared meltwater-supplied river water quality with drinking water guidelines (Li et al. 2016, 2019). For glacier-dominated rivers, pH values often exceed the World Health Organization (WHO) guidelines, and concentrations of Fe, Ni, and Pb also surpass the guidelines in some glacial basins (Table 5.2). pH levels exceeding the guidelines significantly impair water potability, and high Fe concentrations can stain clothing. High doses of Ni can cause kidney failure, lower body mass, and increased mortality at birth in animals, and in humans, high Ni doses may lead to digestive complications, headaches, and weakness. Elevated blood concentrations of Pb can adversely affect physical and mental development.

In glacier-fed rivers, pH levels exceed guidelines at the headwaters of the Yangtze River, while concentrations of B, Al, Mn, Fe, As, and U surpass guidelines at the headwaters of the Brahmaputra and Indrawati/Dudh Koshi rivers (Table 5.2). Elevated B concentrations cause testicular lesions in rats, mice, and dogs. As increases the risk of cancers in the skin, lungs, bladder, and kidneys, and can cause other skin changes such as hyperkeratosis and pigmentation changes. Nephritis is the primary chemically induced effect of U in humans. Given that some harmful elements in meltwater-fed rivers already exceed drinking water guidelines, mountain rivers fed by glaciers may face challenges of water quality deterioration due to glacier melt. This is particularly concerning for rivers that provide drinking water for downstream residents in regions such as Central Asia. Silicate minerals may play an important role in delivering harmful elements downstream (Li et al. 2019). Notably, the concentrations mentioned here were obtained from filtered samples. However, the largest source of some elements may be sediment-bound. Once the dissolved and sediment-bound fractions enter aquatic ecosystems, these elements may further threaten ecological health.

In addition to harmful elements, persistent organic pollutants (POPs) derived primarily from human activities can also be released from glaciers, affecting streamflow water quality. For example, polychlorinated biphenyls (PCBs) and polycyclic aromatic hydrocarbons (PAHs) from Himalayan glaciers have entered surface waters, potentially endangering drinking water and food safety. It is foreseeable that the water environments and ecosystems of rivers, lakes, and coastal zones fed predominantly by glaciers may face severe challenges due to continuous global climate warming and accelerated glacier melting in the future.

5.2.4 Chemical Weathering and Carbon Budget

Terrestrial chemical erosion rates influence atmospheric CO_2 concentrations on a geological scale. Carbonation reactions involving the dissolution and dissociation of CO_2 constitute a significant component of chemical weathering processes in glacial catchments. A strong positive relationship between CO_2 drawdown and discharge was found at Haut Glacier d'Arolla in Switzerland (Sharp et al. 1995). The greatest rates of CO_2 drawdown were associated with high suspended sediment fluxes and waters with low partial pressure of $p(CO_2)$. This highlights the importance of carbonation reactions and suspended sediment transport in major channels at the glacier bed in controlling the rate and magnitude of CO_2 drawdown in glacial meltwater. At high sediment concentrations, these reactions consume protons more rapidly than they can be re-supplied by CO_2 diffusion across the gas-meltwater interface. Consequently, the apparent $p(CO_2)$ of the water decreases, and CO_2 drawdown occurs by steepening the atmosphere-water CO_2 concentration gradient, which increases the rate of CO_2 diffusion into solution.

The link between glacial discharge and CO_2 drawdown suggests a significant role for glacial chemical weathering in carbon cycling over glacial-interglacial scales. If contemporary rates of chemical weathering in alpine catchments apply to glaciers and ice sheets during these cycles, considerable amounts of CO_2 could be removed from the atmosphere, with CO_2 drawdown proportional to glacial discharge during deglacial pulses. Such changes in the global cycling of CO_2 could influence climate change (Sharp et al. 1995). However, much uncertainty surrounds these estimates, as there is no modern analogue for the mid- to high-latitude Quaternary ice sheets. Additionally, the hydrological conditions beneath Quaternary ice sheets, which may impact the magnitude of CO_2 supply and its sequestration through carbonation reactions, remain unclear.

Glacial discharge may act as a sink for atmospheric CO_2, and the CO_2 drawdown figures have been presented as total drawdown, combining both short- and long-term rates. Short-term CO_2 drawdown is associated with the carbonation of carbonates, with CO_2 eventually returned to the atmosphere via reprecipitation of carbonates in the oceans. Conversely, the carbonation of silicates influences long-term CO_2 drawdown, as the CO_2 consumed is not reintroduced to the atmosphere. The correlation between chemical weathering and physical erosion rates for major world rivers supports a coupling between the physical and chemical erosion of silicates (Gaillardet et al. 1999). Thus, regions with the most active physical erosion are also where CO_2 consumption by silicate weathering is most significant. However, other studies suggest that carbonate weathering is more important than silicate weathering in chemically limited systems (Anderson et al. 2000). This indicates the need for further research to resolve these uncertainties and to better understand the rates of chemical weathering under terrestrial ice masses. Such understanding is crucial for evaluating the importance of glaciers and ice sheets in the global cycling of CO_2.

5.3 Permafrost Hydrochemistry

Permafrost degradation has led to a shift in the freshwater system from surface water dominance to groundwater dominance, making mineral-rich groundwater a significant contributor to streamflow. This shift could impact the river transport of organic matter, inorganic nutrients, and major ions downstream. These transport processes are highly complex and exhibit variabilities linked to permafrost conditions, sensitivity to permafrost thaw, modes of permafrost degradation, and the environmental characteristics of watersheds. Future changes in the river transport of these chemical species could have critical implications for primary production and carbon cycling.

5.3.1 Organic Matter

With the accelerated degradation of permafrost, previously frozen organic matter stored in the permafrost will be released as ground ice melts. Currently, there is some consensus on how permafrost degradation affects changes in terrestrial organic carbon and nitrogen.

Permafrost degradation increases DOC concentrations in surface water. In Alaskan permafrost catchments, the highest DOC concentrations were found in areas with the most extensive permafrost coverage, while the lowest concentrations in areas with the least coverage (Fig. 5.9). This relationship can be attributed to variations in soil water levels and the interaction between meltwater and soil organic matter. High permafrost coverage leads to elevated soil water levels and strong interactions between meltwater and shallow organic matter, resulting in high DOC concentrations. Conversely, low permafrost coverage leads to lower soil water levels and stronger interactions between meltwater and deeper mineral matter, resulting in lower DOC concentrations. DOC concentrations also exhibit seasonal variations, being higher in June and July and lower in August and September (Fig. 5.9). These variations are likely due to changes in the freeze–thaw cycle and active layer thickness. Additionally, DOC concentrations are generally higher in permafrost-covered catchments compared to in seasonal frozen soils (Frey and Smith 2005).

Similar to DOC, permafrost degradation also increases dissolved organic nitrogen (DON) concentrations in surface water. The more intense the permafrost degradation, the greater its influence on DON levels. For example, in Alaska, DON concentrations in surface water were higher in catchments with seasonal frozen soils compared to

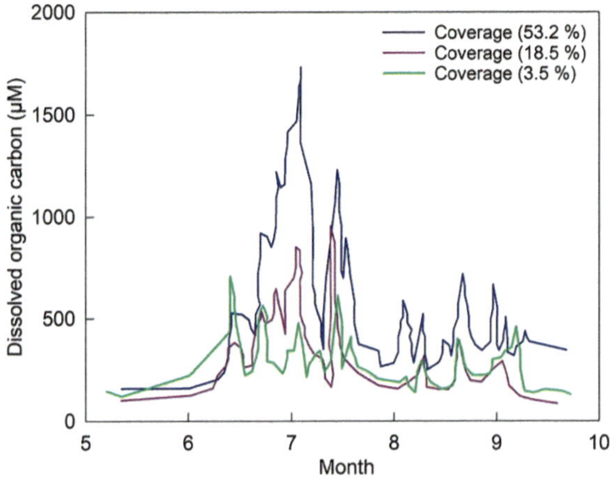

Fig. 5.9 Seasonal changes in concentrations of dissolved organic carbon (DOC) for the high, medium, and low permafrost coverage for the catchments in Alaska (modified from Petrone et al. 2006)

those with permafrost coverage (Jones et al. 2005). Consequently, as continuous permafrost degrades, organic carbon and nitrogen concentrations in surface water are likely to gradually increase in permafrost catchments.

Permafrost degradation may significantly impact the carbon and nitrogen cycles. With climate warming, organic carbon and nitrogen trapped in permafrost are likely to be released, closely linking to global biogeochemical cycles. As temperatures rise and permafrost degradation accelerates, the concentrations of organic carbon and nitrogen in surface water will gradually increase in permafrost catchments. This will lead to higher exports of carbon and nitrogen, further affecting the elemental cycles. Additionally, the thawing of permafrost can release organic matter from peat layers, further increasing organic matter concentrations in surface water. Overall, annual solute fluxes from permafrost catchments are expected to increase in the future.

5.3.2 Inorganic Nutrients

Previous studies have highlighted considerable uncertainty regarding the effects of permafrost degradation on the concentrations and fluxes of inorganic nutrients (e.g., nitrate, phosphate, and silicate) in surface water and the subsequent impact on downstream aquatic ecosystems.

Permafrost degradation may increase nitrate concentrations and exports in surface water. Generally, higher permafrost coverage results in longer water residence times in the soils, promoting the export of dissolved inorganic nitrogen (DIN). As permafrost thaws, water penetrates deeper soil layers, facilitating the export of soil nitrate. Nitrate concentrations in surface water are higher in catchments with seasonal frozen soils at lower latitudes compared to those with permafrost at higher latitudes (Fig. 5.10). In the early 1990s, nitrate concentrations and exports in surface water in some frozen-soil catchments increased significantly, likely due to the emergence of numerous thermokarst phenomena and accelerated permafrost degradation.

Permafrost degradation may increase the concentrations of silicate and phosphate in river water. As permafrost continues to degrade, the active layer's thickness gradually increase, causing water to migrate from the shallow, organic matter-enriched soils to deeper, mineral-enriched soils. This process can elevate silicate and phosphate concentrations in surface water due to increased mineral weathering, a major source of these nutrients. For example, silicate concentrations in river water have been observed to rise in some Siberian catchments as permafrost coverage decreases.

Notably, factors beyond climate warming can also influence permafrost degradation, leading to significant spatial differences in inorganic nutrient concentrations and exports in surface water across permafrost catchments worldwide. Therefore, when examining the impact of permafrost degradation on inorganic nutrient concentrations and exports, it is crucial to consider both the degradation modes of permafrost and the chemical composition and structure of the soils.

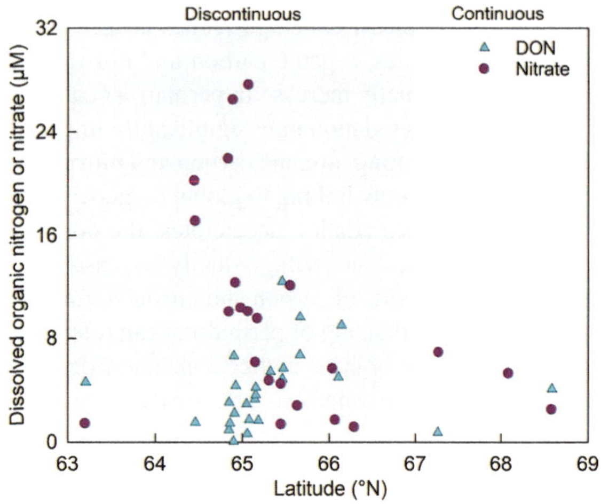

Fig. 5.10 Concentrations of nitrate and dissolved organic nitrogen (DON) in river water spanning the distribution of discontinuous and continuous permafrost in Alaska (modified from Jones et al. 2005)

5.3.3 *Inorganic Ions*

Permafrost degradation may affect the transport of inorganic ions (e.g., Ca^{2+}, Mg^{2+}, and SO_4^{2-}) in river water. The pattern of permafrost degradation influences the concentrations and types of dissolved inorganic ions in surface water. During rapid permafrost degradation, the interaction between water and shallow, organic matter-rich soils weaken, while the interaction between water and deeper, mineral-rich soils strengthen. Consequently, the concentrations of inorganic ions in surface water will change with the rapid decrease in soil water levels.

Permafrost degradation may increase the concentrations of inorganic ions. In some Siberian catchments covered by seasonal frozen soils, the average concentration of total inorganic solutes (TIS) in surface water is significantly higher than in catchments covered by permafrost (Fig. 5.11). This difference is due to the hydrological processes influenced by permafrost, which prevent water from migrating from shallow to deep soils and restrict the contact between deep groundwater and shallow, organic matter-rich soils. In the future, with increased runoff and enhanced dilution, the rising trend of ion concentrations in surface water may slow down.

Permafrost degradation may increase the fluxes of inorganic ions in river water. In some catchments, this degradation has led to higher fluxes of Ca^{2+}, Mg^{2+}, K^+, Na^+, and SO_4^{2-} in surface water. The combination of permafrost degradation and decreased soil water levels can lead to soil oxidation and mineralization during the development of soil and peat, resulting in the release of inorganic ions. These ions tend to be enriched in the shallow soils of permafrost catchments, causing a sudden increase in inorganic ion flux during permafrost degradation. Future changes in runoff will primarily affect inorganic ion concentrations and fluxes. With global warming, hydrological systems in permafrost catchments may shift from being surface water-dominated to groundwater-dominated, impacting both hydrological systems and biogeochemical cycles.

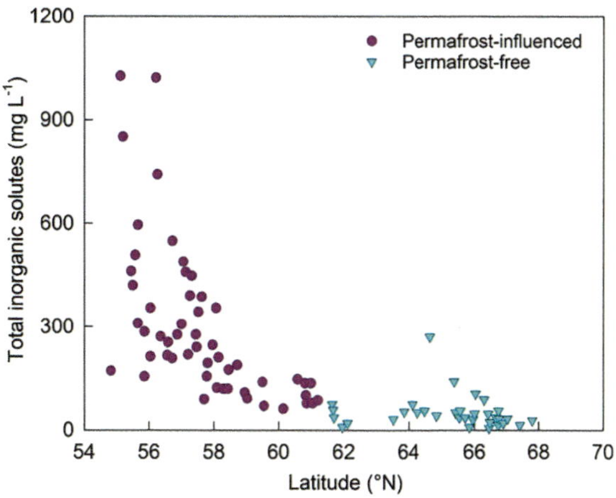

Fig. 5.11 Changes in total inorganic solutes (TIS) in rivers across West Siberia. An increase in TIS is roughly coincident with the southern limit of permafrost (modified from Frey et al. 2007)

5.3.4 Microbial Decomposition and Greenhouse Gases

Permafrost thaw threatens the carbon and nitrogen reservoirs in permafrost soils, exposing these inert pools to microbial decomposition and transformation processes. The mobilization of permafrost carbon and nitrogen can potentially increase the release of greenhouse gases. There is evidence that nitrogenous gases, particularly N_2O, are more potent greenhouse gases than CO_2 and are dominant contributors to ozone depletion in the stratosphere (Ravishankara et al. 2009).

Large quantities of organic carbon in permafrost soils represent a significant reservoir vulnerable to climate change. As the climate warms, microbial breakdown will accelerate, releasing greenhouse gases once permafrost thaws. If released primarily as CO_2 at a constant rate over a century, this amount would be comparable to that from land use change (Schuur et al. 2015). Including CH_4 in the permafrost carbon release would further increase the warming impact. Notably, the emissions of CO_2 and CH_4 from thawing permafrost are unlikely to cause abrupt climate change over a short period. However, a large pulse release of permafrost carbon could lead to catastrophic costs for society. Overall, permafrost carbon emissions are likely to occur over decades and centuries as the permafrost thaws, accelerating climate change beyond current projections. Due to the system's momentum and the continued warming and thawing, permafrost carbon emissions are expected to persist not only during this century but also beyond.

Although permafrost stores large amounts of nitrogen, only a fraction is in bioavailable form. Strong plant-microorganism competition causes nitrogen limitation in permafrost soils, often preventing N_2O production and release. Plant-regulatory effects on the soil nitrogen pool size are significant, with N_2O-emission hotspots occurring in permafrost peatlands. Permafrost thaw could turn these regions into a globally relevant source of N_2O, creating a non-carbon feedback to the global climate system. A recent study reports post-thaw N_2O release from late Pleistocene permafrost deposits, which store a substantial portion of permafrost carbon and nitrogen and are highly vulnerable to thaw. While freshly thawed, unvegetated permafrost deposits in disturbed areas emit little N_2O, emissions increase significantly within a few years after stabilization, drying, and revegetation with grasses. These emissions exceed typical rates from permafrost soils by 1–2 orders of magnitude (Marushchak et al. 2021). Previous studies highlight the importance of extra nitrogen availability from thawing permafrost, which then causes a positive climate feedback.

5.4 Application of Isotope Hydrochemistry to Cryospheric Hydrology

Throughout the water cycle, the stable isotopes (**^{2}H and ^{18}O**) in water exhibit regular variations across different phases and regions due to the processes of mixing and isotopic fractionation. Distinct water bodies, as well as water from various sources, possess unique ratios of **^{2}H and ^{18}O**. Due to their widespread formation in low-temperature environments, the isotopes in the cryosphere's water bodies have their own particularities. By analyzing the spatial distribution and variation of these stable isotopes, it is possible to better understand the hydrological processes and evaluate the volume, quality, and sustainability of water.

5.4.1 The Reconstruction of the Paleoclimate and Paleoenvironment

The water stable isotopes (^{2}H, ^{18}O) are widely employed classical paleoclimatic proxies in numerous studies. The basis for reconstructing past temperature changes using stable isotope records is the correlation between stable isotopes and temperature at both time and spatial scales (δ-T correlation).

Precipitation δD and δ^{18}O are one of the most in-depth and widely used climate proxy indicators in the study of ice core climate records. The related research results have become classic in the study of paleoclimate change, making a huge contribution to a deep understanding of the laws of Earth's climate change (Liu et al. 2022). The Dome C ice core in Antarctica has reconstructed temperature changes over the past 800 ka, making it the longest continuous climate record (which includes the past 8 glacial to interglacial cycles) in the world's ice core at a time scale. Due to its high accumulation rate, the time scale of temperature changes recorded in Greenland ice cores is relatively short. For example, the depths of GRIP and GISP2 ice cores reached 3022 m and 3050 m, respectively. However, only temperature changes over the past 100 ka were recorded. The NEEM ice core in Greenland only recorded climate records over the past ca. 130 ka. However, Greenland ice cores can explore short-term climate change events (such as D-O cycles) based on hydrogen and oxygen isotope records.

Research on ice core isotopes has begun to focus more on deuterium excess (defined as $d_{excess} = \delta D - 8\ \delta^{18}O$) records. Combined with simple Rayleigh fractionation isotope models or atmospheric circulation isotope models, d_{excess} helps to quantitative reconstruction of environmental conditions in water vapor source areas. In recent years, researchers have found that d_{excess} in polar snow and ice is not only controlled by non-equilibrium dynamic fractionation during evaporation, but also by the influence of temperature and water vapor condensation during equilibrium fractionation. Therefore, logarithmic and exponential definitions of excess deuterium have been proposed. The study of d_{excess} records in the Greenland NEEM ice core showed that, compared to linear definitions, the logarithmic definition of excess deuterium in ice cores at high latitudes can better measure non-equilibrium dynamic fractionation (Masson-Delmotte et al. 2008).

In recent years, with the improvement of the accuracy of oxygen isotope testing in waters, accurate determination of $\delta^{17}O$ in ice cores has been achieved, thus defining a new secondary parameter oxygen-17excess ($^{17}O_{excess}$, $^{17}O_{excess} = 10^6 \ln(\delta^{18}O/1000 + 1) - 0.528\ln(\delta^{18}O/1000 + 1)$). Due to the fact that $^{17}O_{excess}$ is not very sensitive to the temperature of the source area and is mainly controlled by the normalized relative humidity of the water vapor source area, it is more advantageous to reconstruct the relative humidity changes in the water vapor source area. In addition, inland regions of polar ice sheets have a significant impact of water vapor supersaturation on $^{17}O_{excess}$ and d_{excess} due to lower temperatures. Therefore, combining $^{17}O_{excess}$ with d_{excess} helps to deepen the understanding of the stable isotope dynamic fractionation process under low-temperature water vapor supersaturation conditions, which is crucial for revealing the climatic significance of isotopes in inland ice cores of polar ice sheets (Liu et al. 2022).

The groundwater aquifer, as a valuable paleoclimatic archive, has gained significant recognition in recent years, which accommodates groundwater recharged over millennia or longer, thereby ideally preserving evidence of climatic evolution (Darling et al. 2023). Furthermore, the integration of radiocarbon dating enables a comprehensive depiction of paleoclimatic variations encompassing temperature and precipitation patterns prevailing during groundwater recharge. Wang et al. (2024) conducted an investigation on the aquifers and performed analysis of stable isotopes and radiocarbon isotopes of groundwater in order to elucidate the paleoclimatic evolution and further verify the effects of paleoclimatic environment on groundwater compositions within the northern Songnen Plain, China. The Cretaceous sandstone confined aquifer (CSA) was designated as the reference aquifer. The ^{14}C results indicate that the CSA preserved groundwater from the late Pleistocene to the modern. These findings, in conjunction with isotope variations, delineated specific climate vents, which were aligning with results derived from conventional proxies. The late Pleistocene-early Holocene showed a climate shift from cold to warm. The middle Holocene was warm and humid. Then the late Holocene entered the Neoglacial, which characterized by a cold climate environment. In confined aquifers, sudden decreases in $\delta^{18}O$ and TDS concentrations and fluctuations in groundwater age occurred along the direction of groundwater runoff (Fig. 5.12), which was not a common phenomenon in natural environments. The stable isotope values of groundwater showed a certain correlation with groundwater age, that is, ancient groundwater had the smallest values of $\delta^{18}O$ and $\delta^{13}C$, while modern groundwater had the opposite. After considering and excluding the effects of latitude and altitude, it is believed that this phenomenon was attributed to the influence of ancient recharge environment and permafrost melting water, and the composition of groundwater was affected accordingly. The hydrogeochemical and isotopic characteristics of groundwater show a certain correlation with regional distribution in both horizontal and vertical directions. Shallow groundwater and spring water are mainly influenced by atmospheric environment and human factors. However, the deep Cretaceous sandstone confined aquifer effectively preserves paleoclimatic environmental signals since the Late Pleistocene, as indicated by the ^{14}C age and stable isotope results of groundwater, indicating that it is a valuable paleoclimatic archive. Compared with various other paleoclimate studies (Fig. 5.13), the paleoclimate reconstruction records obtained from this study showed that the characteristics of the Late Pleistocene Early Holocene were gradually warming of the environment, accompanied by the melting of permafrost. The warm and humid climate prevailed in the mid Holocene. Subsequently, the northern part of the Songnen Plain experienced a relatively cold and humid climate in the late Holocene. Additionally, the recharge gaps also potentially result from arid climate, inadequate samples, and the effects of mixing water.

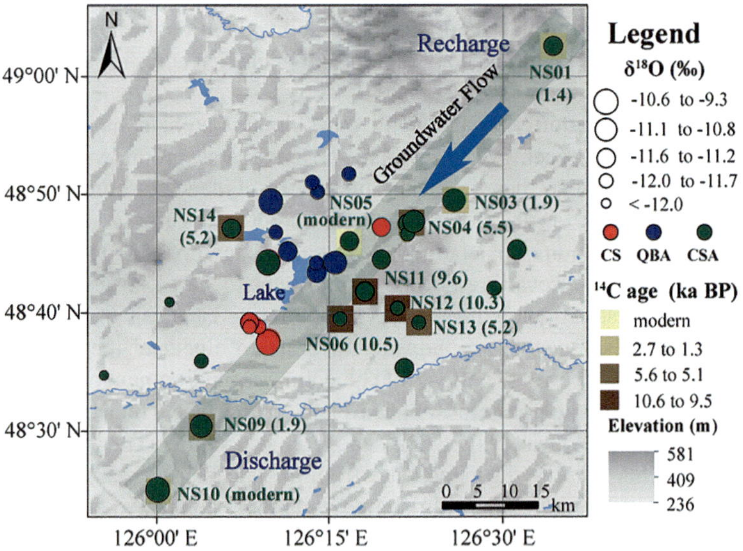

Fig. 5.12 Distribution of TDS concentrations for groundwater from nSP (The groundwater with labels belongs to group CSA-1, and numbers in parentheses are groundwater age, ka BP) (Wang et al. 2024)

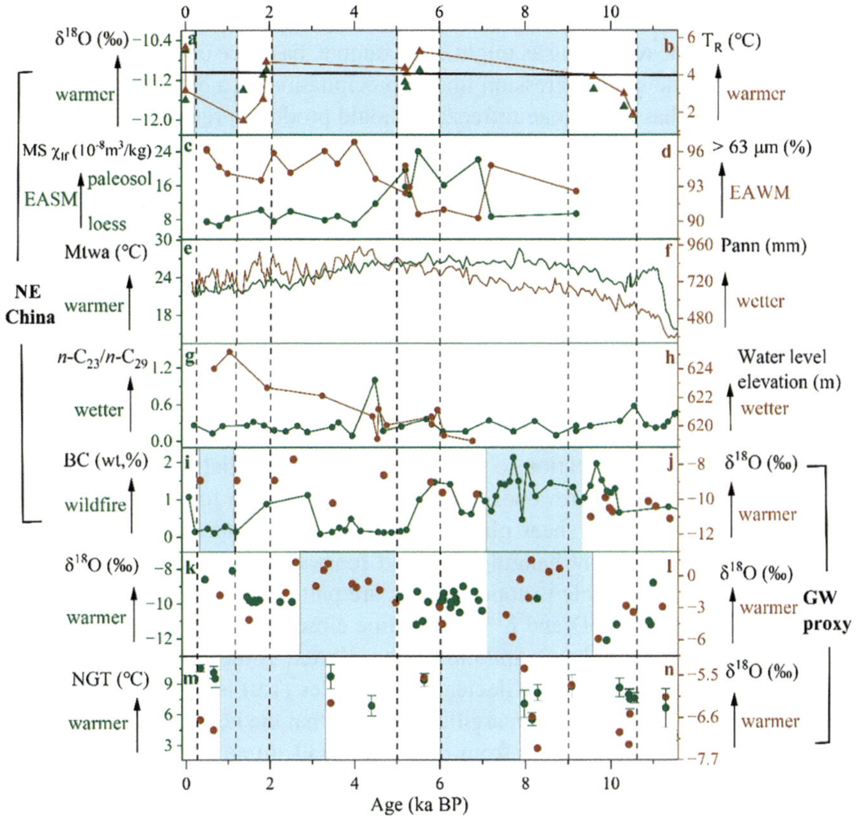

Fig. 5.13 Comparison of the $\delta^{18}O$ and T R records of the study area with the other paleoclimatic records (**a, b**) (Wang et al. 2024). The $\delta^{18}O$ values of the groundwater and T R derived from $\delta^{2}H$ and $\delta^{18}O$ of the study area (**c, d**). The magnetic susceptibility (MS) and the sand-size fraction of >63 μm of Hunshandake Sandy Land, EASM means East Asian summer monsoon, EAWM means East Asian winter monsoon (**e, f**). The mean warmest month temperature (Mtwa) and the mean annual precipitation (Pann) of Sihailongwan Maar Lake sediments (**g**). The n-alkane C_{23}/C_{29} ratios of Hani peat sequence (**h**). The water level elevation of Jinchuan peatland (**i**). The concentration of black carbon (BC) in a sedimentary sequence from Tianchi crater lake (**j, k, l, n**). The $\delta^{18}O$ values of groundwater from North China Plain, northern Ordos Basin, Kuwait and Belgium, respectively (**m**). The noble gas temperature from Belgium groundwater aquifer (The blue rectangular represents the groundwater recharge gap)

5.4.2 *Identification of Glacier Drainage Systems*

Delineating relationships between ice facies, structures and debris entrainment has proven to be an area of emerging interest for many researchers over the past three decades. Because water undergoes isotopic fractionation on freezing, the combined measurements of δD and $\delta^{18}O$ provide the opportunity to distinguish between isotopically unmodified englacial or meteoric ice and isotopically modified basal ice. By plotting values of δD and $\delta^{18}O$ on a co-isotopic diagram, samples of glacier ice that have not undergone refreezing as might, for example, basal ice or superimposed ice, should lie along the same regression line as precipitation on a δD–$\delta^{18}O$ covariance diagram. Ice that has undergone refreezing should produce a regression line with a lower slope, depending on the initial isotope values of the melted ice at the onset of regelation.

Evidence for flow dynamics and deformation at Fox Glacier, a temperate, mid-latitude maritime glacier are evident from time-lapse imagery, surface observations and stable isotope sampling in 2010/2011 (Appleby et al. 2017). The stable isotopes (Fig. 5.14) show contrasts between the debris-bearing shear plane facies and values typical of clean englacial facies. Ice is highly deformed due to topographic constriction within a channelized valley, in addition to a steep, undulating long profile. The sequential development of structures within the glacier consists of: (1) the formation of primary stratification through firnification processes in the accumulation area; (2) folding of primary stratification to produce longitudinal foliation; (3) opening of extensional fractures and crevasses, some of which close to form crevasse traces; (4) arcuate debris-bearing shear planes close to the terminus due to longitudinal compression, with apparent hanging wall and footwall fault blocks, characteristic of thrust faulting. The stable isotope samples are plot on a line with a slope of 6.8 on a co-isotopic plot of δD and $\delta^{18}O$. This line closely approximates the LMWL (7.7) obtained from monthly precipitation data collected by the IAEA at Invercargill. Moreover, the slope of the Fox Glacier névé samples (7.9) is particularly close to the monthly LMWL data at Invercargill, indicating that the névé has not undergone significant isotopic modification from direct snowfall. In addition, all of the Fox Glacier facies sampled either overlap, or are close to the maximum values of the isotope data reported in a prior study of winter snowfall. Comparing the co-isotopic plot slope gradients and regression statistics shows that some slopes are less steep than others, with the shear plane ice at the terminus having the steepest line (7.9), and Victoria Flat englacial facies having the lowest gradient (5.1). Nevertheless, the stable isotope composition of the shear plane ice facies shows slightly more positive values when compared with the nearby englacial ice samples, indicating that freezing of interstitial water occurred as the debris was elevated from the subglacial zone. It accords with ice formation through refreezing of water in close association with a debris-rich bed.

Given the similarities between intraglacial and subglacial hydrological structures and karst systems, chemical ions or isotopes offer advantages in tracing glacial hydrological systems. In tracer experiments, tracers are placed at various locations and times upstream of the glacier, and the spatial structural characteristics of the glacier's drainage channels and their seasonal changes are inferred from the monitored concentration changes at the water outlet. The relationship between tracer propagation velocity (v) and flow rate (Q) can indicate whether the subglacial drainage channel is closed and pressurized or open. The relationship between flow and velocity in open drainage channels is nonlinear and related to the hydraulic radius of the channel. In contrast, the v-Q relationship in a closed pressure drainage channel should be linear,

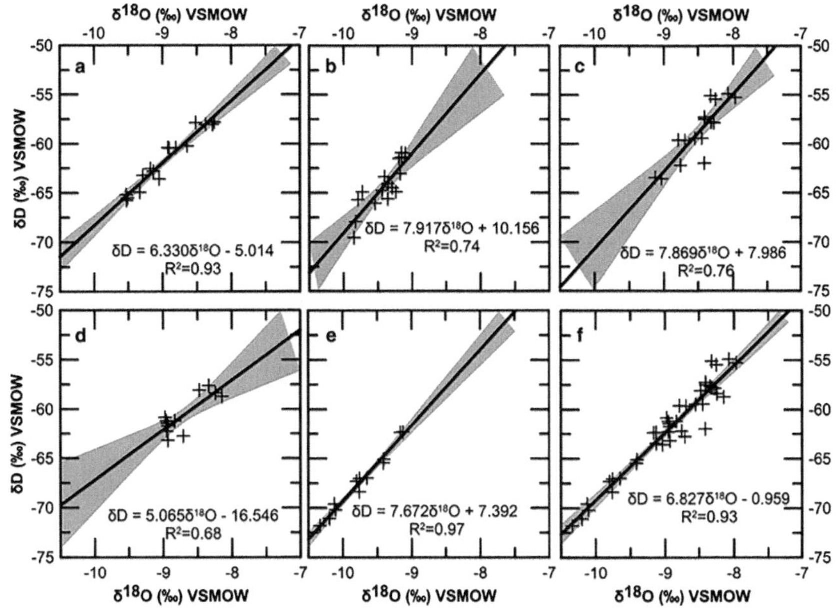

Fig. 5.14 Co-isotopic scatterplots of oxygen and deuterium ($\delta^{18}O$ and δD) from the following ice facies/sampling zones at Fox Glacier (Appleby et al. 2017): **a** basal ice; **b** shear plane ice below the lower icefall; **c** firn from the accumulation zone; **d** englacial ice at Victoria Flat; **e** englacial ice in the lower glacier; and **f** all longitudinal englacial samples (accumulation zone, Victoria Flat and lower glacier). Each plot shows the 95% confidence intervals about the least-squares linear regression model and corresponding R^2 values

with the linear slope related to the channel's hydraulic geometry, indicating the size and blockage of the drainage channel.

Soluble ions in glacial meltwater can be utilized to investigate the changing processes of glacial hydrological systems. For example, the variation range of soluble ions in meltwater was used to identify the variation characteristics of solutes in the Scott Turnablin area of Svalbard. The findings were as follows: (1) a transient high flux from the pre-glacial icing area at the onset of ablation with sodium (Na), a rapid decrease in the concentration of Na^+ and bicarbonate (HCO_3^-), and a slight decrease in sulfate (SO_4^{2-}) concentration; (2) a brief high flux associated with seasonal snowmelt water leaching at the onset of ablation, characterized by a gradual increase in chloride (Cl^-) concentration followed by an exponential decrease; (3) the subsequent gradual dilution process of the meltwater; (4) the increase of solutes from the weathering of suspended matter in the meltwater. However, the application of different ions to investigate the structure of subglacial hydrological systems may yield different results. To explore the hydrological differences between mountain glaciers (De Arora Glacier in Switzerland) and subpolar glaciers (Austria Brogblin Glacier in Svalbard), ions such as nitrate (NO_3^-), bicarbonate (HCO_3^-), and sulfate (SO_4^{2-}) were used. It was assumed that NO_3^- in meltwater associated with sea salt aerosols originated from the atmosphere. However, the NO_3^- concentration in the meltwater of the De Alola Heights Glacier decreased gradually with the ablation season, reflecting the temporary storage of snowmelt water enriched in NO_3^- in a subglacial distributed hydrological system. Conversely, the NO_3^- concentration in the meltwater of the Austrian Brogblin Glacier decreased rapidly and exponentially, which may reflect the seasonal snowmelt water leaching process, with subsequent meltwater primarily traveling through the ice surface and within the ice's hydrological pathways. This allows the meltwater to acquire more solutes through interaction with rim sediment and pre-ice ice. It is evident that soluble ions hold great potential for studying the structure of glacial hydrological systems, and further research is needed to identify ions that can provide more accurate information.

The micro/trace elements in glacial river water are mainly controlled by hydrological and physicochemical effects. For instance, most micro/trace elements in glaciers show an inverse correlation with runoff on daily and seasonal timescales, reflecting the hydrological control of these elements in glacial river water. Meltwater runoff primarily consists of fast flow and delayed flow. Fast flow occurs rapidly within the channel-type hydrological system of the ice wall conduit, limiting the potential for solute acquisition in the meltwater. Delayed flow, on the other hand, occurs more slowly within the distributed hydrological system at the ice-bedrock interface, promoting melting and limiting solute acquisition potential. When the water level of the meltwater runoff is at its lowest, the runoff mainly comes from delayed flow, and the prolonged contact between the meltwater and the sediment beneath the ice introduces more solutes into the meltwater. As runoff increases, meltwater gradually enters directly into the canal system, where solutes are derived from the interaction between the meltwater above the ice and the sediments below. As more meltwater enters the subglacial environment, the delayed flow meltwater becomes diluted, and the interaction time between water and sediment decreases, resulting in lower solute concentrations in the meltwater. It can be observed that certain micro/trace elements (such as boron (B), molybdenum (Mo), and barium (Ba)) in glacial river water can also serve as indicators for subglacial hydrological surveys.

5.4.3 Indicators of the Hydrological Connections in Permafrost Region

Permafrost degradation compels melting permafrost to become an important source of surface runoff, changes the storage of groundwater, and greatly influences the hydrological processes in permafrost regions. Stable isotopes offer valuable information on the connections between changing permafrost (ground ice) and water components. The changes in the stable isotopes of different waterbodies vary over hourly to annual timescales and truly capture the thawing signals and reflect the influence of permafrost (ground ice) on the regional hydrological processes.

By using six years' (2017–2022) observations, Yang et al. (2024) supplied a stable isotope dataset in the Beiluhe Basin in the hinterland of the Qinghai–Xizang Plateau (QXP). The Beiluhe Basin, where the water components related to hydrological cycles are well integrated, is an ideal site to study hydrological effect of permafrost change. The temporal data of stable isotopes ($\delta^{18}O$ and δD) are presented in different water bodies (precipitation, stream water, thermokarst lake, groundwater and ground ice at 17 boreholes and 2 thaw slumps) (Yang et al. 2024). This dataset provides a new basis for understanding the hydrological effects of permafrost degradation on the QXP.

Due to the transitions of moisture sources and the influence of local climate conditions, the stable isotopes of precipitation (in an altitude of 4628 m) exhibit a remarkable seasonal trend (high values in summer and low values in winter). The $\delta^{18}O$ and δD of the local precipitation ranged from −30.44 to 6.20‰ (with amount-weighted average −10.9‰) and from −237.99 to 65.45‰ (with amount-weighted average −72.11‰), respectively.

The $\delta^{18}O$ (ranged from −14.39 to 5.72‰, mean: −5.98‰) and δD (ranged from −104.07 and 22.59‰, mean: −47.96‰) of thermokarst lakes/ponds are more positive than those of precipitation due to strong evaporation and resultant enrichments of heavier isotopes in lake water. Generally, the isotopic contents of lakes/ponds are lower in August and September, which is attributed to the recharges of monsoonal precipitation and isotopic-negative water fed by melting ground ice.

For the streams, the isotope values varied from −13.67 to −7.19‰ ($\delta^{18}O$, mean: −11.07‰) and from −83.76 to −53.26‰ (δD, mean: −73.56‰), respectively. The mean values are equivalent to the average levels of precipitation in the BLH. Compared with thermokarst lakes/ponds, the stable isotopes of streams exhibited relatively stable patterns due to short residence time. However, the stream isotopes also represented seasonal variations, lower values were prevailing in August and September. The temporal changes of stream isotopes are mainly influenced by the seasonal variability of evaporation and differences in the source water, i.e., alternative replenishment of precipitation, melting ground ice, and groundwater.

For supra-permafrost water, the $\delta^{18}O$ and δD ranged from −13.28 to −5.76‰ (mean: −11.16‰) and form −86.75 and −39.04‰ (mean: −74.17‰), respectively. The isotopic peaks of springs lagged behind those of precipitation, indicating replenishments of precipitation via infiltration. By contrast, the stable isotopes of sub-permafrost water are more negative than those of supra-permafrost water, ranging between −12.68 and −11.08‰ (mean: −11.77‰) for $\delta^{18}O$, from −83.75 to −77.73‰ (mean: −80.68‰) for δD.

The ground ice represented more negative isotopes, indicating multi-sources of initial water during ice formation under variable climatic conditions and complex geological contexts on the QXP. Generally, the $\delta^{18}O$ ranging from −15.01 to −8.27‰ (mean: −12.19‰), from −113.67 to −66.38‰ (mean: −94.44‰) for δD, respectively. Specifically, the stable isotopes of ground ice varied between different boreholes. It is attributed to the influences of initial source water and complex ice formation mechanism. For instance, the near-surface ground ice is closely related to the recent precipitation and active layer hydrology (Yang et al. 2023), however, the deep-layer ground ice exhibited complicated formation mechanism, including the various source water (meltwater from glacier, permafrost, and snow; lake water; past precipitation; et al.), climate conditions, and freeze histories (Yang et al. 2020). In addition, the isotopic patterns along depths showed marked differences between boreholes, suggesting influence of lithology on the water migration and freezing fractionation of stable isotopes. Remarkably, the thaw slump ice represented more negative isotopes than those of drilling ground ice, it is due to the considerable differences in the initial source water and freezing processes. The thaw slump ice is considered to replenished by winter snowmelt water via cracks and freezing quickly. However, the isotopic-positive pore ice in these boreholes is suffered isotope fractionation due to freeze–thaw under climate transitions (Yang et al. 2023).

The local meteoric water line (LMWL) is expressed as $\delta D = 7.97\delta^{18}O + 15.26$ ($r^2 = 0.96$) (Fig. 5.15). The $\delta^{18}O$–δD diagrams of lakes, streams, and groundwater, defined as local evaporation line (LELs), were calculated as: $\delta D = 5.88\delta^{18}O$-12.80 ($r^2 = 0.95$, lakes), $\delta D = 4.89\delta^{18}O$-19.41 ($r^2 = 0.83$, streams), $\delta D = 5.69\delta^{18}O$-10.50 ($r^2 = 0.85$) (supra-permafrost water), and $\delta D = 3.54\delta^{18}O$-39.06 ($r^2 = 0.92$) (sub-permafrost water), respectively (Fig. 5.15). The slopes of the LELs are all lower than those of LMWL, and ranging between 4 and 6, indicating strong evaporation. The $\delta^{18}O$–δD relationship for ground ice, defined as the freezing line, is calculated as: $\delta D = 5.36\delta^{18}O$-29.15 ($r^2 = 0.73$), which is significantly different from the LMWL (Fig. 5.15). The lower slope suggests variable freezing rates, kinetic isotopic fractionation during ice formation, as well as the influence of the initial source water of the ground ice at different sites.

All the stable isotopes of stream lie on the LMWL and embrace in the range of supra-permafrost water (Fig. 5.15), in addition, the mean value is close to the amount-weighted average value of annual/summer precipitation, indicating the direct recharge of precipitation and supra-permafrost waters. The LEL of thermokarst lakes/ponds significantly deviated from LMWL, partial of the isotopic dots overlapped with precipitation, groundwater, and ground ice, indicating the hydrological connections between them. The cluster of ground ice is partly overlapped with precipitation, groundwater, lakes, and stream (Fig. 5.15). Some of the isotope dots are more positive than the It is indicative of mutual replenishment relations between them. In addition, the thaw slump ice exhibited more negative isotopes, which is even lower than the amount-weighted average value of winter precipitation, indicating the main recharge of snowmelt water.

5.4.4 Analysis of Water Sources

The sources of river water recharge in cryospheric basins generally include precipitation, meltwater and groundwater (Table 5.3). The recharge ratio of precipitation, meltwater, and groundwater to the river varies significantly across different basins and seasons.

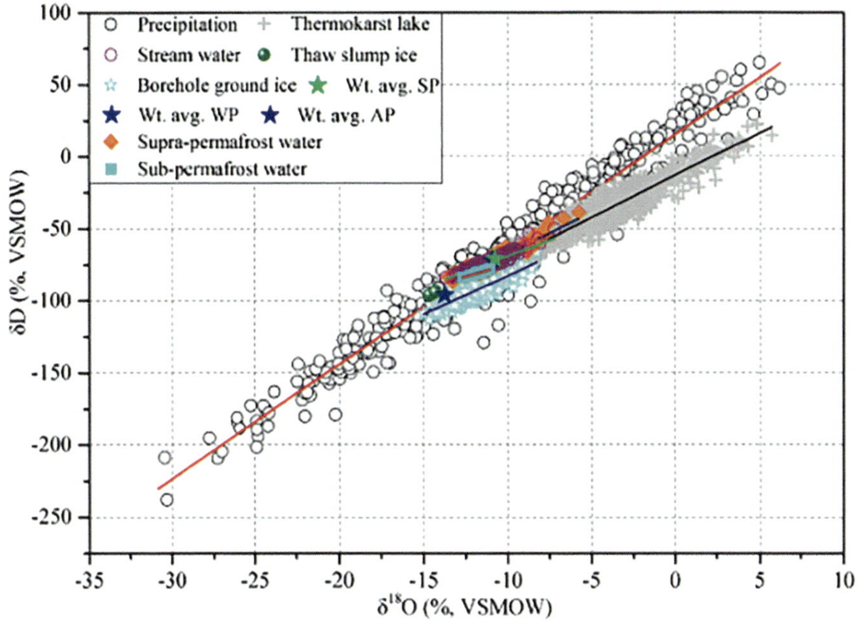

Fig. 5.15 Hydrological connections between different water components (the red line, LMWL; the purple line, LEL of the stream water; the black line, the freezing line of the borehole ground ice)

When different waters merge into a river, the composition of dissolved substances changes significantly. For stable isotopes, the $\delta^{18}O$ and δD of ice and snow in the same area are lower than those of river water and rainwater due to the temperature effect of precipitation. The formation of groundwater through various infiltration pathways can reduce the seasonal variation of isotopic signals, resulting in a relatively uniform isotopic composition. Consequently, there is a distinct isotopic signal difference. Similarly, there are certain differences in the chemical composition of water across different media. For example, the ion content in groundwater is generally much higher than in precipitation due to evaporation and water–rock interactions. Utilizing the stable isotope and hydrochemical marker characteristics of water bodies allows for the determination of the composition and mixing ratio of water bodies in surface runoff, which is of practical significance for elucidating the origin of river water and for the quantitative calculation of the mixing ratio of tributaries in different seasons.

During the mixing process, different surface runoffs, due to the varying complexity of the geographical environments in which they are located, exhibit different rates of mixing. Some mix rapidly, while others do so very slowly. For instance, prior to the confluence of the Liard and Mackenzie rivers in northwestern Canada, the average values of $\delta^{18}O$ were −21.3 and −17.4‰, respectively, with clear isotopic differences. Ten profiles were established to collect samples in the section extending more than 480 km below the confluence point. The results indicated that the river water was not fully mixed until it was at least 30 km away from the confluence point.

Table 5.3 Types of source water and their chemical characteristics

Types	Main forms	Chemical characteristics
Precipitation	Rain and snow directly entering the river channel; Rainwater through slope convergence; Snow (accumulated within 3 days) meltwater	Large seasonal variations in stable isotope and hydrochemical ion composition; Stable isotope values are generally higher in the warm season than in ice and snow meltwater; Ion content is generally low
Meltwater	Glacier ice meltwater; Snow (accumulated over 3 days) meltwater	Small seasonal variations in stable isotope and hydrochemical ion composition; Stable isotope values are generally lower than those in precipitation and groundwater; Ion concentrations are higher than in precipitation
Groundwater	Quaternary pore water; Fissure water; Groundwater in subglacial systems; Permafrost ice meltwater	Small seasonal variations in stable isotope and hydrochemical ion composition; Stable isotope values are generally higher than those of meltwater and most precipitation; Chemical ion content is generally high

In South America, the isotopic and chemical compositions of the Rio Negro and the tributaries of the Rio Solimões are quite distinct. It has been established that the $\delta^{18}O$ value of the Rio Negro River is significantly more negative than that of the Rio Solimões River above Manaus, where the two tributaries meet. There is a transition zone between the isotopic compositions of the two tributaries in the Amazon River at their confluence, suggesting that the mixing between the tributaries is very slow, and it has not been observed to be complete until 120 km below the Manaus confluence. Of course, in sections with narrow river channels, steep slopes, and turbulent water flow, the mixing of river water is more likely to achieve relative uniformity.

The runoff hydrograph separation method using hydrochemical techniques generally involves end-member mixing analysis (EMMA), the most reasonable and valuable method in remote mountain catchments. EMMA includes: (1) application of principal component analysis (PCA) and the Rule of One to determine the suitable dimensionality of the chemistry of stream water; (2) identification of endmembers; and (3) calculation of endmember contributions to streamflow. The sources of river water recharge in cryospheric regions typically include precipitation, meltwater and groundwater. According to the principles of isotopic and water chemical mass conservation, the proportions of different components of river water can be determined:

$$f_1 + f_2 + f_3 = 1 \tag{5.2}$$

$$C_1^1 f_1 + C_2^1 f_2 + C_3^1 f_3 = C_s^1 \tag{5.3}$$

$$C_1^2 f_1 + C_2^2 f_2 + C_3^2 f_3 = C_s^2 \tag{5.4}$$

where f_1, f_2 and f_3 are the proportions of precipitation, meltwater and groundwater in river water, respectively; C_1^1, C_2^1, C_3^1 and C_s^1 are the first kind of isotope/hydrochemical ions in precipitation, meltwater, groundwater and river water, respectively; C_1^2, C_2^2, C_3^2 and C_s^2 are the concentrations of the second isotope/hydrochemical ions in precipitation, meltwater, groundwater and river water, respectively.

The use of stable isotopes for runoff hydrograph separation requires meeting the following assumptions: (1) groundwater and base flow are the same, with a constant isotopic composition; (2) the isotopic content of precipitation and meltwater is temporally and spatially uniform, or its variations can be characterized; (3) there is a significant difference between the isotopic composition of precipitation, meltwater, and runoff; (4) the contribution of soil water to river runoff is negligible, or its isotopic composition is the same as that of groundwater; (5) the contribution of surface storage to the flow hydrograph is negligible.

Uncertainties were also considered for the quantification of endmember contributions. Uncertainty in hydrograph separations can be separated into 'statistical uncertainty' based on the spatial and temporal variability of the endmember values and the laboratory error and into 'model uncertainty' based on the violations of the underlying assumptions. The uncertainties of tracer-based hydrograph separation were usually caused by: (i) tracer analysis and discharge measurement; (ii) intra-storm variability of isotopic and chemical concentrations; (iii) elevation effect of isotopic composition; (iv) dissolution of minerals during runoff formation; and (v) general spatial heterogeneity of tracer concentrations. Studies suggested that the uncertainty in the measurement method was generally less important than that in the temporal and spatial variations of tracer concentrations.

Using the aforementioned analysis methods, the runoff composition of the river at the end of the Laohugou No. 12 glacier watershed in the Qilian Mountains in the 2009 ablation season was estimated. Meltwater accounted for 69.9% of the river water in the glacier basin, with precipitation and groundwater contributing 17.3% and 12.8%, respectively (Fig. 5.16).

Neglecting groundwater recharge to river runoff, the above model can be simplified to two equations, dividing the river runoff by only one isotope or hydrochemical ion concentration. In order to attain knowledge on the diversity of $\delta^{18}O$ variations in Greenlandic rivers, Yde et al. (2016) examined two contrasting glacierised catchments disconnected from the Greenland Ice Sheet (GrIS). At the Mittivakkat

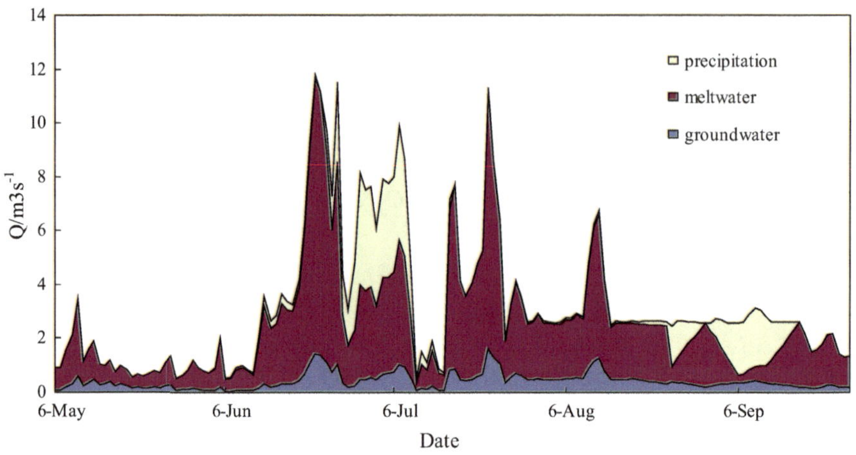

Fig. 5.16 Runoff composition variation in the Laohugou No. 12 Glacier Watershed (Wu et al. 2016)

Gletscher river, diurnal oscillations in $\delta^{18}O$ occurred with a 3 h time lag to the diurnal oscillations in run-off. The mean annual $\delta^{18}O$ was $-14.68 \pm 0.18‰$ during the peak flow period. A hydrograph separation analysis revealed that the ice melt component constituted $82 \pm 5\%$ of the total runoff and dominated the observed variations during peak flow in August 2004. The snowmelt component peaked between 10:00 and 13:00 local time, reflecting the long travel time and an inefficient distributed subglacial drainage network in the upper part of the glacier. At the Kuannersuit Glacier river, the $\delta^{18}O$ characteristics were examined after the major 1995–1998 glacier surge event. The mean annual $\delta^{18}O$ was -19.47 ± 0.55 ‰. Despite large spatial variations in the $\delta^{18}O$ values of glacier ice on the newly formed glacier tongue, there were no diurnal oscillations in the bulk meltwater emanating from the glacier in the post-surge years. This is likely a consequence of a tortuous subglacial drainage system consisting of linked cavities, which formed during the surge event. A comparison of the $\delta^{18}O$ compositions from glacial river water in Greenland shows distinct differences between water draining local glaciers and ice caps (between -23.0 and -13.7‰) and the GrIS (between -29.9 and -23.2‰). It demonstrates that water isotope analyses can be used to obtain important information on water sources and the subglacial drainage system structure that is highly desired for understanding glacier hydrology.

If there are significant differences in isotopic and hydrochemical ion concentrations between soil water, shallow groundwater, and deep groundwater in the region, further division of groundwater is possible. Utilizing hydrogen and oxygen stable isotopes, conservative water chemical parameters (such as EC, Cl, TDS, Si, etc.), and other auxiliary parameters, runoff segmentation from four or more water sources can be conducted. However, it should be noted that the selection of conservative ions should be based on water chemical analysis, and different research areas should be screened separately. In addition to isotopic and hydrochemical parameters for runoff segmentation, dissolved organic carbon (DOC) and excess deuterium can also be considered as auxiliary parameters for runoff segmentation.

5.4.5 Determination of Water Resident Time

Meltwater are crucial water resources in the arid northwest region. After the formation of surface runoff from meltwater, as well as atmospheric precipitation and bedrock fissure spring water in the mountains, these waters enter the piedmont basin from the mountain entrance, where surface water and groundwater frequently interchange. Under natural conditions, approximately 80% of surface water seeps into the thick, coarse-grained alluvial layers of the piedmont, and about 40% of the water in the oasis belt is discharged as surface water in the form of spring water. Within the watershed, human activities such as water diversion projects, groundwater extraction, and agricultural irrigation have altered the recharge conditions of groundwater, complicating the conversion between surface water and groundwater. Determining the transformation time of water bodies is vital for regional water resources management.

Watershed resident time is a fundamental descriptor of watershed characteristics, revealing information about water storage, flow paths, and sources. Quantifying the resident time of water bodies within a basin provides essential parameters for estimating the water resources cycle, studying biogeochemical processes, assessing the impact of human activities on water resources, and tracking pollutant transport. Generally, due to the short resident time of surface water, the term "water resident time" typically refers to the resident time of groundwater, or the "age" of groundwater, which is the time it takes for water to enter an aquifer until it emerges at discharge points such as wells or springs.

The seasonal effects of stable isotopes can be utilized to calculate the average residence time in a watershed. Stable isotopes, deuterium (D) and oxygen-18 (^{18}O), are components of water's molecular structure. They are chemically stable in nature, do not undergo ion exchange or adsorption, and their dynamic behavior mirrors that of water. Their ease of collection and analysis in the field or laboratory make them the preferred choice for evaluating water resident time. The input signals formed by the stable isotopes D and ^{18}O in precipitation are high frequency signals duc to scasonal variations and are attenuated during runoff formation through the distribution of the average residence time within the watershed. This is related to the physical properties and hydraulic parameters of the aquifer medium. Consequently, the isotope output signal at the watershed's runoff outlet exhibits a smaller variation range and is a low-frequency signal. By comparing the amplitudes of the input and output signals of the system, the mean residence time (MRT) of the watershed can be calculated. The isotope transport relationship in the rainfall-runoff system is assumed to be linear, and the system is generalized as a linear lumped-parameter model.

Stable isotopes typically exhibit distinct seasonal variations, which can be represented by wave functions:

$$\delta = \beta_0 + A[\cos(ct - \phi)] \tag{5.5}$$

The above equation is difficult to solve and can be transformed into:

$$\delta = \beta_0 + \beta_{\cos}\cos(ct) + \beta_{\sin}\sin(ct) \tag{5.6}$$

where δ is the isotope output function; β_0 is the average isotope value; $\beta_{\cos}$ and $\beta_{\sin}$ are regression coefficients for the trigonometric function; A is the amplitude of isotopic variation, $A = \sqrt{\beta_{COS}^2 + \beta_{\sin}^2}$; ϕ is the lag phase, $\tan\phi = \left|\frac{\beta_{\sin}}{\beta_{\cos}}\right|$; c is the angular frequency and t is time.

By using the exponential flow model and combining Eq. (5.6), τ is obtained:

$$\tau = c^{-1}\sqrt{f^{-2} - 1} \tag{5.7}$$

where τ is the mean residence time; f is the damping coefficient, $f = B_n/A_n$; B_n and A_n are the amplitude of the isotope input function and the isotope output function, respectively; c is the angular frequency.

Due to the analytical error in isotopes and the limitations of the amplitude of seasonal isotopic variation, the MRT of a watershed estimated by the wave function method should not exceed 5 years. Calculating the MRT of a watershed using the wave function method requires a long-term series of isotopic input and output signals, typically data spanning more than 2 years. However, owing to the constraints of water isotope sampling costs and experimental expenses, the sampling frequency is usually once a week or monthly, resulting in a relatively short sampling time series. Short-term isotopic input and output data can easily lead to increased errors and uncertainties in the calculation of the watershed's MRT. In current research, it is often necessary to extend the time series of isotopic input and output signals in the watershed. The methods to extend the time series of isotopic data mainly include: (1) utilizing the correlation between isotopes and air temperature; (2) directly extending isotopic data based on the fitted isotopic wave function; (3) replacing the lack of precipitation isotope data in the study area with data from nearby stations.

According to the wave function method and the exponential flow model, using the ^{18}O value of the seasonal effect of meltwater above 4150 m in the Laohugou glacier watershed in Xiaochangma River basin in the Qilian Mountains as the input signal, and the ^{18}O value of the seasonal effect of groundwater at 2200 m downstream as the output signal, the mean resident time of water in the Xiaochangma River Basin in the Qilian Mountains is calculated to be 2.17 years (Fig. 5.17). This means that after the meltwater seeps into the exit of Laohugou in the Qilian Mountains, it takes approximately 26 months before it emerges at the surface of the Changma River.

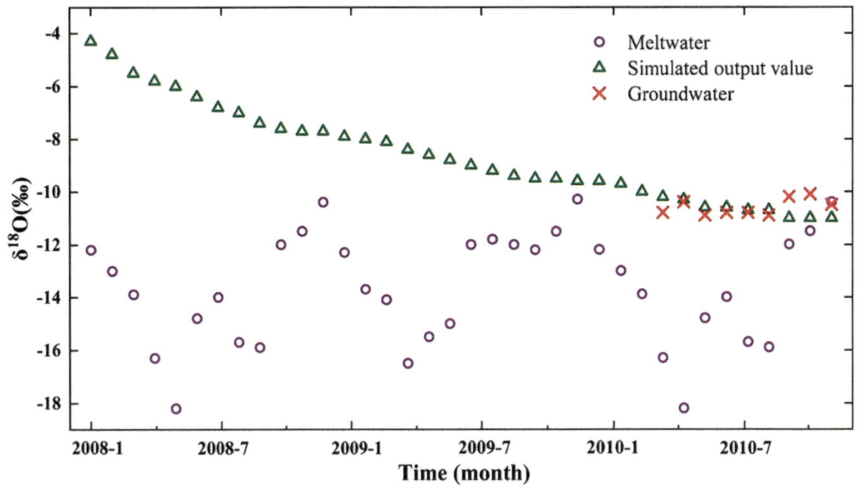

Fig. 5.17 Mean water resident time in the Xiaochangma River Basin, Qilian Mountains

5.4.6 Estimation of Water Evaporation

Evaporative loss potential is often predicted using multi-parametric meteorological models (Penman–Monteith, Priestley–Taylor Equations), energy-budgets (Bowen ratio), simulators, or evaporation pans. Stable isotopes are used to estimate evaporative loss by measuring the $^{18}O/^{16}O$ and $^{2}H/H$ ratios of the water compared to in-and-out flows (e.g., precipitation, atmospheric vapor, rivers, and groundwater) and using isotope mass-balance models to quantify evaporation to inflow (E/I) ratios (Halder et al. 2015). Waters undergoing evaporation exhibit distinctive trends in their $^{18}O/^{16}O$ versus D/H ratios away from the precipitation input, and the characteristic "local evaporation lines" (LEL) are assessed by using air temperature and relative humidity via the Craig–Gordon evaporation model. For meteorological-based evaporation prediction models, crucial information obtainable from stable isotopes to help constrain variables driving water evaporation are rarely used, and conversely, isotope mass-balance models do not incorporate a wide range of hydroclimatic or environmental variables. In both approaches, a full suite of variables of evaporation within the catchment system framework are generally overlooked (e.g., limnicity, solar radiation, wind speed) or are qualitatively acknowledged. One advantage of using stable isotopes is they provide a globally comparative metric (as opposed to often incomparable international hydrometric data) to quantify water evaporation, not only from water surfaces, but also to disentangle the relevance and commonalities of hydroclimatic variables within catchment systems.

Vystavna et al. (2021) have analyzed the stable isotope composition of 1257 global lakes and find that most lakes depend on precipitation and groundwater recharge subsequently altered by catchment and lake evaporation processes. Isotope mass-balance modeling shows that ca. 20% of water inflow in global lakes is lost through evaporation and ca. 10% of lakes in arid and temperate zones experience extreme evaporative losses >40% of the total inflow. Precipitation amount, limnicity, wind speed, relative humidity, and solar radiation are predominant controls on lake isotope composition and evaporation, regardless of the climatic zone. High latitude polar lakes had more negative median $\delta^{18}O_L$ values (−9.8‰) compared to tropical lakes generally (−1.3‰). Lake water undergoing evaporation becomes enriched in the heavy isotopes along the LEL with lower slopes (m = 4–6) that diverge from the Global Meteoric Water Line (GMWL, m = 8). The intersection of each LEL with the GMWL provides an estimate of the lake water origin (input water), whereas the displacement distance from the input value along the LEL reveals the extent of evaporation loss of the lake in relation to the inflow, as expressed by the E/I ratio. The lower intersection of $\Delta_{L\text{-}P}\delta^{18}O$ values (−6.6‰) in arid zone lakes suggests a greater dependency on groundwater storage with isotopic compositions ranging from −6 to −9‰, suggesting recharge from higher elevation or older groundwater recharged under cooler climate conditions, or from cold season recharge bias. The low LEL intersection (−14.6‰) for polar lakes is indicative for water inputs from glacier, ice, snowmelt, or that represents paleoclimate conditions from permafrost melt. The promotion of systematic global isotopic monitoring of Earth's lakes provides a direct and comparative approach to detect the impacts of climatic and catchment-scale changes on water-balance and evaporation trends.

Due to the numerous lakes in cryospheric regions such as the Qinghai-Tibet Plateau, the impact of lake and river water evaporation on the water cycle is substantial and cannot be ignored. In the arid climate and intense sunlight of the Yangtze River's source, water added directly by precipitation or entering lakes and swamps from long-term evaporation of meltwater causes the $\delta^{18}O$ and δD values of lake water to gradually increase. For example, due to long-term evaporation in Qinghai Lake, the average values of $\delta^{18}O$ and δD in the lake water are as high as 2.0‰ and 12.5‰, respectively. In contrast, in some lakes with both river inflow and lake outflow, the increase in lake water $\delta^{18}O$ and δD values is slightly lower. For instance, in Eling Lake, located in the source area of the Yellow River, the $\delta^{18}O$ and δD values of the lake water are −3.1‰ and −32‰, respectively. Lake water subjected to long-term evaporation will inevitably affect the hydrogen and oxygen isotopic composition of the river water it supplies. In the case of the Tuotuo River, the source of the Yangtze River, the average $\delta^{18}O$ value of regional atmospheric precipitation is −11.9‰, the average $\delta^{18}O$ value of snow and ice at its source in the Gladandong Snow Mountain is −12.4‰, and the average $\delta^{18}O$ value of the Tuotuo River water is −10.2‰, demonstrating the influence of evaporative lake water on river water.

Stable isotope methods are valuable tools for estimating lake evaporation. When calculating the evaporation of lake water, it is challenging to determine the amount of water entering the lake, primarily from the surface and underground, due to the large volume of water entering the lake. The characteristic of using the isotope method to calculate lake evaporation is that it is not necessary to measure the volume of water entering the lake, but only the stable isotope ratio. Based on the measured oxygen isotope ratio data and related hydrometeorological data of Qinghai Lake, Zhang and Yao (1997) calculated the average annual evaporation of Qinghai Lake using the stable isotope model to be 877 mm, which is roughly equivalent to the measured evaporation of 904.6 mm during the same period. However, due to the difficulty of isotope sampling, the isotope content of various water bodies is not a constant value and is greatly influenced by factors such as temperature and humidity, which can affect its accuracy. Therefore, evaporation estimates made by stable isotope methods should be compared with those obtained by other methods to ensure accuracy.

References

Anderson SP, Drever JI, Frost CD et al (2000) Chemical weathering in the foreland of a retreating glacier. Geochim Cosmochim Acta 64:1173–1189

Appleby J, Brook M, Horton T et al (2017) Stable isotope (δD–$\delta^{18}O$) relationships of ice facies and glaciological structures within the mid-latitude maritime Fox Glacier, New Zealand. Ann Glaciol 58:155–165

Beusen AHW, Dekkers ALM, Bouwman AF et al (2005) Estimation of global river transport of sediments and associated particulate C, N, and P. Glob Biogeochem Cy 19:GB4S05

Brimblecombe P, Tranter M, Abrahams PW et al (1985) Relocation and preferential elution of acidic solute through the snowpack of a small, remote, high-altitude Scottish catchment. Ann Glaciol 7:141–147

Cermeño P, Falkowski PG, Romero OE et al (2015) Continental erosion and the Cenozoic rise of marine diatoms. Proc Natl Acad Sci USA 112:4239–4244

Dai MH, Yin ZQ, Meng FF et al (2012) Spatial distribution of riverine DOC inputs to the ocean: an updated global synthesis. Curr Opin Environ Sustain 4:170–178

Darling W, Gooddy D, Gulliver et al (2023) Using environmental tracers to evaluate the preservation of palaeoclimate signals in aquifers of the London Basin, UK. J Hydrol 617128972

Frey KE, Siegel DI, Smith LC (2007) Geochemistry of west Siberian streams and their potential response to permafrost degradation. Water Resour Res 43:W03406

Frey KE, Smith LC (2005) Amplified carbon release from vast West Siberian peatlands by 2100. Geophys Res Lett 32:L09401

Gaillardet J, Dupré B, Allègre CJ (1999) Geochemistry of large river suspended sediments: silicate weathering or recycling tracer? Geochim Cosmochim Acta 63:4037–4051

Halder J, Terzer S, Wassenaar L et al (2015) The Global Network of Isotopes in Rivers (GNIR): integration of water isotopes in watershed observation and riverine research. Hydrol Earth Syst Sci 19:3419–3431

Hawkings J, Wadham J, Tranter M et al (2016) The Greenland Ice Sheet as a hot spot of phosphorus weathering and export in the Arctic. Glob Biogeochem Cy 30:191–210

Hawkings JR, Wadham JL, Benning LG et al (2017) Ice sheets as a missing source of silica to the polar oceans. Nat Commun 8:14198

Hawkings JR, Wadham JL, Tranter M et al (2014) Ice sheets as a significant source of highly reactive nanoparticulate iron to the oceans. Nat Commun 5:3929

Hawkings JR, Wadham JL, Tranter M et al (2015) The effect of warming climate on nutrient and solute export from the Greenland Ice Sheet. Geochem Perspect Lett 1:94–104

Hindshaw RS, Rickli J, Leuthold J et al (2014) Identifying weathering sources and processes in an outlet glacier of the Greenland Ice Sheet using Ca and Sr isotope ratios. Geochim Cosmochim Acta 145:50–71

Hodson A, Murnford P, Lister D (2004) Suspended sediment and phosphorus in proglacial rivers: bioavailability and potential impacts upon the P status of ice-marginal receiving waters. Hydrol Process 18:2409–2422

Hood E, Battin TJ, Fellman J et al (2015) Storage and release of organic carbon from glaciers and ice sheets. Nat Geosci 8:91–96

Hudson RO, Golding DL (1998) Snowpack chemistry during snow accumulation and melt in mature subalpine forest and regenerating clear-cut in the southern interior of B. C. Nord Hydrol 29:221–224

Jones JB, Petrone KC, Finlay JC et al (2005) Nitrogen loss from watersheds of interior Alaska underlain with discontinuous permafrost. Geophys Res Lett 32:L02401

Li XY, Qin DH, Jing ZF et al (2013) Diurnal hydrological controls and filtration effects on minor and trace elements in stream water draining Qiyi Glacier, northeastern of Tibetan Plateau. Sci China Earth Sci 56(81):81–92

Li XY, Wang NL, Ding YJ et al (2022) Globally elevated chemical weathering rates beneath glaciers. Nat Commun 13:407

Li XY, Ding YJ, Han TD et al (2019) Seasonal controls of runoff chemistry and chemical weathering at Urumqi Glacier No.1 basin in central Asia. Hydro Process 33(26):3258–3281

Li XY, Ding YJ, Xu JZ et al (2018) Importance of mountain glaciers as a source of dissolved organic carbon. J Geophys Res 123(9):2123–2134

Li XY, He XB, Kang SC et al (2016) Diurnal dynamics of minor and trace elements in stream water draining Dongkemadi Glacier on the Tibetan Plateau and its environmental implications. J Hydrol 541:1104–1118

Liu K, Hou S, Pang H et al (2022) Polar ice core-based climate and environmental research: a review and perspective. Chin J Polar Res 34(4):530–545

Marushchak ME, Kerttula J, Diáková K et al (2021) Thawing Yedoma permafrost is a neglected nitrous oxide source. Nat Commun 12:7107

Masson-Delmotte V, Hou S, Ekaykin A et al (2008) A review of Antarctic surface snow isotopic composition: observations, atmospheric circulation, and isotopic modeling. J Clim 21(13):3359–3387

Mitchell A, Brown GH (2007) Diurnal hydrological-physicochemical controls and sampling methods for trace elements in an Alpine glacial hydrological system. J Hydro 332(1–2):123–143

Petrone KC, Jones JB, Hinzman LD et al (2006) Seasonal export of carbon, nitrogen, and major solutes from Alaskan catchments with discontinuous permafrost. J Geophys Res 111:G02020

Ping C-L, Michaelson GJ, Jorgenson MT et al (2008) High stocks of soil organic carbon in the North American Arctic region. Nat Geosci 1:615–619

Ravishankara AR, Daniel JS, Portmann RW (2009) Nitrous oxide (N_2O): the dominant ozone-depleting substance emitted in the 21st century. Science 326:123–125

Schuur EG, McGuire AD, Schaedel C et al (2015) Climate change and the permafrost carbon feedback. Nature 520:171–179

Schuur EAG, Bockheim J, Canadell JG et al (2008) Vulnerability of permafrost carbon to climate change: implications for the global carbon cycle. Bioscience 58:701–714

Sharp M, Brown GH, Tranter M et al (1995) Comments on the use of chemically based mixing models in glacier hydrology. J Glaciol 41:241–246

Sharpe WE, De Walle DR, Leibfried RT et al (1984) Causes of acidification of four streams on Laurel Hill in Southwestern Pennsylvania. J Environ Qual 13:619–631

Stibal M, Anesio AM, Blues CJD, Tranter M (2009) Phosphatase activity and organic phosphorus turnover on a high Arctic glacier. Biogeosciences 6:913–922

Tranter M, Brown GH, Raiswell R et al (1993) A conceptual model of solute acquisition by Alpine glacial meltwaters. J Glaciol 39:573–581

Vystavna Y, Harjung A, Monteiro L et al (2021) Stable isotopes in global lakes integrate catchment and climatic controls on evaporation. Nat Commun 12:7224

Wang S, Wang S, Huang X et al (2024) Groundwater composition and stable isotopes as proxies to investigate paleoclimate change in permafrost thawing regions. Appl Geochem 169:106036

Wu J, Wu X, Hou D et al (2016) Streamwater hydrograph separation in an alpine glacier area in the Qilian Mountains, Northwestern China. Hydrol Sci J 61(13):2399–2410

Yang Y, Guo X, Wu Q et al (2023) Formation processes of shallow ground ice in permafrost in the Northeastern Qinghai-Tibet Plateau: a stable isotope perspective. Sci Total Environ 863:160967

Yang Y, Wu Q, Guo X et al (2024) First comprehensive stable isotope dataset of diverse water units in a permafrost-dominated catchment on the Qinghai-Xizang Plateau. Earth Syst Sci Data. https://doi.org/10.5194/essd-2024-54

Yang Y, Wu Q, Jiang G et al (2020) Ground ice at depths in the Tianshuihai Lake basin on the western Qinghai-Tibet Plateau: an indication of permafrost evolution. Sci Total Environ 729:138966

Yde J, Knudsen N, Steffensen J et al (2016) Stable oxygen isotope variability in two contrasting glacier river catchments in Greenland. Hydrol Earth Syst Sci 20:1197–1210

Zhang X, Yao T (1997) Estimating evaporation in lakes using stable isotope ratios. J Glaciol Geocryol 19(2):161–166

Chapter 6
Cryospheric River Sediment

Jia Qin and Xiangying Li

Water and sediment transport from rivers to oceans play a crucial role in the global geochemical cycle. Unlike runoff, river sediment discharge is highly sensitive to variations in climate and surface processes, making it a key indicator for quantifying land degradation and the depletion of soil resources. It also serves as a significant pathway in the geochemical cycle. Many of the world's major rivers are either located in or originate from regions influenced by cryospheric elements such as glaciers, permafrost, and snow cover. These cryospheric elements have a unique impact on river sediment, leading to generation processes, characteristics, and effects that differ markedly from those in non-cryospheric rivers.

Glacier melt, snowmelt, and permafrost thaw are major potential sediment sources for rivers within cryospheric regions. Cryospheric erosion can introduce substantial amounts of sediment and nutrients into rivers. However, research on sediment in cryospheric rivers began relatively late, with early studies primarily focused on high-latitude regions such as Siberia in the Northern Hemisphere and the pan-Arctic river basins of Alaska, the United States, and Canada. In China, research on cryospheric river sediment largely began in the 2000s, with a focus on the Qinghai-Tibet Plateau.

This chapter focuses on the sources and physical composition of sediment, the yield and variation of fluvial sediment fluxes, and the impacts of sediment in cryospheric rivers. It aims to systematically introduce the complex origins and temporal-spatial variations of riverine sediment transport within cryospheric regions.

J. Qin (✉)
Northwest Institute of Eco-Environment and Resources, Chinese Academy of Sciences, Lanzhou, China
e-mail: qinjia418@lzb.ac.cn

X. Li
Shaanxi Key Laboratory of Earth Surface System and Environmental Carrying Capacity, Northwest University, Xi'an, China
e-mail: shaanxilxy@163.com

Y. Ding et al., *Cryohydrology*, https://doi.org/10.1007/978-981-96-7360-5_6

By providing a detailed exploration of these aspects, we hope to offer a scientific foundation for the management of ecology and water safety in regions influenced by the cryosphere.

6.1 The Sources and Composition of Sediments in Cryospheric Rivers

6.1.1 Potential Sediment Sources in Cryospheric Rivers

The erosional impact of cryospheric elements, such as glaciers, snow, and permafrost, exerts a significant influence on sediment loads in cryospheric rivers. Key areas of high sediment yield within these regions include alpine barren zones, glacierized areas, snowmelt zones, and permafrost erosion zones. The primary processes contributing to sediment generation in cryospheric river basins are glacial erosion, snowmelt erosion, permafrost thaw slumps, and thermal gullying (Fig. 6.1). Additionally, conventional erosional processes, such as slope wash, mass wasting, and fluvial channel erosion, remain important sources of sediment.

The sediment sources in cryospheric river basins exhibit pronounced seasonal variability. Snowmelt erosion predominantly occurs in the spring when melting snow rapidly mobilizes sediment from the landscape. In the summer, thermal erosion processes, including glacial erosion, thaw slumping, and thermal gullying, become more active, alongside pluvial erosion mechanisms such as slope wash and mass wasting. Fluvial channel erosion, driven by the flow of water within river channels, can occur across spring, summer, and autumn, further contributing to sediment transport throughout the cryospheric river systems.

Glacial erosion is a dominant sediment source in alpine mountainous regions, particularly in cryospheric areas. For instance, in Alaska's Tanana River basin, glacier-melt discharge contributes 26–57% of the total river discharge, yet the sediment load from glacierized regions alone accounts for an overwhelming 76–94% of the total sediment load (Wada et al. 2011). Glaciers transport significant quantities of sediment to the proglacial zone through various mechanisms, including englacial and subglacial streams, as well as direct transport within or on the ice. Supraglacial debris, often derived from valley walls, accumulates along ice margins and terminus areas, forming moraines that are subsequently reworked by mass wasting and glaciofluvial processes (Li et al. 2020a).

Sediment transported through englacial and subglacial pathways is released into the proglacial zone primarily through ablation. Some of this sediment is immediately transported out of the glacial system via glaciofluvial processes, while other portions enter short- or long-term storage within or outside the channel before being remobilized. Storage can occur in both proximal and distal settings within the glacial/proglacial zone. Additionally, sediment may be transported subglacially via deforming bed layers, where entrainment and transfer mechanisms are spatially

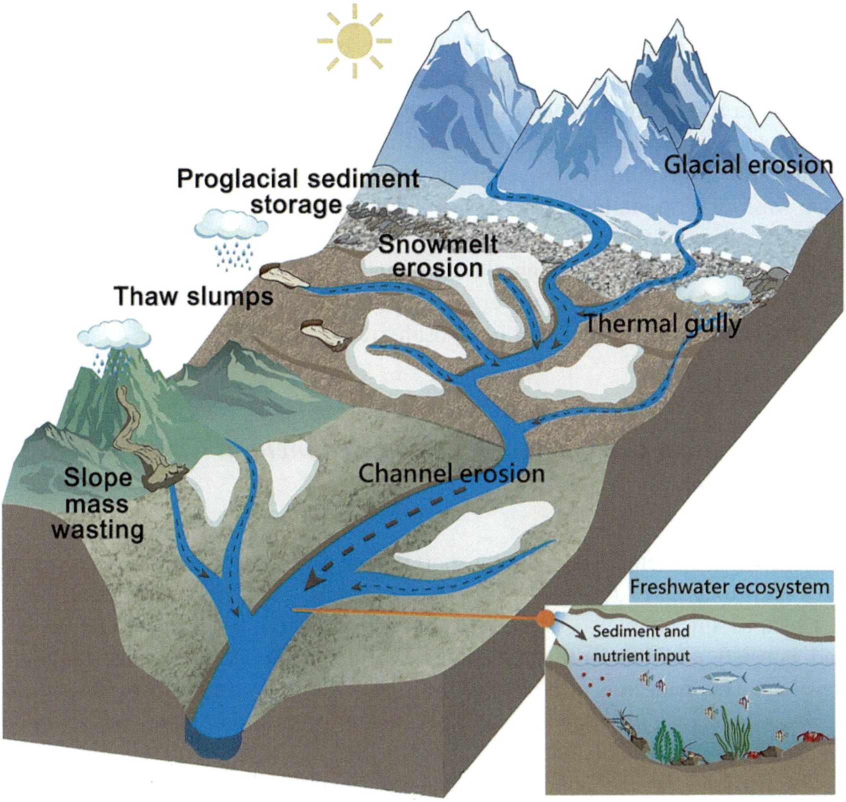

Fig. 6.1 The various erosion processes and sediment sources in cryospheric rivers

and temporally variable, often concentrated in marginal areas away from subglacial streams.

The quantity of sediment transferred by glacial ice to the proglacial zone is influenced by several factors, including lithology, topographical relief, precipitation, and the climate regime, which collectively affect ice flux, thermal conditions, and entrainment processes. These variables contribute to the dynamic nature of sediment transfer in cryospheric regions, making glacial erosion a critical component of sediment generation in these environments.

Permafrost processes are crucial contributors to runoff, sediment production, and transport during the summer months in cryospheric regions. Disturbances in permafrost, such as thaw slumps (illustrated in Fig. 6.1), which are often connected to river networks, can introduce substantial amounts of fine-grained sediment into river channels. Thaw slumps are typically triggered by the exposure and subsequent ablation of ground ice due to extreme thaw events or fluvial-thermal erosion. The resulting slump toes accumulate highly weatherable fine-grained sediment, which is further transported downslope by meltwater, eventually impacting stream systems. In some instances, sediments from these disturbed permafrost slopes may be temporarily stored within river systems, such as in channels, braided bars, or floodplains, which can dampen the immediate impact of increased sediment delivery at the basin outlet.

In addition to permafrost processes, slope wash, mass wasting, and fluvial channel erosion are significant sources of river sediment in cold regions. These processes can transfer large volumes of sediment into rivers, particularly during events of slope wash or slope failure. For instance, major slope failures can drastically alter the morphology of valley walls by shifting interfluves, thereby introducing considerable sediment loads from high elevations into lower elevation fluvial and lacustrine systems. These processes underscore the complex interplay between permafrost dynamics and sediment transport in cryospheric river basins.

6.1.2 Physical Composition of River Sediment in Cryospheric Regions

In natural river systems, sediment transport is generally categorized into two primary types: bed load and suspended load. The suspended load, comprising finer particles such as silt and clay, is typically more abundant in most rivers, while the bed load, consisting of coarser materials like sand, gravel, and pebbles, tends to be less prevalent. This pattern is also observed in rivers within cryospheric regions, where the suspended load often constitutes a significant proportion of the total sediment transported.

The particle size distribution of suspended load in river runoff is detailed in Table 6.1. In glacier-fed rivers, the suspended load primarily consists of medium silt, coarse silt, and fine sand, with fine silt and medium sand following, and clay making up the smallest proportion (Fig. 6.2). During the glacier ablation season, the particle size distribution within the suspended sediment remains relatively stable, indicating a consistent composition of suspended sediment in melt runoff throughout the season.

However, there are notable spatial variations in the suspended load composition across different regions. For instance, in the Gangotri Glacier basin, fine silt makes up about 15% of the suspended load during various ablation seasons (Fig. 6.2). In contrast, in the Dokriani Glacier basin, also situated in the Himalayas, fine silt accounts for only about 10% of the suspended load (Singh et al. 2003). This suggests that while the overall composition of suspended sediment in glacier runoff may remain consistent within a given region or season, it can vary significantly between different locations, even within the same mountain range.

Table 6.1 The particle size of silt and sand size particles in suspended sediment of river runoff

	Clay	Fine silt	Medium silt	Coarse silt	Fine sand	Medium sand
Particle size	<2 μm	2–6 μm	6–20 μm	20–60 μm	60–200 μm	200–600 μm

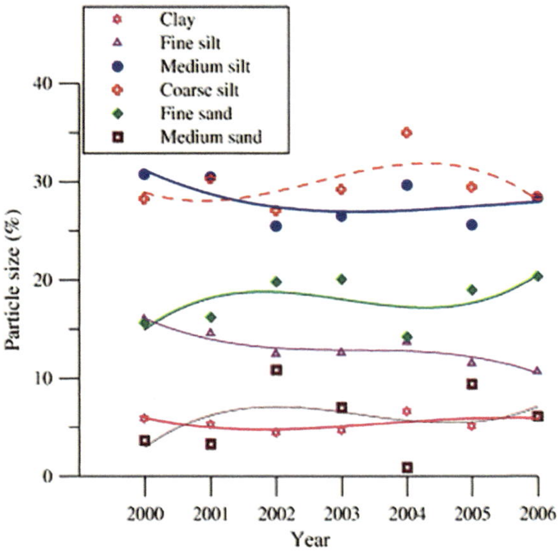

Fig. 6.2 Percentage distribution (symbols) and running mean average (lines) of annual particle size observed in different size ranges during seven consecutive melt seasons near the terminus of the Gangotri Glacier, central Himalayas (Referenced by Haritashya et al. 2010)

In permafrost regions, the particle size distribution of suspended load in river runoff exhibits notable differences compared to that in glacier runoff. In these areas, suspended load is primarily composed of fine silt, medium silt, and coarse silt, with the proportions of sand and clay being minimal, especially in Arctic rivers underlain by permafrost. This distinction highlights the unique sedimentary processes and sources in permafrost-dominated river systems, where fine particulate matter is more prevalent due to the specific erosion and transport mechanisms operating in these cold environments.

6.2 The Fluvial Sediment Fluxes and Their Variations in Cryosphere Rivers

6.2.1 Fluvial Sediment Fluxes of Cryosphere Rivers

Figure 6.3 illustrates the sediment fluxes within the major riverine sediment regions of the Northern Hemisphere, highlighting significant contributions from pan-Arctic, Greenland, and High Mountain Asia (HMA) rivers. The sediment fluxes are approximately 350 Mt/year in the pan-Arctic region, 1100 Mt/year in Greenland, and a substantial 1900 Mt/year in High Mountain Asia (HMA). Li et al. (2021a) compiled data on annual sediment fluxes from 28 monitoring stations across HMA, revealing that all these stations are situated in cold regions where sediment dynamics are significantly influenced by at least one or more cryospheric elements.

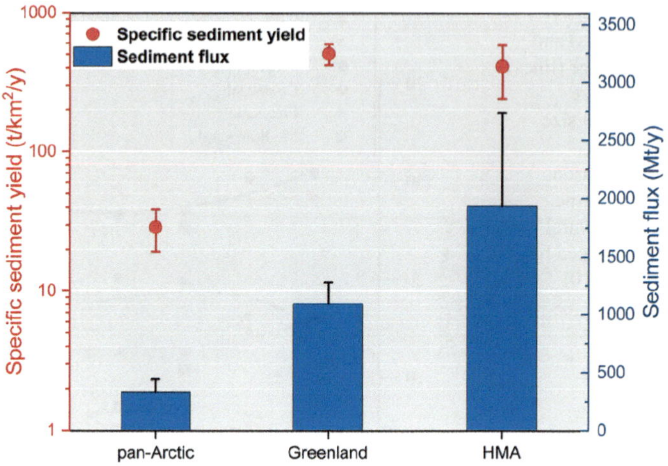

Fig. 6.3 Comparison of present-day specific sediment yields and sediment fluxes in the pan-Arctic, Greenland, and High Mountain Asia (HMA). The pan-Arctic is represented by the eight largest Arctic rivers (Yenisey, Lena, Ob', Pechora, Kolyma, Severnaya Dvina, Yukon, and MacKenzie) (Li et al. 2021a)

The sediment fluxes in these regions are notably high. In the eastern HMA basins, annual sediment fluxes range from 0.45 ± 0.41 Mt/yr at Qilian (Heihe River) to 34.77 ± 20.21 Mt/yr at Daojiebe (Salween), with an average of 8.89 ± 4.88 Mt/yr. In contrast, in the western HMA basins, sediment fluxes vary from 0.13 ± 0.16 Mt/yr at Yingxiongqiao (Wulumuqi River) to 174.12 ± 71.18 Mt/yr at Partab Bridge (Indus River), with an average flux of 21.28 ± 10.79 Mt/yr.

6.2.2 Variations of Fluvial Sediment Fluxes in Cryosphere Rivers

The fluvial sediment fluxes in the pan-Arctic and Andes regions were relatively stable during the period from 1971 to 2010, with a few rivers even showing a decreasing trend (Fig. 6.4). In contrast, observational data from 28 headwater basins in High Mountain Asia (HMA) reveal substantial increases in annual sediment fluxes over the past six decades (Fig. 6.5). These increases have accelerated since the mid-1990s, coinciding with a warmer and wetter climate. The observed sediment fluxes from HMA headwaters show significant rates of increase, averaging 12.99 ± 1.18% per decade.

In the HMA-east basins, sediment flux increases range from 3.67 ± 5.71% per decade at Jimai (Yellow River) to 25.40 ± 4.92% per decade at Daojiebe (Salween River). In comparison, the corresponding rates of runoff increase are only 2.56 ± 2.76% and 2.13 ± 1.30% per decade, respectively. Similarly, in the HMA-west basins,

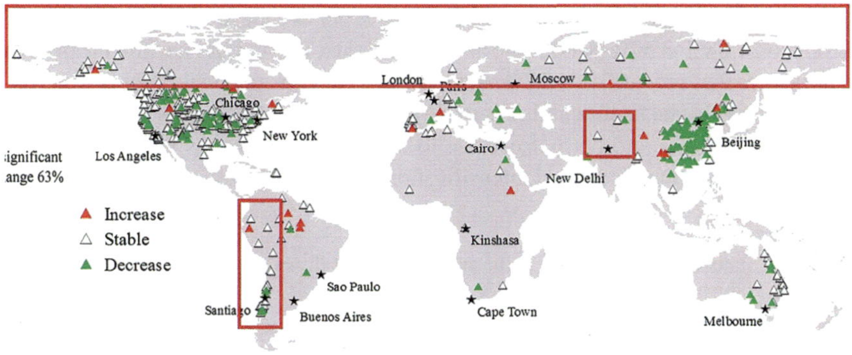

Fig. 6.4 Trends in suspended sediment flux for major large rivers during 1971–2010 in the world. All trends are significant at the 0.05 level (Li et al. 2020b)

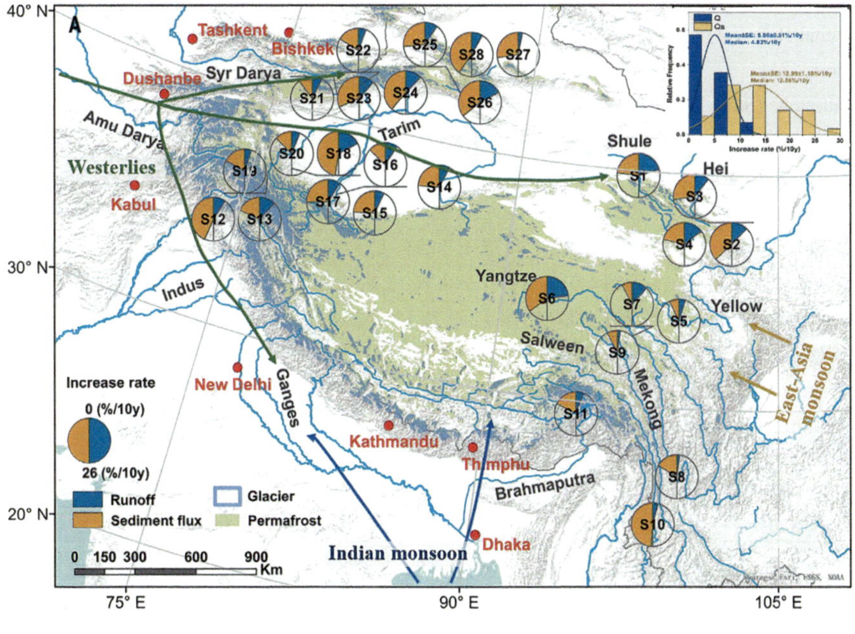

Fig. 6.5 Increasing annual runoff (Q) and fluvial sediment fluxes (Qs) in a warming and wetting High Mountain Asia (HMA). Relative rates of increase of mean annual runoff and sediment flux over the past six decades (1950s–2010s). Stations S1 to S11 represent HMA-east basins (monsoon dominated basins), and stations S12 to S28 represent HMA-west basins (westerly-dominated basins) (Li et al. 2021a)

sediment flux increases range from 5.30 ± 2.04% per decade at Shaliguilanke (Tarim River) to 24.68 ± 8.57% per decade at Yuzimenleke (Tarim River), while the decadal runoff increases are 4.00 ± 1.52% and 9.07 ± 1.67%, respectively (Table 6.2).

The rising fluvial sediment fluxes in HMA are primarily driven by the accelerated melting of glaciers and permafrost due to thermal processes, as well as increased precipitation and the associated intensification of erosion. These findings underscore

Table 6.2 Statistics of the changes in the annual sediment fluxes at the 28 individual measuring stations over the past few decades in HMA

Station	Period	Sediment flux (Mt/yr)	Increase rate (%/10yr)	Change over 1960–2010 (%)	P-value
Changmapu (S1)	1957–2017	3.46 ± 2.47	11.97 ± 5.03	59.85 ± 25.15	0.02
Zhamashike (S2)	1960–2017	1.25 ± 1.17	19.13 ± 7.01	95.65 ± 35.05	0.01
Qilian (S3)	1968–2017	0.45 ± 0.41	14.18 ± 9.05	70.9 ± 45.25	0.12
Buhahekou (S4)	1958–2017	0.47 ± 0.35	11.16 ± 5.49	55.8 ± 27.45	0.05
Jimai (S5)	1956–2015	1.02 ± 0.77	3.67 ± 5.71	18.35 ± 28.55	0.52
Tuotuohe (S6)	1958–2017	0.78 ± 0.58	18.11 ± 5.08	90.55 ± 25.4	<0.01
Zhimenda (S7)	1966–2017	9.42 ± 5.5	3.9 ± 5.47	19.5 ± 27.35	0.48
Shigu (S8)	1958–2016	24.96 ± 11.95	8.56 ± 3.55	42.8 ± 17.75	0.02
Xiangda (S9)	1963–2017	3.69 ± 1.69	3.93 ± 3.93	19.65 ± 19.65	0.32
Daojieba (S10)	1964–2011	34.77 ± 20.21	25.4 ± 4.92	127 ± 24.6	<0.01
Nuxia (S11)	1966–2010	17.47 ± 8.57	8.93 ± 5.6	44.65 ± 28	0.12
Partab Bridge (S12)	1983–2008	174.12 ± 71.18	22.49 ± 10.13	112.45 ± 50.65	0.04
Yugo (S13)	1974–2005	35.66 ± 32.93	9.51 ± 18.17	47.55 ± 90.85	0.60
Qiemo (S14)	1960–2006	4.24 ± 3.27	12.46 ± 8.26	62.3 ± 41.3	0.14
Nunumaimaitilangan (S15)	1960–2009	3.17 ± 1.96	12.63 ± 5.92	63.15 ± 29.6	0.04
Tongguziluoke (S16)	1964–2017	12.4 ± 8.16	6.68 ± 5.78	33.4 ± 28.9	0.25
Wuluwati (S17)	1957–2014	7.3 ± 4.45	13.21 ± 4.54	66.05 ± 22.7	0.01
Yuzimenleke (S18)	1960–2013	3.23 ± 3.35	24.68 ± 8.57	123.4 ± 42.85	0.01
Kulukelangan (S19)	1969–2015	26.5 ± 12.8	8.99 ± 5.14	44.95 ± 25.7	0.09
Kaqun (S20)	1957–2017	31.35 ± 15.4	6.57 ± 3.53	32.85 ± 17.65	0.07
Shaliguilanke (S21)	1960–2015	6.98 ± 1.8	5.3 ± 2.04	26.5 ± 10.2	0.01
Xiehela (S22)	1960–2015	20.15 ± 8.1	8.48 ± 3.18	42.4 ± 15.9	0.01
Xidaqiao (S23)	1958–2017	20.21 ± 11	16.39 ± 3.52	81.95 ± 17.6	<0.01
Heizi (S24)	1960–2009	11.79 ± 5.49	20.2 ± 3.63	101 ± 18.15	<0.01
Dashankou (S25)	1960–2012	1.07 ± 0.52	13.76 ± 3.99	68.8 ± 19.95	<0.01
Keerguti (S26)	1960–2009	0.14 ± 0.1	18.5 ± 6.17	92.5 ± 30.85	<0.01
Yingxiongqiao (S27)	1961–2011	0.13 ± 0.16	14.15 ± 11.49	70.75 ± 57.45	0.22
Kasiwate (S28)	1961–2015	3.35 ± 2.79	20.86 ± 6.61	104.3 ± 33.05	<0.01

the significant impact of climatic changes on sediment dynamics in cryospheric regions, particularly in HMA, where the effects are more pronounced compared to other cold regions such as the pan-Arctic and Andes.

Suspended sediment concentration in river systems typically exhibits spatial and temporal variability. Generally, the concentration is lowest at the surface and increases toward the riverbed. These concentrations also vary on both daily and seasonal time scales.

Daily Variation: The daily variation in sediment concentration is closely tied to the progression of the melting season. During periods of intense ablation, the diurnal variation is pronounced, while during the early and late stages of ablation, the variation is less distinct. For instance, in the Gangotri Glacier Basin in the Himalayas, sediment concentration peaks around 16:00 in August, the peak melting month. In contrast, during the weaker melting period in June, the maximum concentration occurs earlier, around 13:00. Regardless of the strength of ablation, sediment concentration tends to be lower at night (Fig. 6.6a).

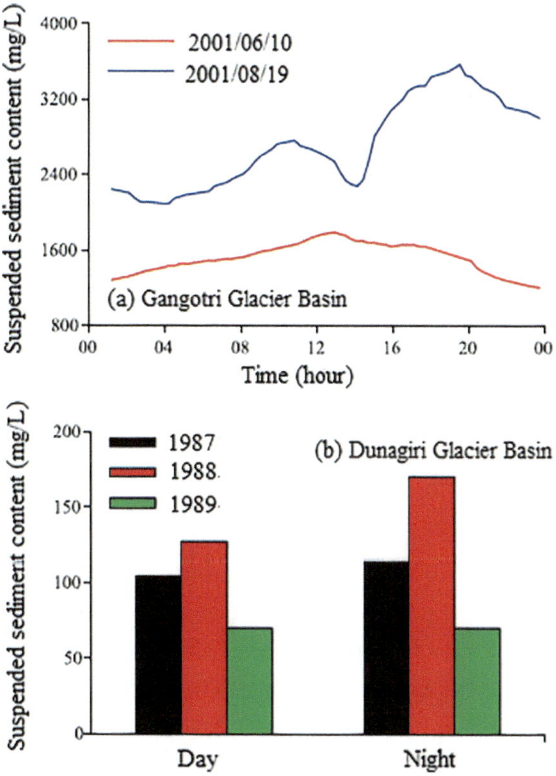

Fig. 6.6 Daily and seasonal variations of suspended sediment content in the Gangotri Glacier Basin (**a**) and the Dunagiri Glacier Basin (**b**) of the Himalayas (Adapted from Singh et al. 2006, and Srivastava et al. 2014)

However, variations can occur between different locations. In the Dunagiri Glacier Basin of the Himalayas, data from 1987 to 1988 show lower sediment concentrations during the day and higher concentrations at night. Interestingly, in 1989, sediment concentrations were nearly equivalent between day and night (Fig. 6.6b). These differences are likely influenced by the erosion effects of atmospheric precipitation and fluctuations in runoff during the sampling period, underscoring the complexity of sediment dynamics in cryospheric river systems.

Seasonal Variation: The seasonal variation in sediment concentration is notably significant in cryospheric river systems. During periods of intense glacier melting, sediment concentration is high, while during periods of weak melting, sediment concentration decreases. For example, in the Gangotri Glacier Basin of the Himalayas, sediment concentrations in 2000 and 2001 peaked in July during periods of strong melting and dropped to their lowest levels in September during weaker melting periods. This trend closely follows the pattern of runoff, with high runoff correlating with high sediment concentration, and low runoff corresponding to lower sediment concentration (Fig. 6.7).

It is also important to note that in June 2000 and May 2001, the maximum sediment concentration was observed earlier in the season. This early peak may be attributed to the erosive effects of snowmelt water and the sudden release of water stored within

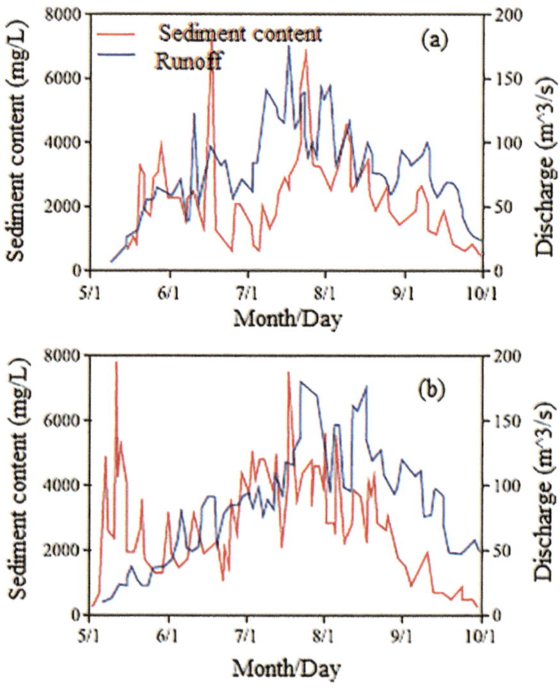

Fig. 6.7 Seasonal variations of sediment content and flow in the meltwater runoff of the Gangotri glacier in the Himalayas (according to Haritashya et al. 2010)

the glacier during the spring flood season. These factors can lead to a sharp increase in sediment load before the main ablation period, highlighting the complex interplay between runoff and sediment transport in cryospheric regions.

6.2.3 Influencing Factors of Riverine Sediment Variations in Cryospheric Regions

Many factors

Various factors influence sediment in cryospheric rivers, including climate (precipitation, temperature, wind, etc.), runoff, glacier melt, snowmelt, permafrost thaw (collapse), and terrain (Fig. 6.8). Additionally, changes in river regimes, underlying surface conditions, and human activities (such as grazing, agriculture, forestry, and infrastructure development) significantly impact riverine sediment in cryospheric regions. Climate factors are the primary drivers of sediment changes in these rivers. Studies have shown that sediment fluxes in the source areas of the 28 major rivers in High Mountain Asia (HMA) are positively correlated with temperature and precipitation (Li et al. 2021a). Increased and more intense precipitation leads to greater erosion, resulting in higher sediment content in rivers. Moreover, in cryospheric regions, the freeze–thaw cycles of permafrost and the thawing of glaciers and snowpacks are critical in influencing river sediment. For instance, in the Qinghai-Tibet Plateau, rainstorms, glacier/permafrost processes, and snowmelt are the three main factors driving the frequency of peak sediment events in rivers. Particularly over the past two decades (since 1995), significant warming has intensified glacial, permafrost, and snowmelt, leading to a notable increase in the frequency of peak sediment events (Fig. 6.9). Accelerated melting of glaciers, snow, and permafrost results in increased erosion, contributing more sediment to rivers.

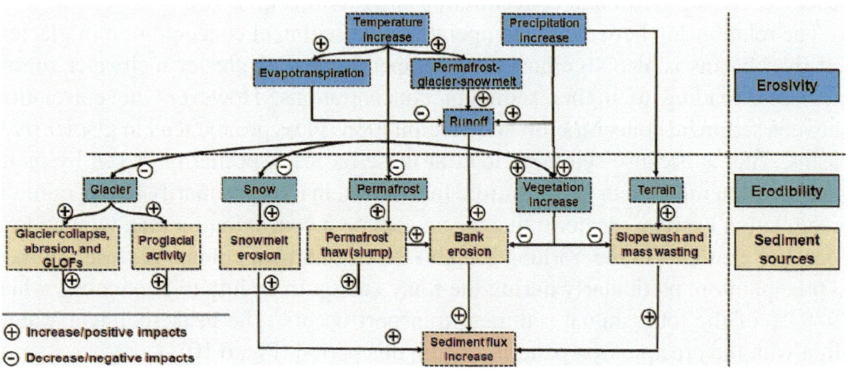

Fig. 6.8 A conceptual model of the responses of sediment flux to climate change in cold environments encompassing erosivity, erodibility and sediment sources

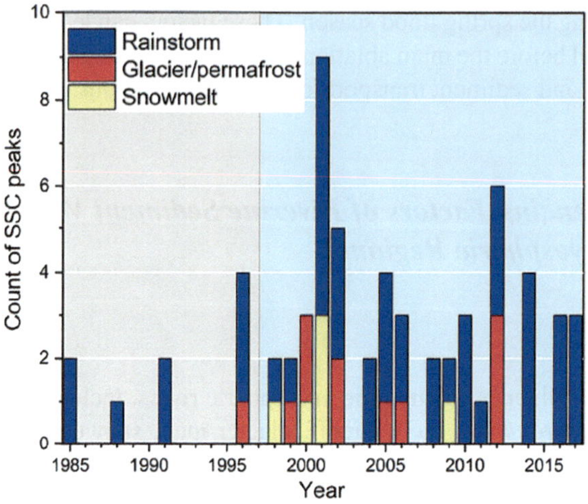

Fig. 6.9 Different drivers of peak suspended sediment concentration (SSC) events from 1985 to 2017. Rainstorm (i.e., rainfall >15 mm, the top one percentile) is the major driver of the SSC peaks (i.e., SSC >2.90 kg/m^3, the top one percentile), followed by glacier/permafrost processes and snowmelt, respectively. Glacier/permafrost melt processes are invoked when SSC peaks occur without rainstorms in the previous 3 days. SSC, suspended sediment concentration (Li et al., 2021a, 2021b)

The relationship between sediment concentration and runoff in cryospheric rivers is closely intertwined. Generally, sediment concentration tends to be proportional to runoff, meaning that when runoff increases, sediment concentration also rises. This is because higher runoff typically results from increased precipitation or accelerated glacier and snowmelt, leading to greater sediment erosion. In a given glacial basin, enhanced glacier ablation and subsequent runoff will intensify erosion and sediment production beneath the glacier. For instance, in the upper reaches of the Yarlung Tsangpo River, which is extensively glaciated, a strong positive correlation has been observed between sediment concentration and runoff (Fig. 6.10).

The relationship between air temperature and sediment concentration in glacier-fed river basins is also significant. As temperatures rise, glacier meltwater runoff increases, leading to higher sediment concentrations. However, the correlation between sediment concentration and precipitation is less pronounced in glacier river basins. This is because sediment load in these basins is primarily driven by meltwater erosion rather than precipitation. In contrast, in rivers primarily fed by rainfall, precipitation exerts a greater influence on sediment concentration than temperature does. For example, in the Yarlung Zangbo River, sediment volume is closely linked to precipitation, particularly during the rainy season from July to September, when 79–93% of the total annual sediment transport occurs. The peak sediment values align with the precipitation patterns during this period (Fig. 6.10).

Sediment load in cryospheric rivers is increasingly influenced by climate change, hydrological shifts, land surface processes, and human activities, leading to rising

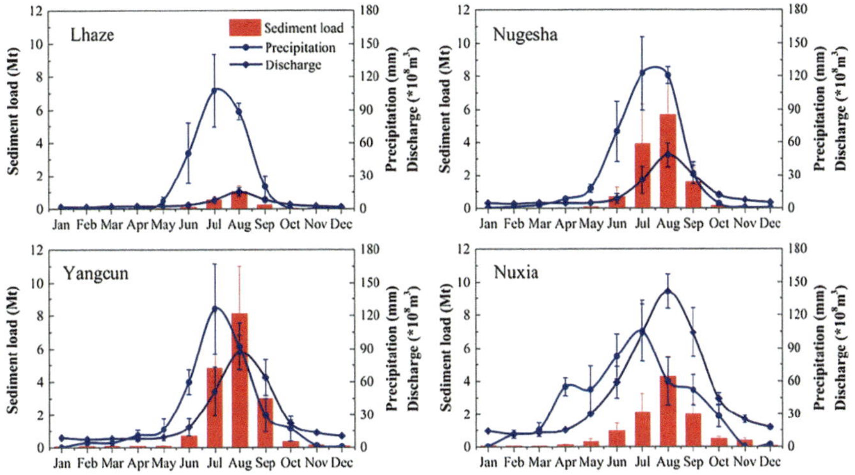

Fig. 6.10 Mean monthly precipitation, discharge, and sediment loads from 2007 to 2009 for the four main stream stations in the upstream and midstream of the Yarlung Tsangpo River

uncertainty in the correlation between these factors and sediment transport. This uncertainty stems primarily from the complex balance between positive and negative feedback mechanisms, which vary spatially. Rising temperatures result in increased glacier and snow melt, as well as permafrost degradation. The increase in meltwater enhances erosion and, when combined with rainfall flood peaks, boosts the sediment transport capacity of river flow. Permafrost freeze–thaw cycles contribute to upper soil erosion, particularly during thawing periods when near-saturated surface soils, combined with the low roughness and permeability of the permafrost table, heighten the risk of severe soil erosion. Changes in precipitation intensity also influence erosion and the sediment transport capacity of slopes, with runoff variations further impacting river sediment transport.

Temperature, precipitation, and glacial meltwater are generally considered positive feedback factors that increase river sediment discharge. In contrast, vegetation acts as a negative feedback factor by reducing soil erosion, intercepting surface runoff, enhancing soil infiltration, and protecting surface soil from erosion. However, under global warming, the interplay between these positive and negative feedbacks becomes increasingly uncertain, leading to greater unpredictability in sediment transport changes in cryospheric rivers.

6.3 The Possible Impacts of River Sediment Changes in Cryospheric Regions

6.3.1 Impact of Sediment on River Ecosystems and the Biogeochemical Cycle

River sediment in cryospheric environments significantly impacts river and lake ecosystems. As a crucial abiotic factor, sediment composition, quantity, and migration rate influence aquatic animals, plants, microorganisms, and surrounding habitats. Changes in sediment concentration and transport can have immediate ecological consequences. For instance, a sharp increase in sediment concentration can diminish water transparency, reduce plankton populations, and threaten the survival of certain fish species. In the Yellow River source area's reservoirs, elevated sediment levels during discharge periods often lead to hypoxia and loss of fish control due to increased sediment volume and the subsequent reduction in available oxygen. Conversely, a marked decrease in sediment concentration can enhance water transparency, promoting aquatic plant growth while simultaneously reducing shelter for zooplankton, which may increase their vulnerability to predation and impact their populations.

Fine-grained sediments, rich in organic and mineral substances, can adversely affect organisms during specific developmental stages. A significant reduction in sediment content may elevate egg and larval mortality rates. Riverbed erosion and sedimentation, driven by river flow and sediment dynamics, produce substantial changes with direct ecological effects categorized into two types: those impacting organisms directly and those altering habitat structures. Erosion can disrupt riverbed structures and harm aquatic flora and fauna, while sedimentation can bury essential gravel substrates that provide refuge for fish fry, thereby affecting their survival rates. Numerous studies indicate that dramatic shifts in riverbed erosion and sedimentation have profound and often unpredictable effects on benthos, fish eggs, and fry.

River biodiversity evolves through the natural succession of organisms over extended periods. Water flow and sediment, as dynamic factors, play a crucial role in shaping this biodiversity. The long-term influence of water and sediment processes creates spatiotemporal heterogeneity in river habitats, forming the foundation for the development and maintenance of river biodiversity. Fine sediment transport and deposition are essential for the movement and deposition of trace elements, carbon, and nutrients within river systems.

The construction of upstream dams impedes sediment transport by trapping a significant amount of sediment in reservoirs, which reduces the sediment content downstream. This alteration leads to coarser sediment particle sizes and a decrease in organic matter and nutrient content downstream. Consequently, the reduction in sediment and the coarsening of particle size following reservoir construction diminish the river system's original nutrient sources, negatively impacting the ecosystem, particularly the floodplain wetland system.

The interaction between rivers and their floodplains drives nutrient circulation and transformation between these areas. Both the volume and composition of sediment particles influence these processes. A decrease in flood frequency or the absence of floods weakens the connection between the river and its floodplain, leading to reduced sediment and nutrient supply. Therefore, changes in floodplain sediment dynamics and reduced flooding disrupt nutrient cycling in river systems and impair the ecological function of river ecotones.

Sediment in cryospheric rivers significantly influences biogeochemical cycles. Glacier ablation releases numerous active elements, which, when transported downstream through glacial meltwater and river sediment, can impact aquatic ecosystems and biogeochemical processes. Essential nutrients such as iron, silicon, and phosphorus originate from glaciers and are critical for plant and algal growth. Organic carbon from glaciers plays a key role in the global carbon cycle. Iron, silicon, and phosphorus are primarily sourced from the chemical weathering of bedrock minerals, while organic carbon is derived from the ice surface's primary productivity, carbon deposition, and surrounding organic matter. Additionally, the degradation of frozen soil releases active elements that can also affect downstream aquatic ecosystems and biogeochemical cycles.

Inland water bodies are significant sources of atmospheric methane, with rivers in permafrost regions contributing notably to global methane emissions. Permafrost degradation enhances methane emissions from river water bodies and may influence the migration and transformation of organic carbon, impacting both ecosystems and the carbon cycle (Li et al. 2017). For example, rivers originating from the Qinghai-Tibet Plateau transport substantial amounts of active elements downstream due to glacier melting and permafrost degradation. The concentration and flux of total organic carbon (TOC), total organic nitrogen (TON), inorganic carbon, and inorganic nitrogen in the rivers from the Yangtze River and Yellow River source areas are notable. The total annual output of TOC from these source areas is approximately 55,400 tons/year, while the total TON output is about 9,900 tons/year. This organic carbon output is comparable to that of some of the world's major river basins and is influenced by seasonal variations in freeze–thaw processes. Active elements transported by cryospheric rivers exhibit pronounced seasonal variation.

6.3.2 Disasters Associated with Cryospheric River Sediment

The impact of sediment in cryospheric rivers extends beyond typical river sediment issues, affecting human survival, living environments, and economies. While common sediment-related disasters include sedimentation and erosion, mountain floods, mudslides, and sediment pollution, cryospheric river sediment presents additional unique challenges. In cryospheric regions, sediment issues notably impact hydropower development, agriculture, animal husbandry, and water quality. Specific problems include reduced river and lake water volumes, riverbank collapses, floods, environmental degradation, and diminished engineering benefits.

Sediment-related disasters in cryospheric regions are characterized by their global and collective nature, gradual and long-term effects, complexity, and management difficulties. In certain river sections, sediment deposition has become a significant source of conflict. For instance, on the Qinghai-Tibet Plateau, side erosion of river channels, expansion of gully banks, and mountain torrents have led to varying degrees of land washout or submersion, affecting cultivated land and agricultural productivity.

Glacier debris flow represents a distinct type of disaster in cryospheric rivers. Rivers fed by modern glaciers or snowmelt are susceptible to heavy rainfall, snowmelt, or glacial lake outburst floods, all of which can transport substantial amounts of sediment and moraine, leading to mudslides. These glacier debris flows can have severe consequences for the ecological environment, infrastructure, and the safety of people and property.

Soil erosion resulting from permafrost degradation has led to the thinning of the cultivated layer, a decline in soil fertility, and significant damage to the ecological environment in cold regions (Zhang et al. 2019). Additionally, many rivers in cryospheric areas, particularly those in mountainous regions with steep terrain, experience high gradients. As water flows downstream into flatter areas, sedimentation occurs. This sedimentation, especially during flood events, can raise the riverbed and narrow the water flow section, thereby affecting the river's flow transport and flood discharge capacity. For instance, sedimentation in the Nianchu River on the Qinghai-Tibet Plateau has been a key factor in the frequent flow interruptions and flooding observed in recent years. Sedimentation in upstream areas also impacts reservoir capacity and regulation, with multiple reservoirs in the Nianchu River Basin facing the risk of sediment accumulation (Qian 2002). Specifically, due to sediment deposition, the actual irrigation area of the Gonggadarando Reservoir is less than one-fifth of the intended area (Cai et al. 1996).

Furthermore, the sediment accumulation on riverbeds contributes to frequent sandstorms in winter and spring. For example, in the Yarlung Zangbo River, reduced water levels in the dry season expose floodplains. Strong winter and spring winds then create sandstorms in the valley, which not only bury roads, farmland, and pastures but also intermittently close Lhasa Gongga Airport for several days each year (Ding 2006; Tan et al. 2007). The temporal and spatial variation of sediment in cryospheric rivers, driven by environmental changes, poses various ecological risks and challenges, which vary depending on the specific physical geography of each river. Therefore, assessing changes in sediment transport and researching early warning systems for sediment disasters in cryospheric rivers are critical scientific and technological tasks that warrant significant attention.

References

Cai X, Chen Q, Jie H (1996) Soil erosion and its countermeasures in the middle reaches of Yarlung Zangbo River in the Tibet Autnomous region. Bull Soil Water Conserv 16(6):48–53

Ding Y (2006) Analysis on cause of formation of wind and sand disaster in Gongga Airport of Tibet and countermeasures for control. For Invent Plan 31(3):122–125
Haritashya UK, Kumar K, Singh P (2010) Particle size characteristics of suspended sediment transported in meltwater from the Gangotri Glacier, central Himalaya—an indicator of subglacial sediment evacuation. Geomorphology 122:140–152
Li X, Ding Y, Han T et al (2017) Seasonal variations of organic carbon and nitrogen in the upper basins of the Yangtze and Yellow Rivers. J Mt Sci 14:1577–1590
Li D, Li Z, Zhou Y, Lu X (2020a) Substantial increases in the water and sediment fluxes in the headwater region of the Tibetan Plateau in response to global warming. Geophys Res Lett 47:e2020GL087745. https://doi.org/10.1029/2020GL087745
Li L, Ni J, Chang F et al (2020b) Global trends in water and sediment fluxes of the world's large rivers. Sci Bull 65:62–69
Li D, Lu X, Overeem I, Walling D, Syvitski J, Kettner AJ, Bookhagen B, Zhou Y, Zhang T (2021a) Exceptional increases in fluvial sediment fluxes in a warmer and wetter High Mountain Asia. Science 374(6567):599–603
Li D, Overeem I, Kettner AJ, Zhou Y, Lu X (2021b) Air temperature regulates erodible landscape, water, and sediment fluxes in the permafrost dominated catchment on the Tibetan Plateau. Water Resour Res 57:e2020WR028193. https://doi.org/10.1029/2020WR028193
Qian C (2002) Soil erosion and ecological reestablishment in the Nianchu River Valley in Tabet. Chin J Ecol 21(5):74–77
Singh K, Haritashy U, Kumar N et al (2006) Hydrological characteristics of the Gangotri Glacier, central Himalayas, India. J Hydrol 327:55–67. https://doi.org/10.1016/j.jhydrol.2005.11.060
Singh P, Ramasatri KS, Kumar N et al (2003) Suspended sediment transport from the Dokriani Glacier in the Garhwal Himalayas. Nord Hydrol 34:221–244
Srivastava D, Kumar A, Verma A et al (2014) Analysis of Climate and Melt-runoff in Dunagiri Glacier of Garhwal Himalaya (India). Water Resour Manage 28(10):3035–3055. https://doi.org/10.1007/s11269-014-0653-8
Tan S, Lin K, Zhang C (2007) Analytical studies on causes of sandstorm formation in the areas of Gongga Airport of Tibet. Yunnan Geogr Environ Res 19(4):139–144
Wada T, Chikita KA, Kim Y, Kudo I (2011) Glacial effects on discharge and sediment load in the Subarctic Tanana River Basin, Alaska. Arctic Antarctic Alp Res 43(4):632–648. https://doi.org/10.1657/1938-4246-43.4.632
Zhang F, Shi X, Zeng C, Wang G, Chen Y, Wang L, Zhang H, Yu Z (2019) Variation and influence of riverine sediment transport from Tibetan Plateau, China. Bull Chin Acad Sci 34(11):1273–1284. https://doi.org/10.16418/j.issn.1000-3045.2019.11.010

Chapter 7
The Roles of the Cryosphere in the Watershed

Junfeng Liu, Rensheng Chen, Xiqiang Wang, Yong Yang, and Yongjian Ding

The cryosphere plays a pivotal role in cryohydrology, contributing to water retention, runoff recharge, and the regulation of water resources. Additionally, the rapid melting of cryospheric elements has significant implications for extreme hydrological events. Glaciers, snow, and permafrost each have distinct hydrological impacts, and their combined characteristics shape the hydrology of various basins.

7.1 Water Conservation in the Watershed

The cryosphere, functioning as a network of "cold islands," plays a pivotal role in altering regional temperature, humidity fields, and circulation patterns. This alteration facilitates the efficient retention and condensation of water vapor, leading to increased precipitation and the formation of a cold, wet microclimate, commonly referred to as the "cold-humidity island effect." This effect is particularly pronounced

J. Liu (✉) · R. Chen · X. Wang · Y. Yang · Y. Ding
Northwest Institute of Eco-Environment and Resources, Chinese Academy of Sciences, Lanzhou, China
e-mail: liujfzyou@lzb.ac.cn

R. Chen
e-mail: crs2008@lzb.ac.cn

X. Wang
e-mail: wangxq@lzb.ac.cn

Y. Yang
e-mail: yy177@lzb.ac.cn

Y. Ding
e-mail: dyj@lzb.ac.cn

Y. Ding et al., *Cryohydrology*, https://doi.org/10.1007/978-981-96-7360-5_7

in mountainous regions, where the cryosphere effectively captures and transforms water vapor into precipitation, thereby significantly enhancing the water supply within cryospheric basins by curbing evapotranspiration.

Despite its significance, the theoretical foundations, systematic observation data, research methodologies, and mathematical models pertaining to the cold-humidity island effect remain underdeveloped. Therefore, this chapter delves into the theoretical principles underlying this effect within the cryosphere, with a specific focus on glaciers. It also presents observational evidence of the cold-humidity island effect, including data from mountain precipitation peaks, temperature and humidity differentials between glacier interiors and exteriors, and cloud distribution patterns in mountainous regions.

7.1.1 Fundamentals of Cold-Humidity Effects in the Cryosphere

High mountain systems, where the cryosphere is located, play a crucial role in intercepting water vapor. This occurs through two primary mechanisms: topographic uplift, which forces vapor to rise along hillsides, and the movement of vapor along prevailing wind directions. The cryosphere, acting as a significant cold source, efficiently condenses this water vapor into precipitation. This process is particularly vital in arid inland river basins, where the cryosphere captures both local and transported water vapor, contributing significantly to the region's water supply.

In the middle and lower reaches of Northwest China's inland rivers, basin precipitation and glacier melt are crucial components of the evapotranspiration process. This process is particularly significant in the region's extensive desert and oasis areas, where moisture evaporates into water vapor. The region's arid climate and high temperatures, contrasted with the cooler conditions in the cold inland mountains, create substantial temperature differentials that drive the transport of water vapor from the desert and oasis areas towards the mountainous regions. Through topographic uplift and these temperature disparities, water vapor from the lower and middle mountains is conveyed to higher elevations. Under suitable meteorological conditions, this vapor condenses due to the alpine cryosphere, leading to precipitation in the mountainous areas. This precipitation subsequently recharges the desert and oasis areas as runoff, establishing an internal cycle of water vapor.

Moreover, global atmospheric circulation patterns, such as westerly winds carrying external water vapor, are intercepted and collected by high mountain ranges. The interaction between external and internal water vapor sources, combined with mountain elevation effects and cryospheric condensation, leads to the formation of high precipitation zones in mountainous terrain (Fig. 7.1).

The alpine cryosphere is characterized by frequent solid precipitation, such as hail and graupel. The formation and development of these precipitation types require strong convective activity, unstable atmospheric stratification, and other specific

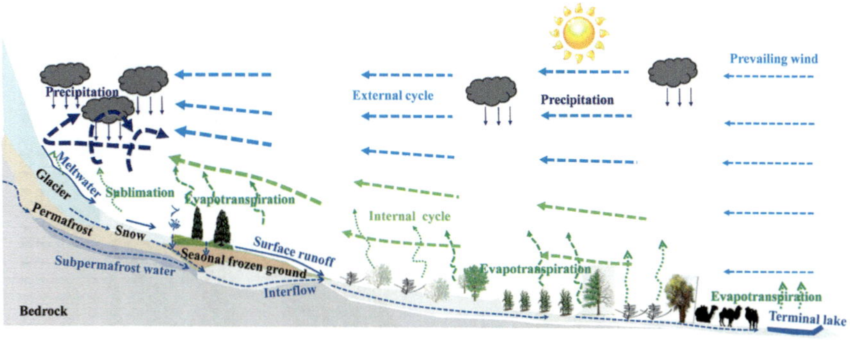

Fig. 7.1 Schematic diagram illustrating water vapor circulation in an arid inland river basin and the interception of precipitation by the alpine cryosphere

conditions. This suggests that vertical atmospheric motion is relatively intense in alpine cryospheric regions, especially near the edges of glaciers, where the glacier's underlying temperature field significantly influences airflow patterns. Figure 7.2 illustrates the wind field distribution at the terminus of the Shiyi Glacier in the Hulu watershed of the Qilian Mountains during two instances in the summer of 2015. The Hulu-5 meteorological station, situated in a high-precipitation area of the Heihe River basin in the Qilian Mountains, recorded notable disturbances and convergence in the wind field (Chen et al. 2018a).

The phenomenon of increasing precipitation with altitude in mountainous regions is often attributed to the dynamic effects of topography. However, in alpine cryospheric areas, this increase is not solely due to topographic influence. The widespread presence of cryospheric elements significantly alters the temperature and humidity fields near the surface, which is the primary driver for the formation of high precipitation zones at elevated altitudes. Numerical simulations have shown that mountains in the arid northwest function as cold islands, creating a pronounced horizontal temperature gradient between the mountains and the surrounding hot and dry desert. This gradient results in strong horizontal turbulent exchanges between the mountainous regions and the desert (Hu 1987). Additionally, the alpine cryosphere and the lower mountain slopes facilitate the exchange of water vapor and heat along the slopes due to this temperature gradient. The movement of turbulent flows modifies the convection of temperature and water vapor within the mountains, generating local circulation patterns. This process also transports warmer air from the desert or lower mountain areas to the cold island, forming an inversion layer with warmer air above and colder air below, which further intensifies turbulent motion (Fig. 7.1). The enhanced turbulence within the cold island causes the air to rise, increasing the hydrothermal gradient between the cold island and the surrounding area, thereby strengthening the cold island effect. As a result, when advection occurs, it encounters strong cold air masses near the cold island, generating more robust updrafts on the upwind slopes. This local circulation is closely related to the extent of the alpine

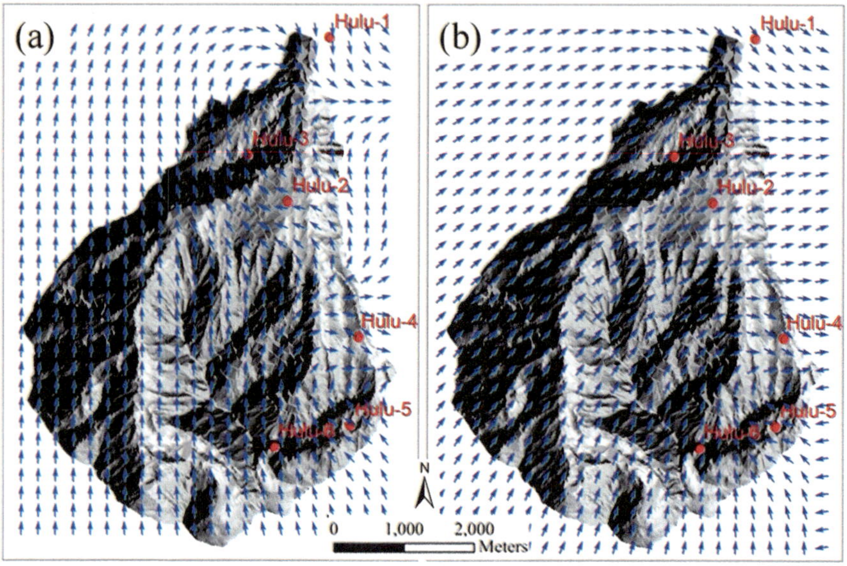

Fig. 7.2 Near-surface wind field distribution in the Hulu watershed of the Qilian Mountains at 8:30 (**a**) and 15:30 (**b**) on June 24, 2015. Data was derived from meteorological stations Hulu-1 to Hulu-6, with wind direction interpolated between these stations. Note that wind direction data beyond the immediate vicinity of the meteorological stations may not accurately represent actual conditions (Chen et al. 2018a)

cryosphere; the larger the cryospheric area, the stronger the resulting local circulation. In alpine cryospheric regions spanning tens of thousands of square kilometers, this leads to intense convection, ultimately creating high-precipitation zones.

In the arid mountainous regions of western China, the interaction between topography and the disturbance of cryospheric temperature and humidity fields significantly enhances the movement and instability of water vapor. This dynamic leads to intensified water vapor condensation, the initiation of precipitation processes, and an increase in precipitation with altitude. The complex topography of high mountain systems means that the influence of cryospheric perturbations varies across different regions. In some areas, water vapor condensation reaches its peak after the cryospheric temperature and humidity fields are disturbed to a certain altitude, resulting in a maximum precipitation altitude. In other regions, the perturbation causes water vapor to continue rising with altitude, leading to progressively increasing precipitation at higher altitude.

The perturbation of cryospheric temperature and humidity fields, known as the cold-humidity island effect, fundamentally alters the classical relationship between precipitation and altitude. In mid-latitude mountains, this effect creates a zone of maximum precipitation at the foot or lower slopes of the mountains. As altitude increases beyond this point, precipitation begins to decrease, resulting in a distinctive spatial distribution of precipitation in high mountain systems where the cryosphere is present. Alpine cryosphere areas receive more precipitation, which in turn increases the cryosphere's storage capacity, thereby strengthening the cold-humidity island effect. This creates a virtuous cycle that plays a crucial role in water conservation.

In summary, in mid-latitude alpine cryosphere regions, cold islands and high humidity centers form due to differences between the temperature and humidity fields in the subsurface and the surrounding environment. These differences enhance horizontal turbulence and intensify internal field turbulence, leading to increased local convection, multiple precipitation processes, and higher precipitation, effectively transforming high mountainous areas into wet islands. When transit airflow passes over these regions, the mountains act as barriers, intensifying airflow activity and generating precipitation. This effect makes alpine regions zones of high precipitation. The higher the elevation and the larger the area covered by snow, glaciers, and permafrost, the stronger the impact on airflow, resulting in increased precipitation. Proximity to water vapor sources further enhances the cold-humidity island effect, leading to extensive glacier distribution and high precipitation around the Tibetan Plateau, while the interior experiences lower precipitation (Shen and Liang 2004).

7.1.2 Observed Cases of Cold-Humidity Effects in the Cryosphere

Understanding the cryosphere's role in water conservation remains limited, with a notable gap in research and quantitative methods to accurately describe this phenomenon. This section delves into the cold-humidity island effect within cryospheric regions by examining several key factors.

1. **Maximum Precipitation Altitude in the Mountains**

On the windward slopes of mountains, water vapor is lifted due to topography, resulting in precipitation that typically increases with elevation. However, once the mountain reaches a certain height, the amount of available water vapor decreases, leading to a decline in precipitation beyond this elevation. The point at which precipitation reaches its peak is referred to as the maximum precipitation altitude.

Classical meteorological theories, supported by model simulations and initial observations, suggest that in mid-latitude mountainous regions, the maximum precipitation altitude is generally found at the foot or on the lower slopes of the mountains. This pattern holds true for most low and mid-altitude mountainous areas globally.

In China, studies have shown that precipitation continues to increase with elevation in mountain ranges below 2600 m, reaching its peak at the summit. However, in higher-altitude mountain ranges, the maximum precipitation altitude is typically located on the mountainsides. For instance, in Huangshan Mountain, the maximum precipitation altitude is approximately 1400 m above sea level, while in the Qinling Mountains, it ranges from 2000 to 2500 m. This variation illustrates the influence of altitude on precipitation patterns and the complex interactions between topography and atmospheric processes in different mountain regions.

Traditional meteorological theories and models concerning maximum precipitation height do not fully apply to the high mountain systems of western China, largely due to the cold-humidity island effect created by the cryosphere. This unique effect alters the expected precipitation patterns, leading to variations in maximum precipitation height that differ from classical models.

For example, on the northern slope of the Tianshan Mountain, the maximum precipitation height has been estimated to be around 2100 m (Kang et al. 2000) or 2500 m (Fu 1983) above sea level. Similarly, on the northern slope of the Qilian Mountain, the maximum precipitation height is found at approximately 2800 m (Tang 1985) or 3000 m (Ding and Kang 1985) above sea level, positioned on the mountainside.

Snow and glacier surveys, as well as studies on the relationship between precipitation and altitude conducted in the 1950s and 1970s, particularly in regions like the northern slopes of Mount Everest, the Tian Shan, the Karakoram Mountains, the Caucasus Mountains, and the Pamir Plateau, indicated that the highest precipitation altitudes in these high mountain areas are typically found near alpine glacier regions. Direct evidence includes the presence of very thick snow accumulation in glacier basins. Subsequent observations and studies on the distribution of maximum precipitation levels have revealed that in the high mountains of western China, precipitation generally increases with elevation, without a distinct peak in maximum precipitation levels near glaciers.

For example, gradient observations conducted in 1987 in the Urumqi River source area of the Tianshan Mountains identified two maximum precipitation heights: one at the mountain base, approximately 1800 m in elevation (annual precipitation of 583.2 mm), and another in the grain-snow basin of the Urumqi River source No. 1 glacier at around 4030 m (annual precipitation of 650.2 mm; Yang et al. 1991). Observations from August 10, 2007, to September 12, 2008, on the Qiyi Glacier on the northern slope of the Qilian Mountains revealed a significant maximum precipitation height between 4500 and 4700 m (Wang et al. 2009). Additionally, observations from 2014 to 2016 in the Zhangye-Hulu watershed, located in the upper reaches of the Heihe River in the Qilian Mountains, showed that the maximum precipitation height on the north slope was at the terminus of the Shiyi Glacier, at an altitude of approximately 4200 m (Fig. 7.3; Chen et al. 2018a). These findings were based on measurements from six weighing gauges deployed within the elevation range of 2980 to 4484 m in the Hulu watershed (approximately 20 km^2) and observations from the Zhangye meteorological station, located along the same profile.

Recent observations from the Hulu watershed profile between 2019 and 2020 supported these findings, revealing a clear linear correlation between annual precipitation and altitude, with a slight peak in precipitation around 4150 m (Fig. 7.5a; Chen et al. 2023). Despite the extensive monitoring and broad elevation range, the results were insufficient to fully characterize the elevation gradient distribution of precipitation in mountainous areas due to the spatial and temporal variability of precipitation in high mountain regions. For instance, monitoring data from the Shule River basin in the Qilian Mountains indicated a maximum precipitation altitude of approximately 3400 m above sea level with a limited number of gauges (Fig. 7.4a). However, when

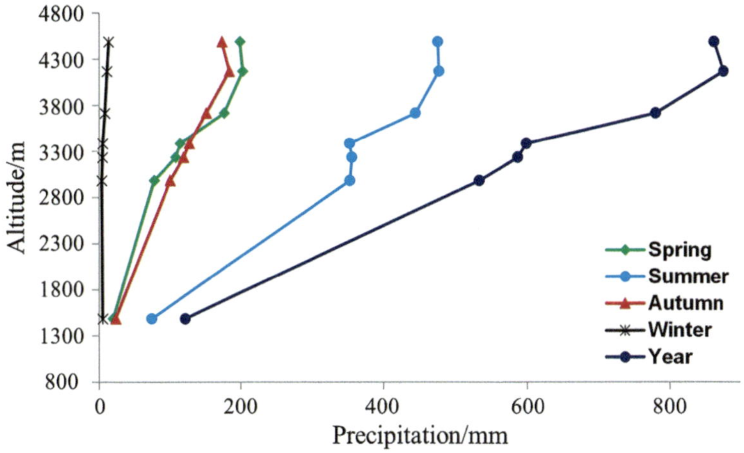

Fig. 7.3 Variation of precipitation with altitude along the Zhangye-Hulu profile in the Qilian Mountains, based on data from 2014 to 2016 (Chen et al. 2018a)

observations were analyzed in more detail, precipitation increased with elevation, suggesting that no distinct maximum precipitation altitude exists (Fig. 7.4b).

Observation results from 46 stations in and around the Qilian Mountains between 2019 and 2020 were consistent with the findings from the Shule River monitoring, showing a significant linear correlation between annual precipitation and altitude (Fig. 7.5b; Chen et al. 2023). This data was collected from six TRwS stations in the Hulu watershed basin, 13T-200B gauges in the Qilian Mountains Precipitation Observation Network, and 27 Chinese Standard Precipitation Gauges from the National Meteorological Station. However, given that the data in Fig. 7.4a, b were collected in different years, with substantial variations in annual precipitation, it is possible that

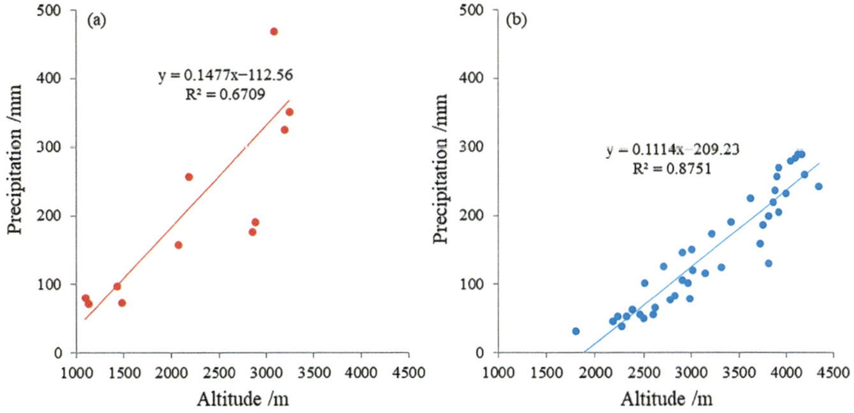

Fig. 7.4 Variation of precipitation with altitude during the summer season in the Shule River basin of the Qilian Mountains, before and after intensified observation

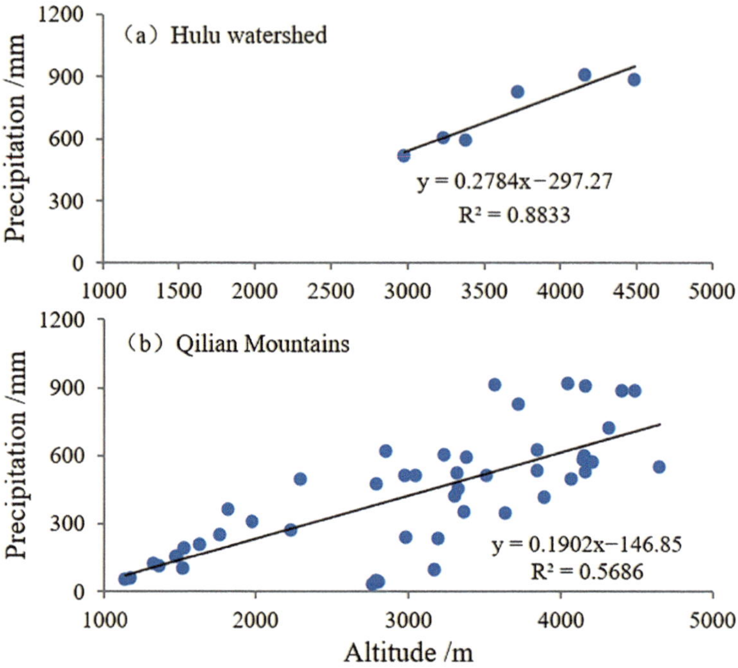

Fig. 7.5 Annual precipitation versus elevation during 2019–2020 for **a** the Hulu watershed and **b** the Qilian Mountains (Chen et al. 2023)

the observed differences in the spatial distribution of precipitation were influenced by interannual variability.

In summary, maximum precipitation altitudes are generally observed near glaciers in high mountain snowy regions. For instance, research on the No. 1 Glacier at the Urumqi River source on the north slope of the Tianshan Mountains and the Qiyi Glacier in the Qilian Mountains supports this pattern. Similar findings have been reported in snow and glacier studies across the northern slopes of Mount Everest, the Tian Shan, the Karakoram Mountains, the Caucasus Mountains, and the Pamir Plateau. Notably, the maximum precipitation altitude of approximately 4200 m found in the Zhangye-Hulu watershed profile on the north slope of the Qilian Mountains is situated at the terminus of the Shiyi Glacier. This altitude aligns with the average elevation of glacier termini on the southern slopes of the Shule River in the Qilian Mountains.

It is important to recognize that the formation of maximum precipitation altitudes cannot be solely attributed to the presence of glaciers. The observed association of maximum precipitation altitudes with glacier areas is largely a result of the limited and non-systematic precipitation observations in high-altitude mountainous regions. Most available precipitation data from these areas are derived from glacier studies. Theoretically, it is the sufficiently high mountain systems, influenced by the cold-humidity island effect, that contribute to the development of the cryosphere.

The presence of the alpine cryosphere, in turn, supports the formation of maximum precipitation altitudes in these mountainous regions. The location of glaciers and the surrounding topographic features likely play a significant role, with large glaciers having a particularly notable impact. Additionally, the interplay between mountain system distribution, atmospheric circulation, and mountain topography also influences precipitation patterns, warranting further investigation and analysis.

2. **Differences in Temperature and Humidity Fields Inside and Outside the Glacier**

Limited comparative studies of air temperature and relative humidity inside and outside glaciers reveal notable differences. Specifically, air temperatures on glacier surfaces are significantly lower, and relative humidity is higher, indicating a colder and wetter microclimate within glacier areas. Figure 7.6 illustrates the comparison of monthly mean air temperature and relative humidity inside and outside the glacier from November 2007 to June 2008 at Koxkar Glacier in the Tianshan Mountains. The two observation sites differ in elevation by 65 m. During the observation period, the average temperature inside the glacier was 1.2 °C lower than outside, a discrepancy significantly greater than the typical lapse rate. Additionally, relative humidity was 1.7% higher inside the glacier. It is important to note that the observation sites are in close proximity, with the off-glacier site still situated on bedrock within the glacier cover. This observation suggests a pronounced cold and wet island effect within the glacier. Similar findings were observed at two neighboring sites inside and outside the Qiyi Glacier in the Qilian Mountains (Wang et al. 2009). At sites farther apart, the temperature difference between inside and outside the glacier was even more marked (Shen and Liang 2004).

During the summer months, from June to August, the intense melting of snow and glaciers over large areas can lead to the cooling of the near-surface air layer above glaciers, resulting in the formation of an inversion layer. In the Russian mountain glacier system, this glacial cooling effect can cause the temperature of the near-surface air layer, which is 70–100 m thick, to drop by an average of 1–2 °C, with variations between 2–4 °C depending on the glacier's size and distribution density. This cooling effect, driven by the height of the glacier and the temperature difference over the moraine, can be quantified using the thermocline value (δT), derived from the average temperature difference between these areas.

In addition to the cooling effect of the low-level air, which is influenced by the thickness and duration of the ablation inversion layer, periodic weather patterns and the horizontal flow of warm and cold air across the glacier area also play a role. This is reflected in the diurnal variations of the thermocline value, which shows a gradual increase with the ice surface temperature. Observations on the Qiyi Glacier indicated that thermocline values varied significantly between day and night across different temperature intervals. In June and July, the average thermocline values were higher during the day than at night (0.97 and 0.56 °C in June; 1.33 and 0.78 °C in July). However, the opposite pattern was observed in August, with thermocline values of 0.76 °C during the day and 1.12 °C at night. This reversal may be attributed to the

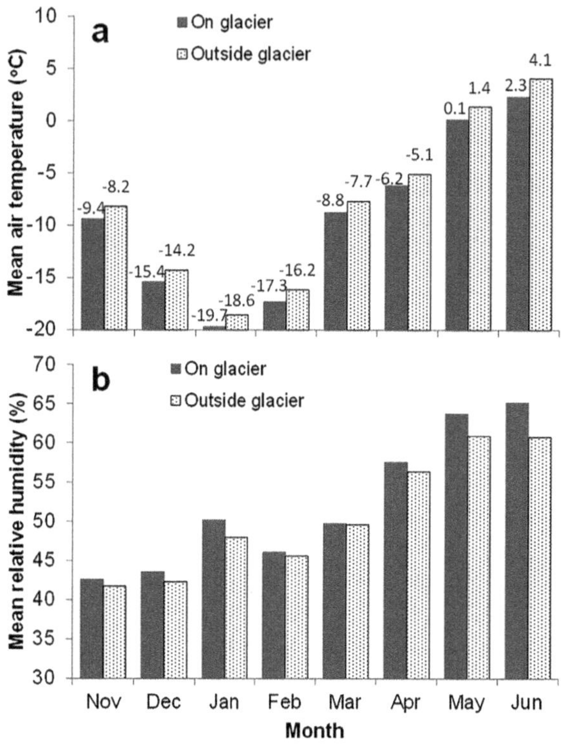

Fig. 7.6 Comparison of monthly mean air temperature (**a**) and relative humidity (**b**) on and outside the Koxkar Glacier in the Tianshan Mountains (Chen et al. 2018a)

melting of the bare ice surface, the thickening of the strong inversion layer, and the enhanced downward long-wave radiation from the lower atmosphere (Bai 1989).

As the inversion layer thickens with rising temperatures, the duration of the inversion extends, and horizontal airflow intensifies. This results in a linear and gradual increase in the thermocline value as the ice surface temperature rises. Glaciers of a certain scale exhibit a strong correlation with the cooling effect of ablation, driven by specific energy exchange processes. On average, the thermocline value is less influenced by temperature and horizontal airflow; however, it is well correlated with the characteristic parameter L_{max}, which describes the effect and scale of glaciers (defined as the maximum length of the glacier along its longest axis).

Based on data from 12 glaciers in China, it has been observed that in any glacial climate area, the relationship between δ_T and L_{max} follows a logarithmic trend, as described by an empirical regression equation (Bai 1989):

$$\log \delta_T = 0.4208 \log L_{\max} - 0.118, \; r = 0.9452 \tag{7.1}$$

3. **Cloud Distribution in the Mountains**

The retention of water vapor by mountain glaciers theoretically leads to a more uniform distribution of clouds within and surrounding mountain systems, a phenomenon influenced by the cryosphere's cold-humidity island effect. Figure 7.7 illustrates the spatial distribution of clouds over the Qilian and Tianshan Mountains, highlighting a clear correlation between cloud patterns and the mountainous terrain, which aligns closely with the distribution of snow and glaciers. These observations provide further evidence supporting the existence and impact of the cold-humidity island effect in mountainous cryospheric regions.

Satellite imagery captures the frequent occurrence of convective clouds with bubble-like structures during the summer months over the high Tibetan plateau. These clouds are predominantly centered around mountain ranges, forming numerous wet islands within glacial regions (Ye and Gao 1979). This underscores the significant role that glaciers play in fostering moist microclimates in mountainous areas.

Fig. 7.7 Examples of clouds distributed along mountain systems in the **a** Qilian Mountains and **b** Tianshan Mountains

7.2 Basin Runoff Recharge Effect

Mountainous regions, due to their significant precipitation, serve as reservoirs of glacial and snow water resources, which are critical in sustaining the river systems that flow from these areas. The cryosphere of High Asia, encompassing regions such as the Tibetan Plateau, Tianshan Mountains, and Altay Mountains, plays a pivotal role in replenishing major rivers like the Yangtze, Yellow, Indus, Ganges, Tarim, Yenisey, and Yili. These rivers, nourished predominantly by the meltwater from the High Asia Cryosphere, are vital water sources for the semi-arid regions of China. The recharge effect of the cryosphere on basin runoff is profound, making it a central topic of discussion in this section.

7.2.1 Cryosphere Water Reserves

Approximately 89% of the Earth's surface freshwater is stored in the form of solid ice in Antarctica. Excluding Antarctica, the remaining six continents account for only 11% of the surface freshwater, of which glaciers represent 10.16%. This means that glaciers make up about 92.84% of the land surface freshwater, with lakes storing 6.65% and river channels accounting for just 0.08%.

Globally, the land surface (excluding the ice sheets of Antarctica and Greenland) contains a significant number of glaciers, covering approximately 7.26×10^8 km^2 and storing around 14.94×10^4 km^3 of water (Table 7.1). The maximum area covered by snow globally is 47×10^6 km^2 (1966–present, NOAA Snow Data), which represents about 31.5% of the global land surface. Notably, 98% of this snow cover is found in the Northern Hemisphere (Fig. 7.8), where, in August, it spans approximately 1.9×10^6 km^2, expanding to 45.2×10^6 km^2 in January—nearly one and a half times the land area of the Northern Hemisphere. The maximum snow water storage in the Northern Hemisphere is estimated at about 3×10^{15} kg (Foster and Chang 1993), equivalent to roughly 65 mm of snow water equivalent over the maximum snow-covered area. In contrast, the Southern Hemisphere has minimal snow cover on land outside of Antarctica.

Permafrost contains significant quantities of ground ice, which can be categorized as structural ice, cave ice, and buried ice. Throughout the development phases of permafrost, water from various sources transforms into ground ice within the soil, making it a critical sink for regional water resources. During periods of stable permafrost, ground ice functions as storage; however, during phases of permafrost degradation, it melts, releasing water and acting as a source. The presence of ground ice profoundly influences the distribution and regulation of regional water resources and hydrological processes.

According to the 1998 Northern Hemisphere Permafrost Distribution Map by the International Permafrost Association, ice-rich permafrost (with ground ice content exceeding 20%) predominantly occurs in high latitudes, covering approximately

Table 7.1 Regional distribution of glaciers worldwide (based on calculations from IPCC AR5 2013)

Region	Number of glaciers	Area (10^3 km^2)	Proportion of regional area (%)	Glacial water resources[a] (10^4 km^3)
The Arctic Islands	4035	98,655.7	13.5	2.563
Alaska	23,112	89,267	12.3	1.983
USA and Canada	25,733	159,094.4	21.9	3.857
Europe	5259	3183.7	0.5	0.018
Asian	71,431	123,587.8	17.0	1.095
Southern hemisphere	21,607	33,076.4	5.3	0.518
Greenland	13,880	87,125.9	12.0	1.41
Antarctic Region	3274	132,267.4	18.2	3.491
Total	168,331	726,258.3		14.94

[a] The ocean area is calculated as 362.5×10^6 km^2

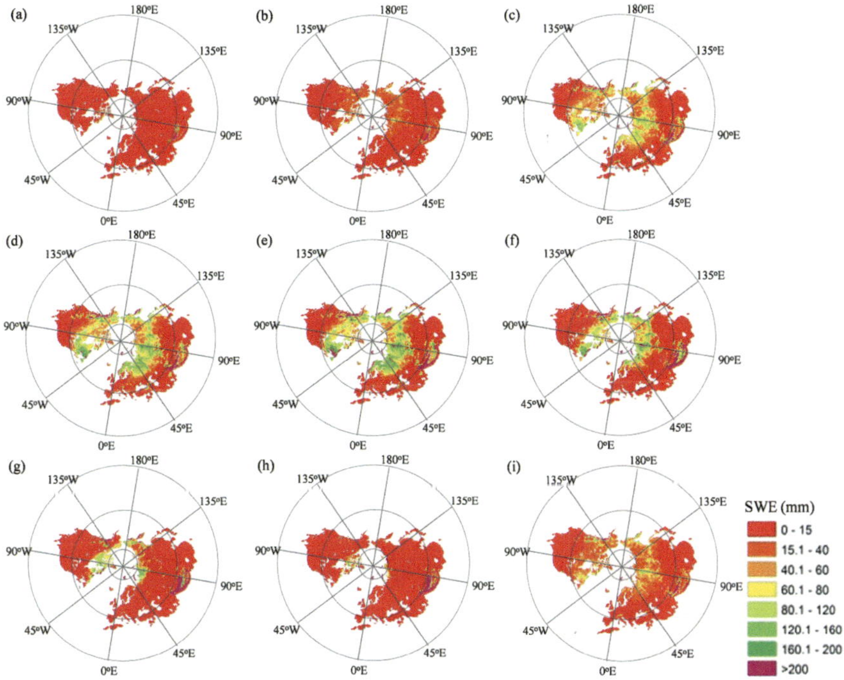

Fig. 7.8 Monthly distribution of Snow Water Equivalent (SWE) in the Northern Hemisphere. Figs. a, b, c, d, e and f are SWE in Jan., Feb., Mar., Apr., May. and Jun. Figs. g, h and i are SWE in Oct., Nov. and Dec.. The data is derived from the National Weather Centre of Canada, utilizing SWE records since 1999

2.02% of the Northern Hemisphere land surface and 8.57% of its permafrost area. Ice-poor permafrost (with less than 10% subsurface ice) is mainly found in mountainous regions and high altitudes, covering about 15.8% of the land surface and 66.5% of the permafrost area in the Northern Hemisphere.

These estimates assume a sediment thickness of 10 m for thick sediment cover and 5 m for thin sediment cover, which likely underestimates the actual thickness of Circum-Arctic sediment. Zhang et al. (2008) consider thick sedimentary cover to have a thickness of 20 m and thin sedimentary cover to have a thickness of 10 m, with ice contents ranging from 20 to 30% and 10% to 20%, respectively. Excluding glaciers, ice sheets, submarine permafrost, and lake permafrost, the total volume of ground ice in the Northern Hemisphere is estimated to range from 11.37×10^6 to 36.55×10^6 km^3, which is equivalent to a potential sea-level change of 2.7–8.8 cm (Table 7.2).

The calculation method mentioned above likely underestimates the ice content in regions with high ground ice concentrations, as it only accounts for the ice content within a 20-m thick permafrost layer. To achieve more accurate assessments, further studies are necessary to refine these estimates, taking into consideration the full thickness of the permafrost and potential variations in ice content beyond the currently assumed limits.

Table 7.2 Calculated ground ice content for each permafrost type in the Northern Hemisphere, based on cover thickness and content hypotheses

Ice content for each permafrost type	Thick sediment cover						Thin sediment cover (10 m)				Total	
	20–30%		10–20%		0–10%		10–20%		0–10%		/10^6 km^3	
	High	Low	High	Low	High	Low	High	Low	High	Low	High	Low
Continuous permafrost (>90%)	9	0.54	5.24	2.36	0.76	0	4.32	1.94	5.7	0	25.02	9.7
Discontinuous permafrost (50–90%)	0.43	0.24	3.14	0.88	0.68	0	1.35	0.38	2.12	0	7.72	1.5
Sporadic (10–50%)	0.33	0.07	0.62	0.06	0.58	0	0.32	0.04	1.34	0	3.19	0.17
Isolated Patches (0–10%)	0.2	0	0.02	0	0.12	0	0	0	0.28	0	0.62	0
Total / 10^6 km^3	9.96	5.71	9.02	3.3	2.14	0	5.99	2.36	9.44	0	36.55	11.37

Data source Zhang et al. (2008)

7.2.2 *Glacier Contribution to Runoff*

At the basin scale, the contribution of glacier meltwater to basin runoff is influenced by several factors, including the number, size, shape, area ratio, and ice storage capacity of glaciers. In mountainous regions characterized by arid climates and significant glacier coverage, glacier meltwater can play a crucial role in replenishing basin runoff. For example, research by Gao et al. (2010) demonstrates that glacier meltwater accounts for up to 50% of the runoff in the Tianshan and Kunlun Mountains of China. Similarly, in the West Corridor of China, the Shule River benefits from glacier runoff contributing up to 30% of its total runoff.

In contrast, major rivers originating from the Tibetan Plateau exhibit lower contributions of glacier runoff, typically around 10%, due to the region's abundant precipitation. The general trend shows an increasing proportion of glacier meltwater runoff from the periphery towards the interior of the Tibetan Plateau, which correlates with increased aridity and greater glacier coverage. In the Austrian Alps, glacier runoff recharge rates vary between 1.5 and 9%, depending on glacier extent (Lambrecht and Mayer 2009).

In the Canadian Rockies, studies reveal that annual glacier runoff contributions to rivers like the Edmonton and Calgary average around 3%, with peak contributions rising to as much as 10% from July to September, particularly in headwaters near glaciers (Comeau et al. 2009). In some glaciated rivers of the Gulf of Alaska, glacier runoff can constitute up to 47% of total recharge rates (Neal et al. 2010).

As temperatures rise, glaciers in China are melting more rapidly and shrinking in size, leading to significant changes in glacier meltwater resources within major river systems (see Table 7.3). Overall, glacier meltwater has shown a noticeable increasing trend, particularly since the 1990s. However, despite the ongoing global warming and glacier retreat, some glaciers have already begun to experience a decline in meltwater output. It is anticipated that the future trajectory of glacier meltwater resources will follow a pattern of initial increase followed by a decrease. The timing of these inflection points in glacier runoff is likely to be influenced by factors such as glacier type, size, shape, and geographic location.

Simulation studies conducted across 56 major glacier basins worldwide suggest that approximately half of these basins will experience a peak in glacier runoff followed by a decline (Huss and Hock 2018). The remaining basins have already reached their peak meltwater output. Basins with these inflection points typically feature extensive glacier coverage and large glacier areas. Intra-annual variability in glacier runoff is typically characterized by increased runoff in early summer, followed by a decrease in late summer. Regionally, Central Asia and the Andes have observed declines in glacier runoff exceeding 10% (Huss and Hock 2018).

Table 7.3 Glaciers and glacial meltwater runoff in the mountainous regions of western China (adapted from Kang et al. 2000)

Mountains	Glacier area (km^2)	Glacial meltwater runoff (10^8 m^3)	Percentage of national glacial meltwater runoff (%)
Qilian Mountains	1930.51	11.32	1.9
Altai Mountains[a]	296.75	3.86	0.6
Tianshan Mountains	9224.8	96.3	15.9
Pamir	2696.11	15.35	2.5
Karakorum Mountains	6262.21	38.47	6.4
Unlun Mountains	12,267.19	61.87	10.2
Himalayas	8417.65	76.6	12.7
Qiangtang Plateau	1802.12	9.29	1.5
Gangdis Mountains	1759.52	9.41	1.6
Nyingchi Tanggula Mountains	10,700.43	213.27	35.3
Hengduan Mountains	1579.49	49.94	8.3
Tanggula Mountains	2213.4	17.59	2.9
Alpine Mountain	275	1.39	0.2
Total	59,425.18	604.65	100

[a] Including the glacier with an area of 16.84 km^2 in the Mustang Ridge

7.2.3 *Snowmelt Runoff Contribution*

Seasonal runoff recharge is essential for rivers that rely on snowmelt, with its impact varying across different river basins. In regions such as Northeast China and the Altai Mountains, where snow cover is extensive, rivers are predominantly recharged by snowmelt. In these areas, snowmelt is concentrated between March and June, resulting in peak annual runoff from April to June. Around the Tibetan Plateau and in the Tianshan Mountains, watersheds are replenished by a combination of snowmelt, glacier meltwater, and precipitation, due to the abundant presence of snow and glaciers. Consequently, runoff in these regions primarily occurs in spring and summer. During spring, minor flood peaks are driven by snowmelt recharge, while the main contributions to peak river flows in July and August come from glacial meltwater and summer precipitation. Despite the varying characteristics of snowmelt runoff across different river basins, it plays a crucial role in alleviating spring water shortages, thereby supporting agricultural and ecological water demands during this critical period.

The contribution of snowmelt runoff varies significantly across different regions. For instance, studies in Czech mountain watersheds have reported snowmelt runoff contribution rates ranging from 12 to 37% (Jenicek and Ledvinka 2020). In the western United States, snowmelt accounts for 53% of the total runoff (Li et al. 2017), while in the inland river basins of western China, it contributes approximately 20–25% of runoff. In the Tianshan Mountains of Xinjiang, snowmelt can constitute between 22 to 49% of runoff during snow ablation periods (Chen et al. 2019a, b). Comparative studies indicate that snowmelt runoff generates higher runoff rates compared to rainfall runoff. For example, in the western United States, although snowfall constitutes 37% of precipitation, snowmelt runoff contributes to more than half of the total runoff. Similarly, in the Czech Republic, where snowfall averages 20%, snowmelt contributes to an average of 26% of total runoff.

Projections for future changes in snowmelt runoff vary by region. Climate simulations using regional climate models (RCMs) for the United States and southern Canada suggest that by the end of the twenty-first century, under different scenarios (RCP2.6, RCP4.5, RCP8.5), snow water equivalent (SWE), snowmelt, and runoff could decrease by 72%, 73%, and 45%, respectively. Conversely, in Alaska and northern Canada, SWE, snowmelt, and runoff potentials are projected to increase by 8, 53, and 41% (Cho et al. 2021). From 1951 to 2017, significant increases in snowmelt were observed on the Tibetan Plateau, whereas decreases were noted in northern, central, and southeastern China. Overall, China experienced a decreasing trend in snowmelt, although this trend was statistically insignificant (Yang et al. 2022). Under different climate scenarios (RCP2.6, RCP4.5, RCP8.5), China is projected to experience substantial declines in snowmelt from 2006 to 2099. The projected decreases in snowmelt for the near future (2011–2040), mid-future (2041–2070), and far future (2071–2099) are estimated at 10.4% (15.8%, 13.9%), 12.0% (17.9%, 21.1%), and 11.7% (24.8%, 36.5%), respectively, compared to the reference period (1981–2010).

7.3 Regulation of Basin Water Resources

The cryosphere plays a vital role in regulating water resources within watersheds, functioning as a key water source on both annual and seasonal scales. Glaciers, through their melt processes, regulate river runoff across different temporal frameworks. The freeze–thaw cycles of frozen ground, including seasonally frozen soils and the active layer above permafrost, significantly influence the annual runoff dynamics within a watershed. Moreover, the presence of ground ice can affect runoff and water resources over extended periods. Snow cover also contributes to the redistribution of annual runoff, further influencing water availability.

The regulation of runoff and water resources in cryospheric basins is particularly crucial for maintaining a stable water supply during extreme hydrological events, such as during dry years and months, especially in arid regions. In these regions, surface water predominantly originates from alpine mountains within the cryosphere. The cryosphere's ability to regulate runoff ensures a consistent water supply for oases in arid areas, thereby supporting their stability and development.

7.3.1 Annual and Seasonal Regulation Effect of Glaciers on Water Resources

Glaciers act as natural solid reservoirs, playing a crucial role in regulating water resources on both annual and seasonal scales. Positioned as cold and wet islands, glaciers are typically located in high-precipitation zones, enabling them to receive substantial annual recharge from precipitation, which is stored in solid form. The relatively low air temperatures in glacier regions during precipitation events, influenced by the shielding, reflection, and absorption of solar radiation by clouds, help to moderate the ablation of glaciers to some extent. In years with increased rainfall, higher precipitation levels lead to greater glacier accumulation, which in turn reduces the output of glacier meltwater. Furthermore, the lower air temperatures associated with precipitation events in glacier regions further curtail the melting of glaciers.

In contrast, reduced precipitation and higher air temperatures during dry months and years can accelerate glacier melting and diminish accumulation. As a result, the water stored by glaciers during wetter periods is gradually released during drier times. Elevated air temperatures expedite glacier melting, leading to a greater proportion of meltwater being generated during these dry periods. Additionally, the continuous movement of water from accumulation areas to melting areas, facilitated by glacier dynamics, ensures the persistence of glacier mass over time.

In large continental or marine glaciers, extensive networks of interglacial and subglacial channels often develop. These channels serve as reservoirs, storing rainfall and meltwater from glaciers and snow. During dry periods, these channels can release stored water, potentially leading to flood events, thereby functioning as another mechanism for runoff regulation. For example, the Koxkar Glacier in the Tianshan Mountains, which spans approximately 80 km^2, features a complex system of ice channels. Figure 7.9 depicts the intra-annual variation of glacial runoff and meltwater from the Koxkar Glacier. The runoff process of glacier meltwater lags behind the peak of the glacier melting process by approximately two months, likely due to the storage capacity of the ice channels, which retain summer meltwater and gradually release it from October to November of the following year.

The regulatory and stabilizing impact of glaciers on watershed runoff is highly dependent on their distribution within a basin. Research suggests that glaciers significantly stabilize runoff when they cover more than 5% of the basin area. In rivers where glacier recharge contributes over 30% of the total flow, the ratio of the annual runoff variation coefficient to the annual precipitation variation coefficient is typically less than 0.5. In contrast, this ratio exceeds 1.0 in rivers that lack glacier contributions (Fig. 7.10). This indicates that rivers with substantial glacier meltwater input are less susceptible to the extremes of droughts or floods, emphasizing the critical multi-annual runoff regulation effect that glaciers provide.

It is important to note that even in watersheds where glacier meltwater represents a small fraction of the overall runoff, the glacier's role in regulating runoff remains significant. For instance, in the Upper Heihe River Basin of the Qilian Mountains, warmer and drier climates have led to increased runoff, with glacier meltwater making

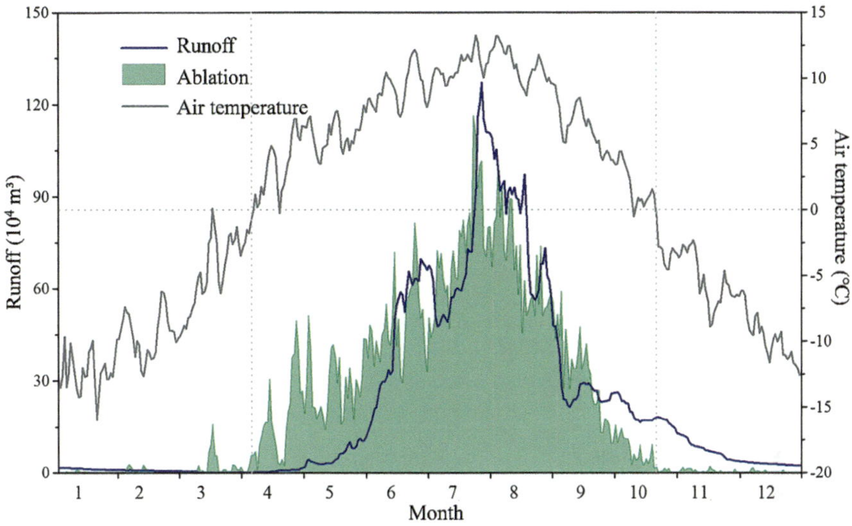

Fig. 7.9 Comparison between daily average ablation amount and terminal runoff at the Koxkar glacier terminus in the Tianshan Mountains (Han et al. 2015)

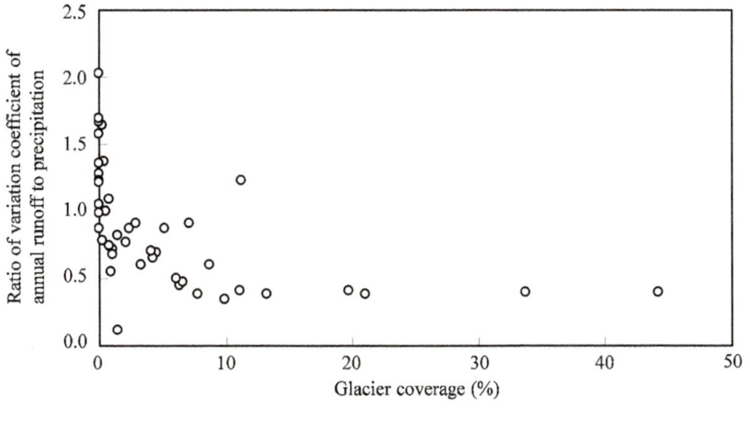

Fig. 7.10 Relationship between coverage of glacier and the ratio of variation coefficient of annual runoff to precipitation among different watersheds in China (Ding et al. 2017)

up a larger share of the basin's total runoff. Although the multi-annual average glacier runoff constitutes only about 3.5% of the total river runoff in this basin, this figure rises to approximately 5.0% during dry years and can peak at 16% during particularly dry months (Fig. 7.11). This highlights the essential role that glaciers play in moderating both intra-annual and multi-annual runoff dynamics, ensuring a more stable water supply even in challenging climatic conditions.

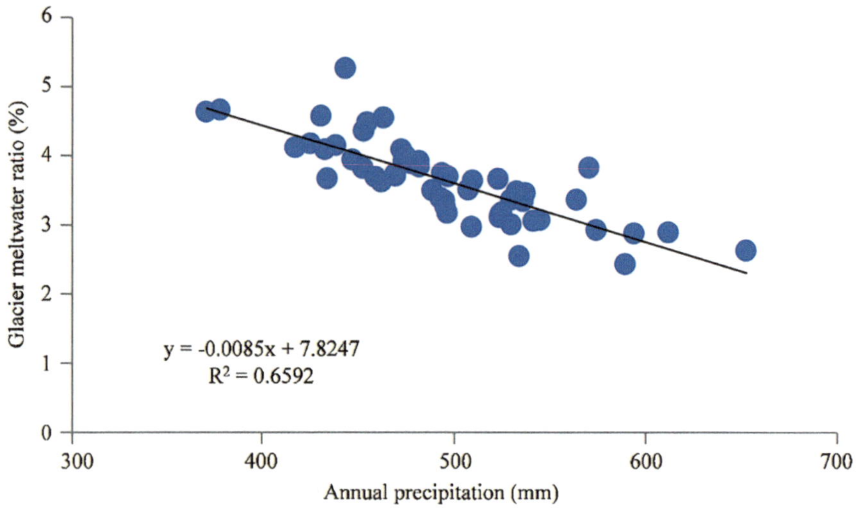

Fig. 7.11 Relationship between annual precipitation and glacier meltwater ratio (the percentage of glacier runoff contribution to the total annual flow) in the Upper Heihe River Basin

Over the past few decades, global glaciers have experienced significant retreat due to climate warming, which may undermine their capacity to regulate both intra-annual and multi-annual runoff. For example, in the Aksu River basin, the glacier area decreased by 8.9% in 2000 and by 13.2% in 2007 compared to 1990 levels. This reduction in glacier coverage was accompanied by an increase in the annual runoff variation coefficient by 2.4% (approximately 0.004) and 3.2% (0.005), respectively, indicating a weakening of the glacier's regulatory influence on runoff stability (Zhao et al. 2015).

7.3.2 Seasonal Regulation of Snow on Water Resources

Snow plays a crucial role in regulating the seasonal distribution of runoff within a basin by acting as both an internal regulator and a temporary storage medium for water. As snow accumulates during the winter months, it effectively stores water, which is then gradually released as snowmelt during warmer periods. This slow release, facilitated by the gradual drainage through the snowpack, results in a more moderated runoff hydrograph compared to direct precipitation. This moderation alters the timing and intensity of runoff, smoothing out peaks and reducing the likelihood of extreme runoff events. Additionally, in high-altitude areas within the basin, increased snowfall during the summer months further influences the basin's overall runoff dynamics, contributing to a more sustained and regulated water supply throughout the year.

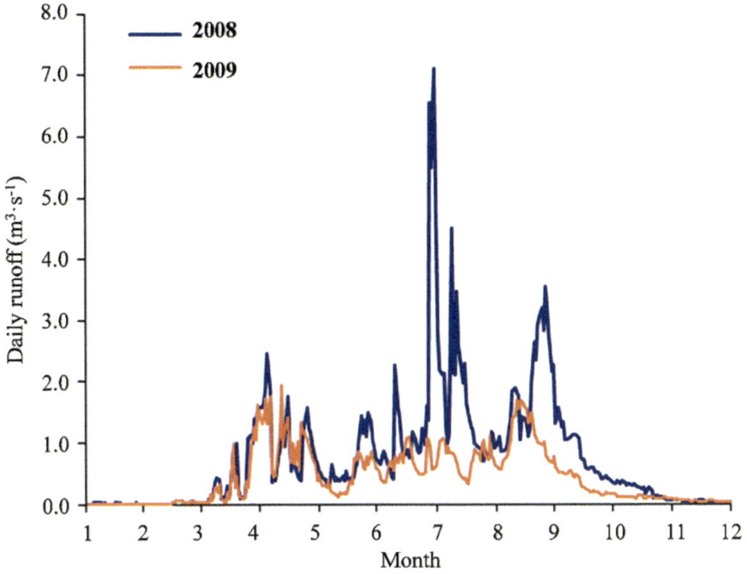

Fig. 7.12 Difference of daily runoff hydrograph from February to December in rainy and dry years in Binggou watershed of the Qilian Mountains

Snowmelt plays a critical role in supplementing low-flow conditions during spring, leading to peak runoff during this season. In basins where precipitation peaks from June to September, runoff patterns often exhibit double or multi-peak characteristics annually. This is due to the combined effects of spring floods driven by snowmelt, along with peak precipitation and glacier meltwater during the summer. Figures 7.12 and 7.13 illustrate these runoff dynamics in the Binggou watershed of the Qilian Mountains and the Kayeltes River Watershed in the Altai Mountains.

In the Binggou watershed, snowmelt contributes a relatively small but consistent amount to the overall runoff. Snow cover begins to melt at the end of March and is largely depleted by the end of May. The primary contribution of snowmelt to river runoff occurs from April to May, a period characterized by minimal precipitation in the Qilian Mountains. Consequently, the hydrographs from April to May during the rainy year of 2008 closely resemble those from the dry year of 2009, underscoring the stable and reliable contribution of snowmelt to river runoff during this time.

In basins that predominantly rely on snowmelt, such as the Kayaltes River Basin in the Altai Mountains, the amount of snowmelt and its contribution to the annual runoff differs markedly from that observed in the Binggou Basin of the Qilian Mountains (Fig. 7.12). Snowmelt in the Kayaltes River Basin typically begins in mid-April and continues until mid to late June. During this period, hydrological patterns generally show consistent trends across different years, with peak annual flows typically occurring from late May to early June. However, the magnitude of daily average flows can vary significantly between years.

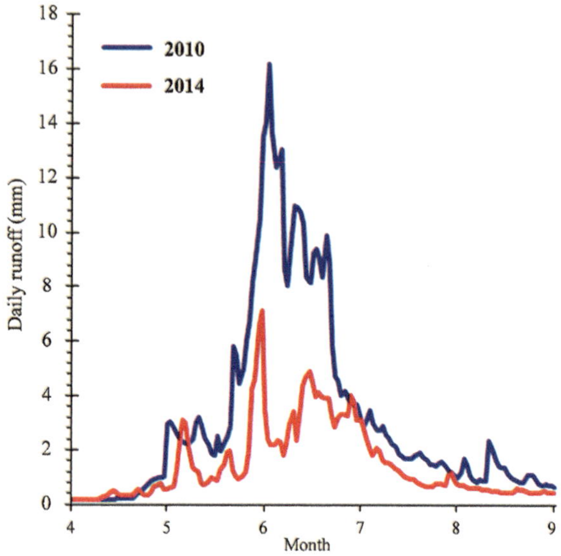

Fig. 7.13 Difference of daily runoff hydrograph from April to September in rainy and dry years in Kayaltes River Basin of the Altay Mountains

For instance, in 2010, the peak flow on June 2 reached a maximum daily average of 442.0 m^3/s, whereas in 2014, the peak flow occurred on May 31, with a maximum daily average of only 194.0 m^3/s, which is less than half of the 2010 value. This disparity likely results from differences in winter snow accumulation. Specifically, precipitation (mainly in the form of snow) from October to March was nearly three times higher from 2009 to 2010 (304.7 mm) compared to the period from 2013 to 2014 (137.3 mm).

Overall, while the regulatory effect of snow cover on watershed runoff might seem modest in snowmelt-dominated rivers, it plays a crucial role in ensuring the consistent generation of runoff within these basins, highlighting the importance of snow accumulation in sustaining water resources throughout the year.

Simulation results from the Upper Heihe River Basin in the Qilian Mountains from 1960 to 2013 indicate that peak snowmelt generally corresponds to normal basin years (Fig. 7.14a), with the highest snowmelt runoff not occurring during rainy years. Generally, larger watershed runoff corresponds to a smaller proportion of snowmelt runoff (Fig. 7.14b), illustrating the multi-annual runoff regulation effect of snow cover.

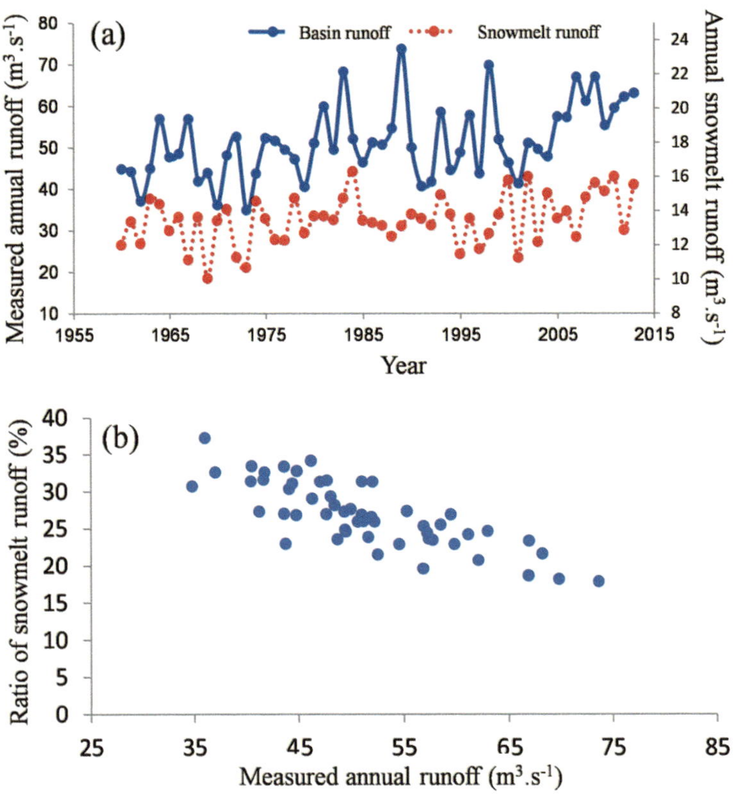

Fig. 7.14 Relationship between snowmelt runoff and watershed runoff in the Upper Heihe River Basin of the Qilian Mountains (Chen et al. 2018b)

7.3.3 Intra-annual and Multi-annual Regulation of Permafrost on Water Resources

The regulation of water resources by permafrost is influenced by both intra-annual and multi-annual changes, particularly under shifting climate conditions. The thickness of the active layer above permafrost undergoes substantial variation, directly impacting the thermal and hydrological dynamics of permafrost regions. These changes inevitably affect the annual hydrological processes within these areas.

In spring, the ground typically remains frozen, with low hydraulic conductivity and permeability in the frozen layers, which impedes the infiltration of liquid water. This condition affects groundwater storage and disrupts the balance between surface water and groundwater flow (Walvoord and Kurylyk 2016). As air temperatures rise, increased rainfall and snowmelt contribute to peak runoff in late spring to early summer. During early summer, the permafrost begins to thaw, causing the active layer above it to thicken. This thickening reduces the depth of seasonal freezing and lowers the freezing front, which initially leads to a decrease in surface runoff from its peak.

As summer progresses, additional precipitation and recharge from glacier and snowmelt contribute to a secondary peak in surface runoff. When the active layer reaches its maximum thickness, the increased infiltration capacity enhances basin water storage and evapotranspiration. This process can mitigate flood peaks during this period. In winter, the active layer gradually refreezes, limiting groundwater recharge and significantly reducing winter runoff. With the combination of lower precipitation and diminished meltwater recharge, winter runoff declines rapidly.

The degradation of permafrost has significant implications for runoff generation and concentration processes within basins, particularly through the melting of ground ice, which serves as a regulator of both intra-annual and inter-annual runoff dynamics. As permafrost degrades, marked by the thickening of the active layer, there is an increase in deep soil flow and an expansion of groundwater storage capacity. This can lead to higher winter runoff and slower recession rates in autumn hydrographs, effectively modulating the distribution of runoff throughout the year and across multiple years (Fig. 7.15).

Figure 7.15 illustrates comparative hydrographs from the Heihe River and Shule River, highlighting the changes before and after 1985 when air temperatures in the Qilian Mountains began to rise significantly. The degradation of permafrost during this period contributed to increased low-flow conditions in these basins and a slower decline in water levels within annual hydrographs. Notably, despite a similar increase in annual precipitation (approximately 27.0 mm) after 1985 compared to the period before, the Shule River Catchment, with a higher proportion of glacial runoff (>20%), experienced greater increases in summer and winter runoff compared to the Heihe River Catchment.

Additionally, permafrost regions contain substantial amounts of ground ice. With ongoing climate warming, the gradual melting of this ground ice contributes to river runoff, providing a regulatory function over multiple years. This slow release of water from ground ice melt helps sustain river flows, particularly during periods of low precipitation, highlighting the critical role of permafrost in long-term hydrological stability.

7.4 Integrated Role of Cryosphere in Basin Scale

The hydrological processes within the cryosphere, encompassing glacial, snow, and permafrost dynamics, each play crucial roles at different times within a basin's hydrological cycle (Fig. 7.16). Among these processes, glacier ablation is particularly significant, especially in arid regions where glacier runoff acts as a stabilizing force, mitigating extreme fluctuations in water availability. Recent global glacier retreat has led to substantial alterations in basin water resources due to increased glacier meltwater. Understanding the impact of glacier changes on runoff is essential for the effective assessment, utilization, and management of basin water resources.

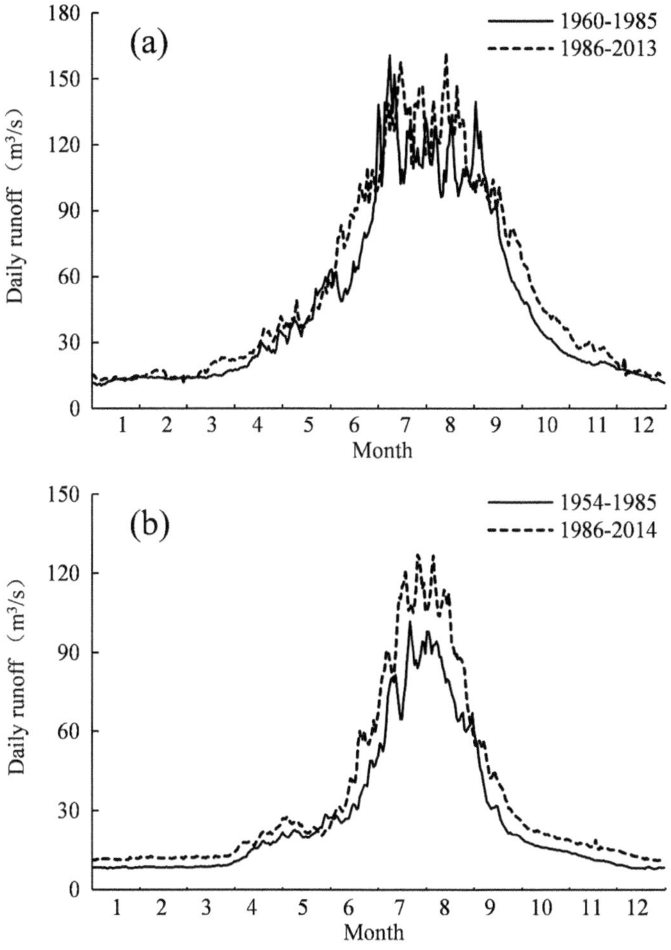

Fig. 7.15 Daily runoff hydrograph comparison before and after 1985 in the **a** upper reaches of Heihe River basin and **b** upper reaches of Shule River (Chen et al. 2018b)

Snow, another vital component of the cryosphere, undergoes pronounced seasonal variations that influence the land-air energy balance through changes in surface albedo. Snowmelt runoff serves as a primary freshwater source during the spring in arid and semi-arid regions, playing a critical role in shaping the basin water cycle. Permafrost, acting as a unique regional water barrier with limited permeability, forms a fundamental basis for maintaining water balance in permafrost zones. With its characteristics of low permeability, high water storage capacity, and seasonal freeze–thaw processes that facilitate water redistribution, permafrost alters the hydraulic connections between surface water bodies and groundwater.

For instance, the annual runoff curve of the Shule River (refer to Fig. 7.16) illustrates the temporal distribution of runoff sources: glacier runoff peaks from June

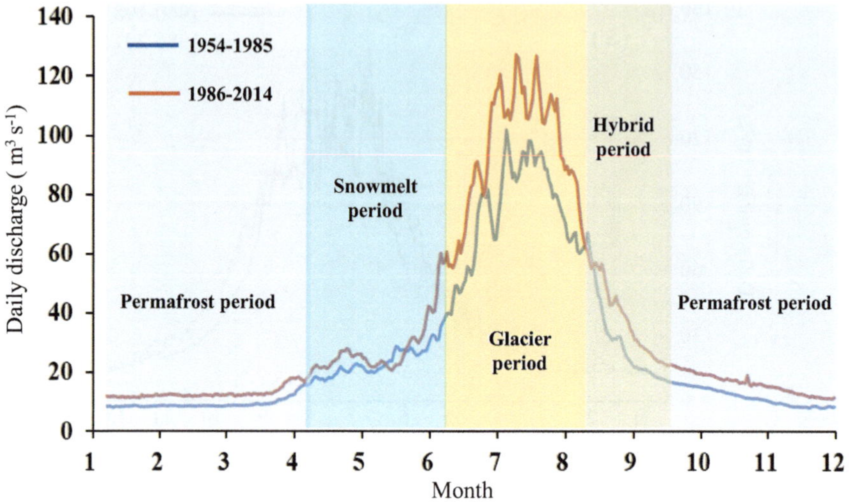

Fig. 7.16 Influence of glacier hydrology, snow hydrology and permafrost hydrology on intra-annual variability of runoff in the Shule River

to August, snowmelt runoff prevails from April to mid-June, and permafrost runoff dominates from mid-September to early April. Historical data from 1954 to 1985, showing lower winter runoff due to thinner active layers, contrasts with the period from 1986 to 2014, during which warmer temperatures and deeper active layers led to increased winter runoff and glacier meltwater, thereby significantly raising peak flows.

Cryospheric basins are characterized by diverse land cover types, including alpine desert, alpine meadow, scrub, swamp, grassland, and forest. These basins often experience seasonal snow cover, especially during colder seasons or in high mountain areas, and some also contain glaciers. Permafrost is prevalent throughout these basins, playing a critical role in hydrological processes. Meltwater from snow and glaciers, as well as rainfall, infiltrates the surface and penetrates permafrost through these various land cover types, driving key processes such as runoff generation, infiltration, evaporation, and subsurface drainage.

In the cryospheric watersheds of western China, alpine grasslands, meadows, and deserts are the dominant land cover types, with some regions also featuring alpine scrub and coniferous forests (Fig. 7.17). In contrast, the mountainous inland river basins of northwestern China present a distinct vertical vegetation belt spectrum (Fig. 7.18). Each of these typical land cover types exhibits unique water balance characteristics, shaped by the regional and local climate conditions in which they are found. These diverse ecosystems contribute to the overall hydrological dynamics of cryospheric basins, influencing how water is stored, transported, and ultimately utilized within these environments.

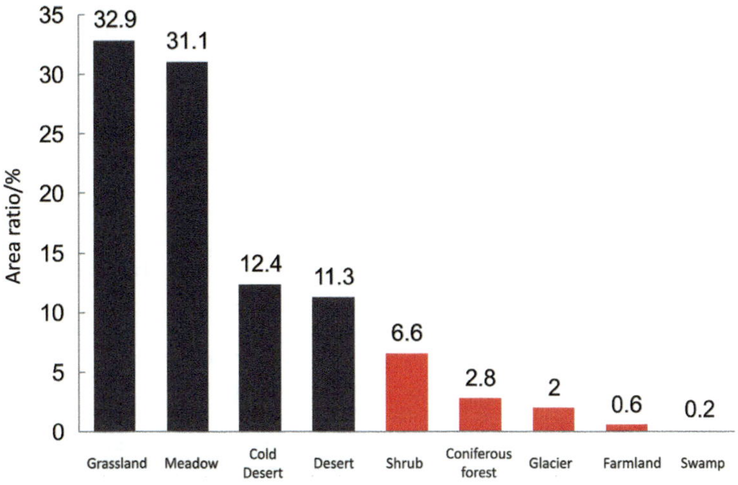

Fig. 7.17 Typical land cover types and their area ratios in the cryosphere of western China

Fig. 7.18 Typical vertical vegetation belt in inland mountainous region (upper reaches of the Heihe River in Qilian Mountains)

Based on the variation in hydrological processes across different elevations, the region can be categorized into four distinct elevation zones: the tundra zone, grassland and forest zone, oasis zone, and desert zone (Fig. 7.19).

1. **Tundra Zone**: At the highest elevations, the tundra zone, dominated by the cryosphere, serves as a crucial water storage area for the watershed. Annual precipitation in this zone ranges from 500–700 mm, largely contributing to the accumulation of snow and glacier ice that supports downstream water flows.

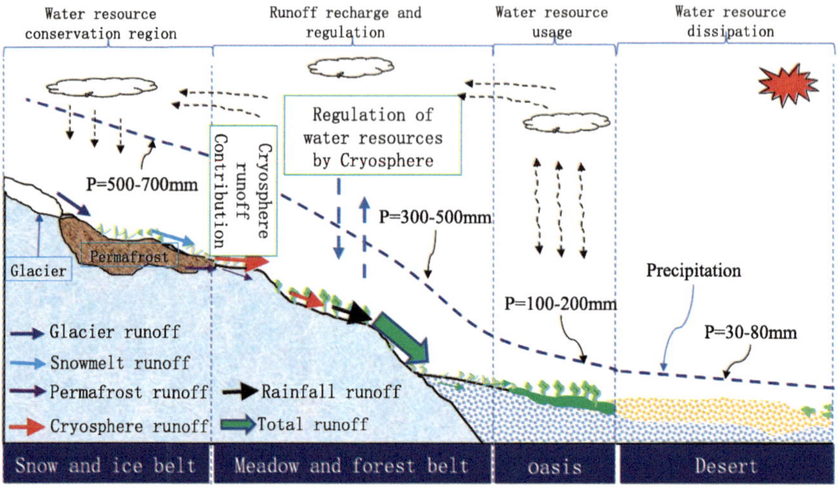

Fig. 7.19 Watershed hydrological belt function in arid region

2. **Grassland and Forest Zone**: As elevation decreases, the landscape transitions into the grassland and forest zone, where precipitation increases in proportion to rainfall, totaling between 300–500 mm annually. This zone benefits from both cryospheric runoff and direct rainfall, with cryospheric runoff playing a significant role in regulating peak flows and maintaining baseflows, thereby supporting the region's hydrological stability.
3. **Oasis Zone**: Moving further down in altitude, the oasis zone experiences 100–200 mm of rainfall annually. This zone is vital for human activities and major water resource utilization, making it highly sensitive to variations in cryospheric runoff. The stability of cryosphere-sourced water, as well as the occurrence of droughts and floods, significantly impacts the livelihoods and production activities of the local population.
4. **Desert Zone**: At the lowest altitudes lies the desert zone, which receives less than 100 mm of rainfall per year. This zone primarily relies on runoff from the mountain cryosphere and occasional rainfall for its water supply, acting as the main area for water dissipation in the region.

These elevation zones illustrate the critical role that cryospheric and rainfall runoff play in sustaining the hydrological balance across the region, from the high-altitude tundra to the arid desert lowlands.

7.5 Extreme Hydrological Events in the Watershed

The hydrological functions of the cryosphere within a watershed extend beyond beneficial roles like water conservation, runoff replenishment, and regulation of water resources; they also encompass the potential for significant hazards, including floods and debris flows—collectively referred to as extreme hydrological events. At the basin scale, these events are primarily associated with glacier and snowmelt floods, glacial lake outburst floods (GLOFs), and glacier-induced debris flows.

7.5.1 Glacier and Snowmelt Flood

Glacier and snowmelt floods are hydrological events primarily driven by the melting of glaciers and snow. When glacier meltwater is the predominant source, these events are referred to as glacier melt floods. Conversely, when snowmelt is the primary contributor, they are termed snowmelt floods. Additionally, floods that result from the combined effects of heavy rainfall, glacier melt, and snowmelt are categorized into two types: rain-on-snow floods and rainfall and glacier melt floods. The characteristics, geographic distribution, and response to climate warming of these four types of floods are discussed in detail in Table 7.4 (Shen et al. 2013).

Rivers typically do not rely solely on glacier meltwater or snowmelt for their flow; rather, they are replenished by a combination of these sources along with rainfall. Glacier and snowmelt floods are seasonal phenomena closely linked to climate variability. As temperatures rise above 0 °C, glaciers and snow begin to melt, transforming into liquid water. Intense solar radiation, particularly in regions with extensive glacier coverage and thick early snowpack, accelerates the rate of melt. The resulting water from glaciers and snow either flows directly into rivers as surface runoff or infiltrates the ground as shallow groundwater, contributing to spring and summer floods.

Glacier and snowmelt floods are prevalent in mid to high latitudes and high mountain regions around the world, including the Caucasus in Russia, Central Asia, the Alps in Europe, and the western mountains of North America. In China, glacier melt floods are most commonly found in the Manas River Basin on the northern slope of the central Tianshan Mountains, the Muzart and Tailan Rivers on the southern slope of the western Tianshan Mountains, the Karakash River in the western Kunlun Mountains, the Yarkand River in the Karakoram Mountains, the Changma and Dang Rivers in the western Qilian Mountains, and several tributaries of the Yarlung Zangbo River on the northern slope of the Himalayas. Snowmelt floods are primarily observed in the Altai Mountains of Xinjiang, and some rivers in Northeast China also experience glacier and snowmelt floods.

1. **Snowmelt Flood**

Snowmelt flood, commonly referred to as snow flood, is caused by the rapid melting of accumulated snow primarily in middle and high latitudes and high mountain

Table 7.4 Key characteristics and distribution of glacier and snow melt floods and their response to climate warming (Shen et al. 2013)

Flood type	Key characteristics	Distribution	Response to climate warming
Snowmelt floods	Rapid snowmelt often coincides with warming temperatures and high solar radiation	Altai Mountains (Xinjiang, China), Northeast China, Central Asia, Russia, Western North America	Earlier onset and potential increase in flood intensity as temperatures rise, leading to quicker and more extensive snowmelt
Glacier melt floods	High runoff, with peak flows coinciding with maximum glacier melt rates	High mountain regions: Tianshan, Kunlun, Karakoram, Qilian, Himalayas, Alps, Caucasus, North America	Increased frequency and magnitude expected due to accelerated glacier melting driven by rising temperatures
Rain-on-snow floods	Characterized by rapid runoff when rain falls on snow-covered ground, leading to potential significant flooding	High-altitude regions prone to late-season snow, such as mountainous areas worldwide	Increased occurrence likely due to warmer winters, leading to more rain events during the snow season, which could enhance the frequency and severity of these floods
Rainfall and glacier melt floods	High runoff resulting from the combined effects of rainfall and glacier melt, potentially leading to severe floods	Regions with high precipitation and glacier presence, particularly in areas similar to those affected by glacier melt floods	Elevated flood risks due to the dual effects of enhanced precipitation and accelerated glacier melt under warmer climate conditions

regions during spring and summer. It can be categorized into flat-type and mountain-type floods. Due to the substantial early-season snowfall and rapid temperature rise, snowmelt accelerates, leading to flooding. Unlike typical floods, snowmelt floods often involve a significant amount of ice, causing extensive damage wherever they occur. In regions with long, severe winters and deep snow cover, temperatures rising above 0 °C in spring and summer trigger snowmelt and subsequent flooding. Snowmelt floods are closely tied to accumulated snow, exhibiting distinct temporal and spatial distribution characteristics (Ablimitjan et al. 2015). Key factors influencing the size and progression of snowmelt floods include snow cover area, depth, density, water-holding capacity, surface freezing depth, snowmelt heat, terrain, landform, orientation, climate, and land use in snowy areas, each influencing one another.

Spring snowmelt floods typically occur from April to June (Fig. 7.20), with peak runoffs typically observed from May to June. In rivers located at similar latitudes, snowmelt floods in plains regions generally precede those in mountainous areas. In China's Altai Mountains, for instance, snowmelt floods typically occur from April to May, concluding by June at the latest.

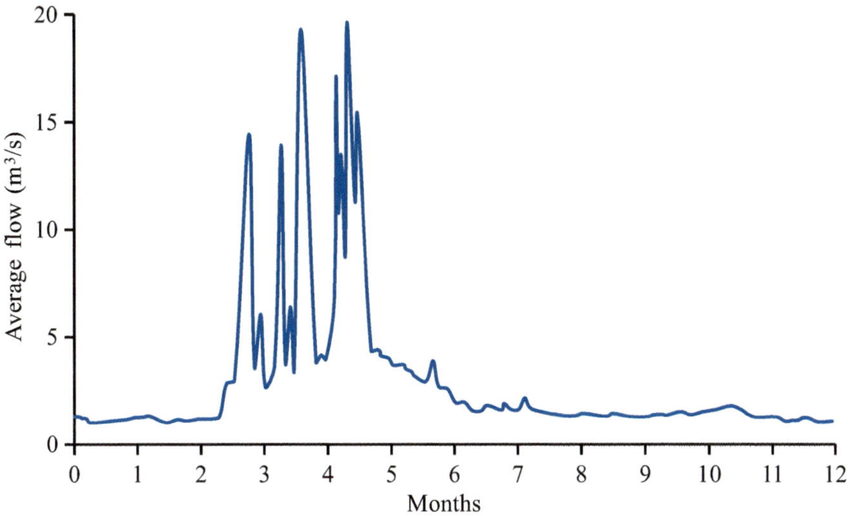

Fig. 7.20 Snowmelt flood flow process line. This figure depicts the snowmelt flood process observed at the Wushi Water Station on the Shalaimiele River in 1970

The snowmelt flood is characterized by distinct dynamics that differentiate it from rainfall-induced floods, primarily due to the interplay of snow accumulation, thermal conditions, and topographical influences. These characteristics are as follows:

1. **Temperature-Driven Dynamics**: The progression of snowmelt floods closely mirrors temperature changes. Initially, as temperatures begin to rise, the response in flood dynamics is gradual. However, as thermal energy accumulates and temperatures reach a critical threshold, the flood process accelerates, often outpacing the rate of temperature increase. Despite this, the flood process generally lags behind the thermal process, with the flood peak occurring after the peak in temperature (Fig. 7.21).
2. **Diurnal Variations**: Snowmelt floods exhibit significant diurnal fluctuations, typically featuring a single daily peak followed by a trough. The flood peak usually occurs in the afternoon, coinciding with the warmest part of the day, while a decline in flow is observed at night. The timing of these peaks and troughs can vary across different rivers, depending on the distance between the hydrological observation stations and the snow-covered areas.
3. **Interannual Variability**: Although snowmelt floods occur early in the spring, they are heavily influenced by the variability of spring temperatures. The rate at which temperatures rise can differ significantly from year to year, leading to substantial interannual variation in the timing of spring floods.

2. **Glacier Melt Flood**

Glacier melt floods are driven by the runoff from both glacier melt and the melting of overlying snow, with glacier melt being the primary contributor (Ding et al.

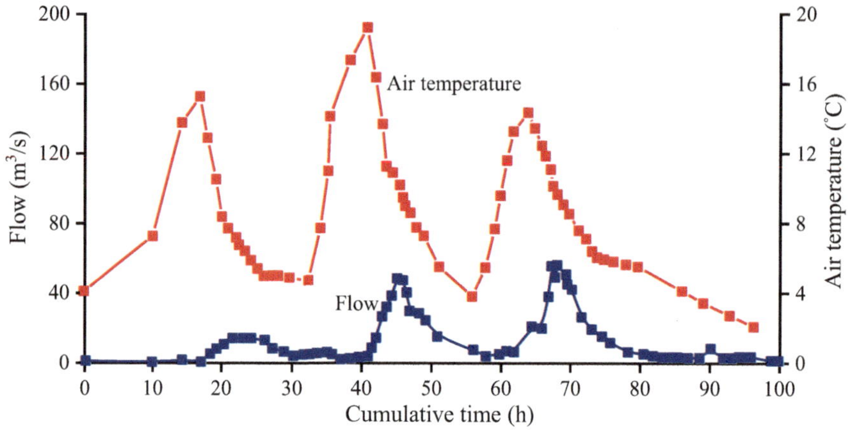

Fig. 7.21 Hourly process of daily runoff temperature of Juntanghu river basin from March 25 to 29, 2000 (Kui 2006)

2017). These floods can be categorized into two distinct types based on their runoff characteristics:

1. **Seasonal Glacier Melt Floods**: These floods occur annually as part of the normal glacier melting cycle. The flood process exhibits a gradual and steady rise, typically without significant fluctuations. The glacier melt area largely determines the flood's peak, volume, and pattern. These floods generally have a broad, single-peaked hydrograph, with the peak usually occurring between July and August (Fig. 7.22).
2. **Extreme Glacier Melt Floods**: Triggered by extreme weather conditions, such as sudden and exceptionally high temperatures, these floods result in rapid and abnormal glacier melting. The runoff in extreme glacier melt floods rises and falls sharply, directly correlating with the temperature fluctuations.

Glacier melt floods typically occur after prolonged periods of high summer temperatures and are characterized by several key aspects:

1. **Temperature Synchronization**: Glacier melt floods are strongly correlated with temperature changes, while their relationship with precipitation is less synchronized.
2. **Factors Influencing Flood Peaks**: The peaks and volumes of these floods are closely tied to the magnitude of temperature increases, as well as factors such as glacier area, snow reserves, and summer snowfall. Additional influences include solar radiation, the duration of temperature rise, thermal conditions, and the regulatory effects of glaciers, snow, and river runoff.
3. **Diurnal Variation**: Similar to snowmelt floods, glacier melt floods exhibit noticeable diurnal variations, though the extent of these variations depends on the mountainous origins of different rivers.

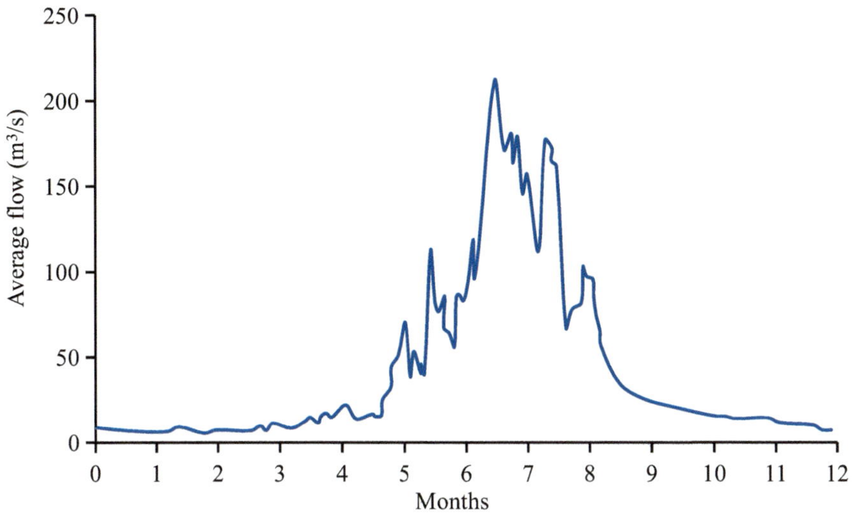

Fig. 7.22 Glacier melt flood runoff process line (measured data from Hongshanzui Station of Manas River in 1974)

4. **Interannual Stability**: Glacier melt floods display minimal interannual variability. For example, the ratio of maximum to minimum flood volumes in the major rivers of Xinjiang ranges from 1.54 to 2.80.
5. **Peak Characteristics**: The characteristics of glacier melt flood peaks reflect the duration of high temperatures in mountainous regions. Rapid temperature increases lead to intense glacier melting and steep flood rises, while sudden drops in temperature during precipitation events reduce meltwater, causing a rapid recession of floods. Conversely, sustained high temperatures result in a slower recession of floodwaters. For instance, rivers originating in the Tianshan Mountains typically experience flood durations of 4 to 10 days, with steep rising and receding slopes.

3. **Rain-on-Snow Flood**

In late spring and early summer in the Northern Hemisphere, heavy rainstorms—particularly those with relatively high temperatures—can accelerate the snowmelt process. This accelerated snowmelt, coupled with the intense rainfall, disrupts the natural regulation and storage capacity of snow, leading to a rapid convergence of meltwater. This combination often results in a snowmelt flood, which can merge with rainstorm runoff to create a Rain-on-snow flood.

The Rain-on-snow flood is characterized by the following:

1. **Absence of Diurnal Variation**: Unlike typical snowmelt floods that exhibit diurnal variations, Rain-on-snow floods generally lack such fluctuations. The rainstorm disrupts the usual temperature-dependent snowmelt patterns, eliminating the typical peaks and valleys in the flood process.

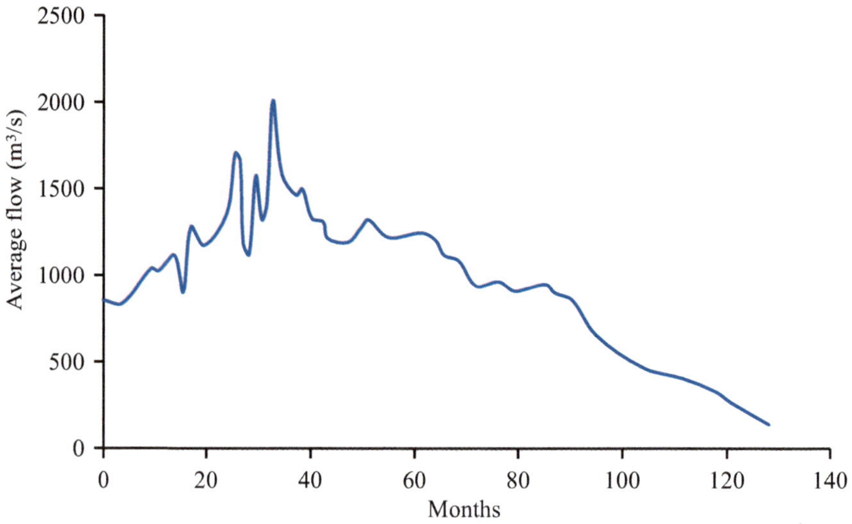

Fig. 7.23 Rainstorm snowmelt flood process recorded at Kaqun station of Yerqiang River from 0:00 on August 1 to 8:00 on August 6, 1999

2. **Broad Base and Prolonged Duration**: The flood process line is typically broad and extended due to the prolonged effects of both rain and snowmelt. The flood peak is primarily determined by the intensity and volume of the rainfall. Higher rainfall intensity and greater volume lead to steeper and higher flood peaks (Fig. 7.23). The shape of the flood process line can vary significantly at the outlet sections of different rivers, depending on factors such as snow cover, distribution, temperature fluctuations, and the patterns and intensity of rainfall.

Rain-on-snow floods account for a significant proportion of flood disasters in Northwest China. Statistics from 19 major rivers in Xinjiang indicate that 37% of the flood disasters in 2010 involved Rain-on-snow floods. Among the four types of snow and glacier floods, mixed flood disasters in the Tianshan, Altay, and Tacheng regions were the most prevalent, accounting for 56% of the total, the highest proportion among the categories.

4. **Rainfall + Glacier Melt Mixed Flood**

The primary distinction between rainfall + glacier melt mixed floods and Rain-on-snow floods lies in their origins. Rainfall + glacier melt mixed floods occur in high mountainous regions where glaciers (including snow in glacial areas) are present. During the summer, prolonged drought and early high temperatures can cause glaciers and snow to completely melt, resulting in meltwater that flows down to the middle and lower mountain belts. When a large-scale precipitation event, such as a heavy rainstorm, occurs, the runoff from the rainstorm combines with the meltwater, forming a mixed flood type known as a rainfall + glacier melt flood.

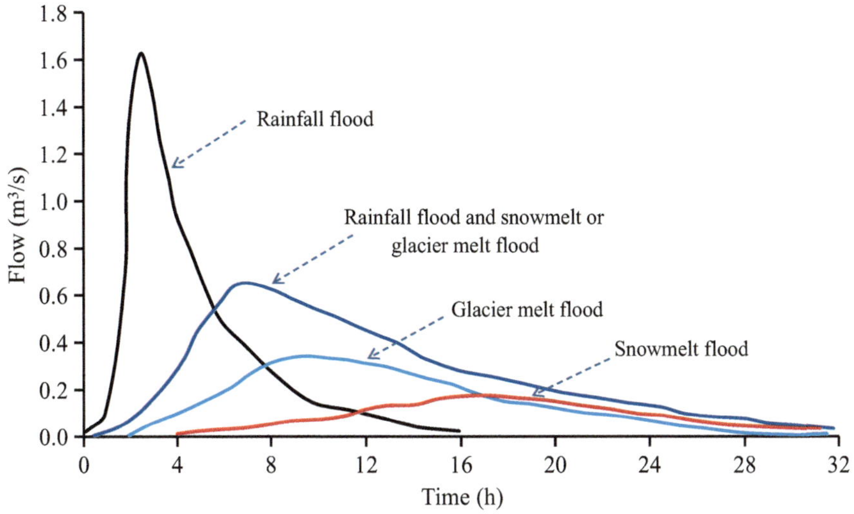

Fig. 7.24 Flood process of different types in mountainous area

Compared to Rain-on-snow floods, rainfall + glacier melt mixed floods typically occur later in the year; Rain-on-snow floods generally happen in the spring or early summer, whereas rainfall + glacier melt mixed floods predominantly occur in midsummer. The source areas for rainfall + glacier melt mixed floods are at higher altitudes. Additionally, the peak flood volume of these mixed floods depends on pre-rainfall temperatures and rainfall intensity. Higher temperatures sustained over a longer period before rainfall lead to a broader base in the flood process line, while larger and more intense rainfall events result in steeper flood peaks.

As rivers flow from high mountains to plains, passing through different regions, various types of floods often overlap and intersect. Consequently, in alpine mountain rivers, most floods are mixed types, such as Rain-on-snow floods and rainfall + glacier melt floods (Fig. 7.24). Despite the distinct processes of these different flood types, accurately attributing the contribution of rainfall remains challenging due to the sparse network of meteorological and hydrological stations in mountainous regions and limited precipitation data.

7.5.2 Glacial Lake Outburst Floods

A glacial lake is a natural lake formed by the accumulation of water in depressions, typically located in glacial regions. These lakes, which are directly or indirectly connected to glaciers, are among the most dynamic components of the cryosphere. Glacial lakes primarily receive their water supply from glacier and snowmelt, and they can be categorized into moraine lakes, glacial lakes, and glacier surface lakes. Compared to typical natural lakes, glacial lakes exhibit several distinct characteristics:

1. They are relatively small, typically ranging in size from 10^{-3} to 10^2 km^2.
2. They experience significant fluctuations in area and water volume both within and between years.
3. They have a short lifespan, ranging from less than a year to several decades.
4. They are directly or indirectly linked to glaciers and are highly sensitive to climate change.

A Glacier Lake Outburst Flood (GLOF) refers to a natural disaster in glacial regions where a sudden release of water from a glacial lake leads to flooding or debris flow, posing a threat to lives, property, and the natural and social ecological environment. GLOFs can originate from various types of glacial lakes, including moraine-dammed lakes, ice-dammed lakes, and internal meltwater lakes. Among these, moraine lake outburst floods and glacial lake outburst floods are the most common.

GLOFs occur worldwide, with glacial regions in the Himalayas, Andes, Central Asia, Alps, and North America being particularly prone to these disasters (Fig. 7.25). Since the 1930s, recorded GLOF events in the Hindu Kush-Himalayas have shown an increasing trend. By 2010, there had been over 32 recorded GLOF events, averaging 0.46 events per year. Particularly since the mid-1960s, there has been an average of one GLOF event per year. Among these, moraine lake outburst floods are particularly impactful due to their large scale and wide-ranging effects. In lakes of similar size, moraine lake outburst floods can result in peak flows 2 to 10 times larger than those from other types of glacial lake outbursts. Consequently, research into moraine lake outburst disasters has garnered significant attention.

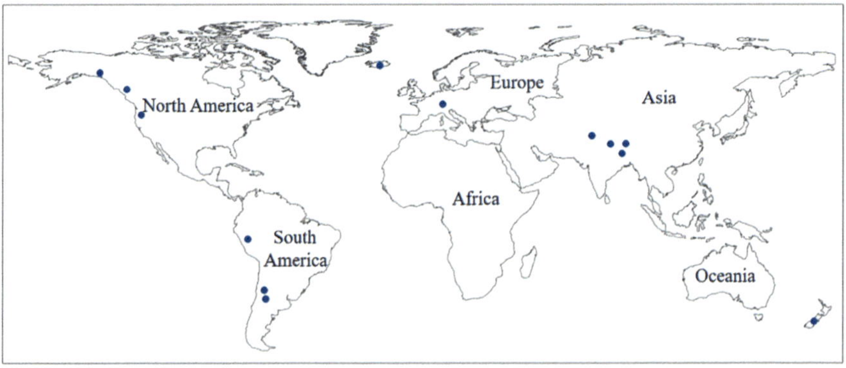

Fig. 7.25 Distribution of glacial moraine lakes with record outburst in the world (Ding et al. 2017)

7.5.3 *Glacial Debris Flow*

Glacial debris flow is a distinct type of flood characterized by the presence of mud, sand, and stones, typically formed at the margins of modern glaciers and snowfields. These debris flows result from erosion by meltwater or are triggered during glacial lake outburst floods. They generally occur during the summer and autumn months, when meltwater is abundant, and temperatures are higher. Unlike debris flows caused by rainstorms, glacial debris flows are often larger in scale and longer in duration. These events can occur under various weather conditions, including sunny, cloudy, or rainy days.

Currently, there is no universally accepted classification for glacial debris flows among scholars worldwide. However, they are widely recognized as a subclass of debris flow, and their classification can be approached from three primary perspectives:

1. **Classification based on the physical properties of the debris flow fluid**: This includes factors such as flow state, solid particle content, unit weight, viscosity, velocity, and energy.
2. **Classification based on practical applications**: This considers intensity, size, deposition of particulate matter, impact on human life and production, and potential losses.
3. **Classification based on the causes of debris flow**: This involves the water source, the intensity of solid particulate matter recharge, geological and geomorphological background, and human activities influencing the occurrence.

However, the criteria and characteristics of the first two classification methods often overlap with those used for general debris flow classification (Shi et al. 2000). This section primarily focuses on differentiating glacial debris flow types based on their water sources. The formation of glacial debris flow is complex, and its classification can involve multiple factors such as water source, mode of solid material supply, and geological and geomorphological conditions. Based on the primary water sources, glacial debris flows can be categorized into six types:

1. Glacial meltwater type
2. Snowmelt type
3. Glacier and snow avalanche type
4. Moraine-dammed lake outburst
5. Glacier-dammed lake outburst
6. Glacier and snowmelt mixed with rainfall type

For further details, refer to Table 7.5.

Table 7.5 Genetic classification of glacial Debris flows

Type	Main water supply source	Description
Glacial melt type	Debris flow caused by the rapid melt of glaciers (especially marine glaciers)	Debris flows primarily driven by meltwater generated from the melting of glaciers. These occur typically during warmer periods when glacial melting is at its peak
Snowmelt type	Flood caused by sudden melt of snow	Debris flows triggered by the melting of accumulated snow, often observed during the spring and early summer when temperatures rise, leading to rapid snowmelt
Glacier and snow avalanche type	The rapid melt of glacier and snow form by an snow and glacier avalanche	Debris flows resulting from the combined effects of glacier and snow avalanches. These events occur when a large mass of snow and ice detaches and cascades down, picking up debris along its path
	Outburst after glacier avalanche or snow avalanche blocks river channel	The scale and frequency are generally smaller than other types of glacier debris flow, but it has greater sudden occurrence
Moraine-dammed lake outburst type	Outburst of moraine blocking lake	These debris flows are caused by the sudden release of water from a moraine-dammed lake, which can rapidly erode the moraine and create a large-scale debris flow
Glacial-dammed lake outburst	Outburst of glacier blocked lake	Similar to moraine-dammed lake outbursts, but involving lakes that are dammed by glaciers. The sudden release of water due to glacial melting or structural failure results in a debris flow
Glacier and snowmelt mixed with rainfall type	Strong water flow composed of rainfall and rapid melt of glacier and snow	Debris flows that are triggered by a combination of glacier meltwater, snowmelt, and rainfall. This type is common during periods of heavy rain, which accelerates the melting process and contributes additional runoff

References

Ablimitjan A, Chen CY, Yusup A et al (2015) The temporal and spatial distribution features of snowmelt flood events in Xinjiang from 2001 to 2012. J Glaciol Geocryol 37(1):226–232

Bai CY (1989) A study of relationship between climate and mountain glaciers. J Glaciol Geocryol 11(4):287–297

Chen R, Han C, Liu J et al (2018a) Maximum precipitation altitude on the northern flank of the Qilian Mountains, Northwest China. Hydrol Res 49(5):1696–1710

Chen R, Wang G, Yang Y et al (2018b) Effects of cryospheric change on alpine hydrology: combining a model with observations in the upper reaches of the Hei River, China. J Geophys Res Atmos 123:3414–3442

Chen RS, Zhang SQ, Yang Y et al (2019a) Influence of cryosphere change on runoff in cold regions of Western China. Science Press, Beijing

Chen H, Chen Y, Li W, Li Z (2019b) Quantifying the contributions of snow/glacier meltwater to river runoff in the Tianshan Mountains, Central Asia. Glob Plan Change 174:47–57

Chen RS, Han CT, Wang L et al (2023) Impact of the alpine precipitation measurements on the precipitation in 2019 and 2020 in the Qilian Mountains. J Glaciol Geocryol 45(2):1–12

Cho E, McCrary RR, Jacobs JM (2021) Future changes in snowpack, snowmelt, and runoff potential extremes over North America. Geophys Res Lett 48(122):e2021GL094985

Comeau LEL, Pietroniro A, Demuth MN (2009) Glacier contribution to the North and South Saskatchewan Rivers. Hydrol Process 23:2640–2653

Ding YJ, Zhang SQ, Chen RS (2017) Introduction to cold-zone hydrology. Science Press, Beijing

Ding L, Kang X (1985) Climatic conditions of glacier development in the Qilian Mountains and their influence on glacier characteristics. In: Proceedings of the Institute of glaciation and permafrost, Chinese Academy of Sciences, Lanzhou, China, no 5, pp 9–15

Foster JL, Chang ATC (1993) Snow cover. In: Gumey RJ, Foster JL, Parkinson CL (eds) Atlas of satellite observations related to global change. Cambridge University Press, Cambridge

Fu BP (1983) Mountain climate. Science Press, Beijing

Gao X, Zhang SQ, Ye BS et al (2010) Glacier runoff change in the upper stream of Yarkant River and its impact on river runoff during 1961–2006. Scientia Sinica (Terrae) 40(5):654–665

Han HD, Ding YJ, Liu SY et al (2015) Regimes of runoff components on the debris-covered Koxkar glacier in western China. J Mt Sci 12(2):313–329

Hu YQ (1987) A result of numerical simulating of the strong Cold Island. Plateau Meteorol 6(1):1–8

Huss M, Hock R (2018) Global-scale hydrological response to future glacier mass loss. Nat Clim Change 8:1135–1140

IPCC (2013) Climate change 2013: the physical science basis. Cambridge University Press, Cambridge, United Kingdom and New York, NY, USA

Jenicek M, Ledvinka O (2020) Importance of snowmelt contribution to seasonal runoff and summer lowflows in Czechia. Hydrol Earth Syst Sci 24:3475–3491

Kang ES, Yang ZN, Lai ZM et al (2000) Meltwater runoff and river runoff in mountainous areas. In: Shi Yafeng. Glaciers and Environment in China-Present, Past and Future. Beijing: Chinese Science Press, 190–205

Kui JB (2006) Snowmelt floods of the Juntanghu River in Xinjiang Region. J Glaciol Geocryol 28(4):530–534

Lambrecht A, Mayer C (2009) Temporal variability of the non-steady contribution from glaciers to water discharge in western Austria. J Hydrol 376:353–361

Li D, Wrzesien ML, Durand M, Adam J, Lettenmaier DP (2017) How much runoff originates as snow in the western United States, and how will that change in the future? Geophys Res Lett 44:6163–6172. https://doi.org/10.1002/2017GL073551

Neal EG, Hood E, Smikrud K (2010) Contribution of glacier runoff to freshwater discharge into the Gulf of Alaska. Geophys Res Lett 37:L06404

Shen YP, Su HC, Wang GY et al (2013) The responses of glaciers and snow cover to climate change in Xinjiang(II): hazards effects. 35(6):1355–1370

Shen YP, Liang H (2004) High precipitation in glacial region of high mountains in high Asia: possible cause. J Glaciol Geocryol 26(6):806–809
Shi Y, Huang M, Yao T et al (2000) Glaciers and environment in China—present, past and future. Science Press, Beijing
Tang MC (1985) The distribution of precipitation in mountain Qilian (Nanshan). Acta Geogr Sin 40(4):323–332
Walvoord MA, Kurylyk BL (2016) Hydrologic impacts of thawing permafrost-a review. Vadose Zone J 15(6)
Wang NL, He JQ, Jiang X et al (2009) Study on the zone of maximum precipitation in the north slopes of the Central Qilian Mountains. J Glaciol Geocryol 31(3):395–403
Yang D, Shi Y, Kang E et al (1991) Results of solid precipitation measurement intercomparison in the alpine area of Urumqi River basin. Chin Sci Bull 36:1105–1109
Yang Y, Chen R, Liu G, Liu Z, Wang X (2022) Trends and variability in snowmelt in China under climate change. Hydrol Earth Syst Sci 26:305–329
Ye D, Gao Y (1979) Meteorology of the Qinghai-Tibetan Plateau. Science Press, Beijing
Zhang T, Barry RG, Knowles K et al (2008) Statistics and characteristics of permafrost and ground-ice distribution in the Northern Hemisphere. Polar Geogr 31(1–2):47–68
Zhao Q, Zhang S, Ding Y et al (2015) Modeling hydrologic response to climate change and shrinking glaciers in the highty glacierized Kunma Like River catchment, central Tian Shan Mountains. J Hydrometeorol 16:2383–2402

Chapter 8
Global Cryospheric Water Cycle

Donghui Shangguan, Shengxia Wang, and Yongjian Ding

Throughout Earth's environmental history, there has been a reciprocal feedback relationship between the cryosphere and the ocean. If the climate continues to warm, significant melting of land ice will occur, leading to rising sea levels due to the influx of meltwater. Conversely, if the climate cools, a substantial amount of ocean water would freeze, forming ice on land and causing sea levels to fall. During the conversion between ocean water and land ice, changes in water volume and ice mass will alter ocean circulation, ultimately impacting the global water cycle.

8.1 Cryosphere and Freshwater in the Ocean

The cryosphere is more extensively developed at higher latitudes than at lower ones. Cryospheric meltwater, a form of freshwater, flows into the ocean and creates marine freshwater reservoirs. There is a dynamic balance between this marine freshwater, cryospheric meltwater, sea ice, and ocean water. However, continuous climate change can disrupt this balance, leading to significant and far-reaching changes in ocean circulation, marine ecosystems, and the global climate.

D. Shangguan · Y. Ding
Northwest Institute of Eco-Environment and Resources, Chinese Academy of Sciences, Lanzhou, China
e-mail: dhguan@lzb.ac.cn

Y. Ding
e-mail: dyj@lzb.ac.cn

S. Wang (✉)
Lanzhou University of Finance and Economics, Lanzhou, China
e-mail: wangshxia@lzb.ac.cn

Y. Ding et al., *Cryohydrology*, https://doi.org/10.1007/978-981-96-7360-5_8

8.1.1 Marine Freshwater at High Latitudes

Marine freshwater at high latitudes is primarily replenished by atmospheric precipitation, river runoff, and the melting of land ice. Significant amounts of freshwater from the melting cryosphere can accumulate in deep ocean basins, where they can reside for extended periods, forming marine freshwater reservoirs. Given the considerable variation in the residence time of this freshwater, the cryosphere plays a crucial role in the freshwater budget of polar oceans.

Marine freshwater in the Arctic and Antarctic is stored in two distinct "reservoirs": the "sea ice reservoir" and the "liquid freshwater reservoir." In the sea ice reservoir, the salinity of sea ice varies with its age. In the liquid freshwater reservoir, the volume is typically defined by vertically integrating the difference between the base salinity and the actual salinity, which can be interpreted as the height of a freshwater column that would result from diluting the base salinity water column to the actual salinity. Freshwater is exchanged between these two reservoirs through the processes of sea ice freezing and melting. Specifically, less sea ice leads to increased melting, resulting in more freshwater entering the ocean and expanding the capacity of the liquid freshwater reservoir. It is important to note that, over long time scales, external forces acting on these two reservoirs are generally more significant than the exchange of freshwater between them.

In addition to sea ice, liquid freshwater reservoirs are also influenced by rainfall and other cryospheric elements. Figure 8.1 illustrates the multi-year average freshwater balance in the Arctic and Antarctic. The freshwater circulation in these regions primarily involves exchanges between the atmosphere, ocean, land, and sea ice. Atmospheric precipitation has a notable impact on liquid freshwater reservoirs, with annual contributions of 4,900 km^3 to Arctic waters and 2,800 km^3 to Antarctic waters (Fig. 8.1). However, despite the significant replenishment from precipitation, the net input to freshwater reservoirs is relatively limited. This limitation is due to the high annual evaporation rates of 3,500 km^3 in the Arctic and 2,500 km^3 in the Antarctic, combined with the mixing of freshwater with ocean water.

Land surface runoff is also a crucial component of freshwater input at high latitudes. In the Arctic, this runoff primarily consists of snowmelt but also includes meltwater from glaciers and ice sheets. Sea ice and snow cover play significant roles in the freshwater cycle in the Arctic. In contrast, Antarctica does not receive direct freshwater input from land runoff; instead, its freshwater contribution comes solely from outlet glaciers flowing into the sea.

In summary, within the freshwater budget of the Arctic and Antarctic seas, the liquid freshwater reservoir is dominant, followed by the sea ice reservoir. The melting and freezing of sea ice play the most critical roles in the freshwater cycle. For instance, approximately 1.7×10^4–1.8×10^4 km^3 of freshwater is involved in the annual freshwater cycle in the Arctic and Antarctic, a volume significantly larger than that contributed by snowmelt and the precipitation-evaporation process (Fig. 8.1).

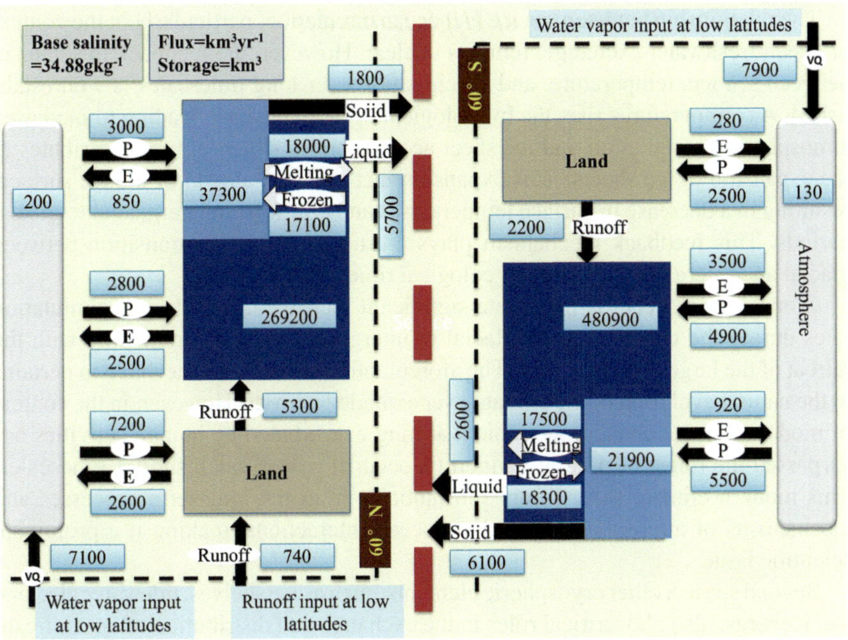

Fig. 8.1 Average freshwater inflow and outflow at 60° to 90° north and south latitudes from 1961 to 1990 (Lehner et al. 2011; Ding and Zhang 2015). The values next to the arrows indicate flux (km^3/year), while the values in the boxes represent reserves (km^3). The base salinity unit is 34.88 g/kg. E: Evaporation; P: Precipitation; VQ: Water vapor input

8.1.2 *High-Latitude Solid–Liquid Freshwater Exchange*

Section 8.1.1 analyzed the composition of freshwater in the ocean from the perspective of dynamic equilibrium. This section focuses on the exchange and variation of solid–liquid freshwater. If the climate continues to warm, changes in the cryosphere will disrupt the dynamic balance between marine freshwater, cryospheric meltwater, sea ice, and ocean water, further complicating the relationships among the cryosphere, oceanic freshwater, ocean circulation, and climate change. For example, since the mid-1960s, Arctic sea ice has significantly decreased, and liquid freshwater in the ocean has tended to decline, indicating that the outflow of liquid freshwater from the Arctic may be exceeding its inflow. This trend is partly due to the strengthening of the North Atlantic Oscillation (NAO) during this period, which not only accelerated warming at high latitudes but also increased sea ice export through the Fram Strait, intensified atmospheric circulation over Northern Europe, and strengthened ocean currents into the Atlantic. Consequently, the outflow of liquid freshwater from the Arctic increased, expanding and intensifying the cold and freshwater effects of sea ice.

The relationship between sea ice and ocean circulation, particularly in the context of ocean freshwater exchange, remains unclear. However, the feedback mechanism between sea ice, temperature, and precipitation over long timescales is well established. As temperatures rise, the hydrological cycle intensifies, leading to increased atmospheric precipitation and ice sheet accumulation, which in turn contributes to the expansion of ice sheets. This expansion increases the albedo of the ice surface, resulting in a decrease in surface temperatures and a shift from interglacial to glacial periods. This feedback mechanism plays a crucial role in the transition between glacial and interglacial cycles on geological timescales.

Climate proxy records indicate a significant increase in ice sheet accumulation rates during the transition from glacial to interglacial periods, coinciding with the retreat of the largest sea ice areas. The aforementioned feedback mechanism pertains to the natural evolution of the climate over extended periods. However, in the context of modern climate change, the rapid warming exacerbated by human activities has surpassed the range of natural variability, occurring over much shorter timescales. This modern climate shift differs substantially from the long-term processes and mechanisms of atmosphere-cryosphere-ocean interactions, making it a prominent scientific issue.

Beyond sea ice, other cryospheric elements, such as ice shelves, tidewater glaciers, and icebergs, also play critical roles in the exchange and distribution of ocean freshwater. For instance, icebergs formed from calving ice shelves or tidewater glaciers, along with large ice floes breaking away from sea ice, can be transported far into the ocean. As the climate continues to warm, the influx of meltwater and large calving ice masses from the Greenland and Antarctic ice sheets is expected to significantly impact ocean circulation, the global water cycle, and the climate. Therefore, the exchange and redistribution of marine freshwater at high latitudes, driven by cryospheric changes, are of considerable concern in the global water cycle.

8.2 Cryosphere and Global Water Cycle

8.2.1 Interaction Between the Cryosphere and Atmosphere

The cryosphere, along with the atmosphere, hydrosphere, lithosphere, and biosphere, forms a complex and interconnected climate system. Each of these subsystems plays a crucial role in the global water cycle, with intricate links binding them together. The oceans are central to the water cycle, containing approximately 97% of Earth's water, while the cryosphere holds about 70% of the planet's freshwater. The interactions between the cryosphere, atmosphere, and oceans are pivotal in shaping global climate, ocean circulation, and the global water cycle. These interactions occur during the conversion of cryospheric solid freshwater (such as ice and snow) into liquid freshwater in the oceans, and subsequently, the transformation of cryospheric liquid freshwater into marine brine through energy exchange and mass changes.

The interaction between the cryosphere and the atmosphere is a vital component of the climate system. At the snow/ice-atmosphere interface, there is a constant exchange of mass and energy that significantly influences atmospheric processes. The dynamics and thermodynamics of the atmosphere-sea ice and atmosphere-snow/ice boundary layers are governed by these ice-gas interactions. This includes not only the exchange of energy, heat, and mass but also the effects of wind on the movement of sea ice and snow. Additionally, the cryosphere plays a role in the biogeochemical cycles by mediating the exchange of gases such as carbon dioxide and dimethyl sulfide between the atmosphere and the ice. These processes underscore the importance of the cryosphere in regulating climate and maintaining the balance of Earth's environmental systems.

The cryosphere is highly sensitive to climate change and serves as a critical indicator of such changes. Variations within the cryosphere can lead to shifts in the energy and water dynamics on local, regional, or global scales. Any alteration in the energy-water balance within the cryosphere triggers a cascade of interactions with the atmosphere, oceans, hydrological systems, environments, and ecosystems. Elements of the cryosphere, such as snow cover, sea ice, and ice sheets, play pivotal roles in driving climate anomalies and serve as essential predictors of climate trends. As climate change progresses, the relationship between the cryosphere and the hydrosphere becomes increasingly interdependent. On a global scale, the retreat of the cryosphere intensifies the hydrological cycle, contributing to sea-level rise. Additionally, the influx of cold, fresh water from the cryosphere into the oceans can alter thermohaline circulation, further influencing climate change.

The interaction between the cryosphere and the atmosphere significantly impacts the oceans, primarily through changes in sea level, freshwater flux, ocean currents, and thermohaline circulation (Fig. 8.2). Among the various components of the cryosphere, sea ice has garnered considerable attention due to its pronounced short- and medium-term effects on climate, particularly its interactions with the atmosphere and the ocean. The extent, thickness, structure, and composition of sea ice are closely linked to climate change, as they influence global climate by altering heat and water vapor exchange fluxes.

The effects of sea ice on the climate system are primarily observed in the following aspects:

1. **Albedo Effect**: As the lower boundary of the atmosphere in polar regions, sea ice has a much higher albedo compared to the ocean surface, reflecting most of the solar radiation back into the atmosphere. This limits the energy available to the atmosphere in polar regions.
2. **Thermal Insulation**: Serving as the upper boundary of the ocean in polar areas, sea ice acts as a barrier, blocking or even isolating heat transfer between the atmosphere and the ocean.
3. **Impact on Thermohaline Circulation**: The processes of freezing and thawing of sea ice influence the formation and intensity of thermohaline circulation, which is crucial for global ocean currents.

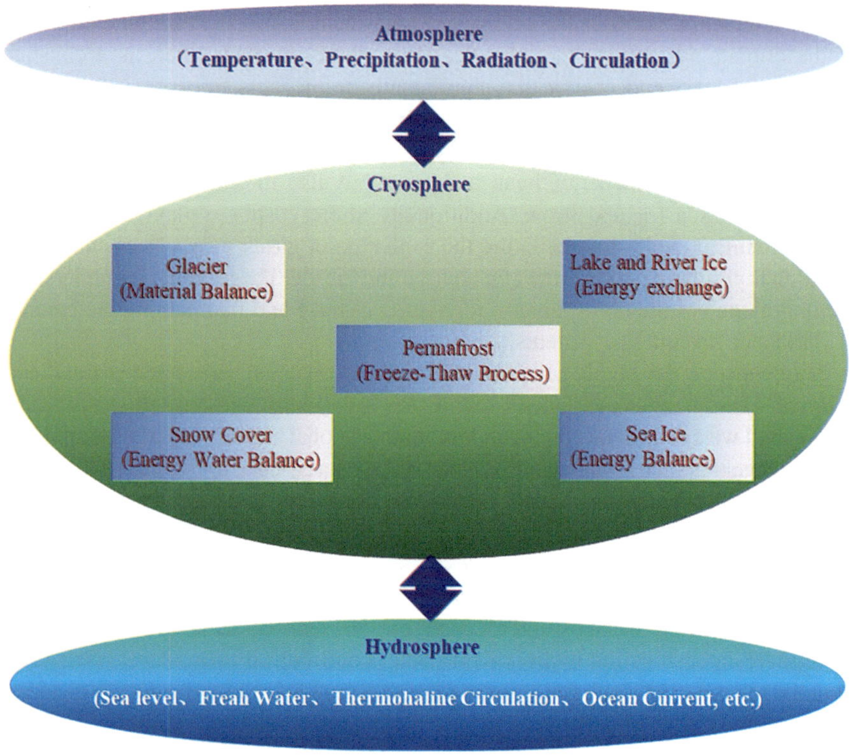

Fig. 8.2 Schematic diagram of cryosphere-atmosphere-ocean interaction

4. **Temperature Regulation**: The absorption and release of heat during the freezing and thawing of sea ice smooth out regional temperature extremes and delay seasonal temperature changes.

Under the influence of climate change, these processes—particularly the freezing and thawing of sea ice—affect thermohaline circulation, sea level changes, and freshwater fluxes in the oceans by altering seawater temperature and salinity.

8.2.2 Cryospheric Role in the Global Water Cycle

Ocean circulation is a fundamental component of the global water cycle, playing a crucial role in the distribution of heat and freshwater across the Earth's oceans. It consists of two primary elements: surface currents, which are primarily wind-driven, and deep currents, which are driven by variations in water temperature and salinity, a process collectively known as thermohaline circulation. This circulation is fundamentally governed by differences in water density caused by these temperature and salinity variations.

The study of ocean circulation on a global scale has long been a central challenge for oceanographers and climatologists. As early as a century ago, it was proposed that ocean circulation functions to transport solar energy from the equator toward the poles. However, the significance of this concept gained substantial recognition in the 1980s, marked by the introduction of the ocean conveyor belt model. This conceptual model, grounded in the mechanisms driving thermohaline circulation, provided a framework for understanding the large-scale movement of water masses across the global oceans (Fig. 8.3).

Further refinements to this model were made by Schmitz in 1995, who proposed a hierarchical structure of thermohaline circulation based on the exchange of water between ocean basins. Schmitz's work also included estimates of the transport fluxes associated with different layers of circulation. The ocean conveyor belt model, along with its hierarchical structure, has since become a classical representation of global thermohaline circulation, serving as a vital tool for understanding the complexities of oceanic heat and mass transport.

Ocean thermohaline circulation is a key driver of the global water cycle, functioning through four primary components: the sinking circulation at high latitudes, the transmission of water from the deep ocean at high latitudes to lower latitudes, the upwelling at low latitudes (i.e., reverse circulation), and the surface circulation that transports water from low to high latitudes (Fig. 8.4). The principal mechanism

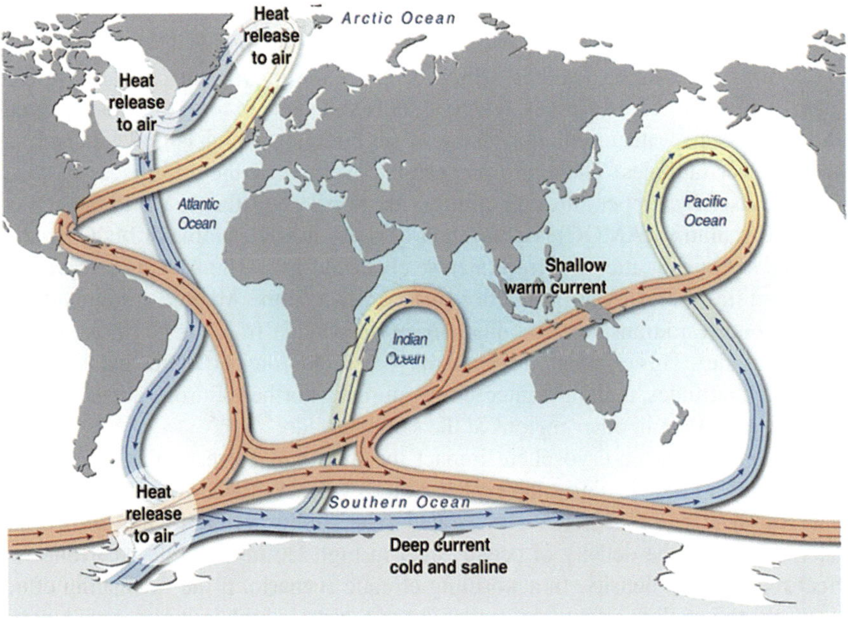

Fig. 8.3 Conceptual model of global Ocean conveyor belt (Broecker 1987, 1991)

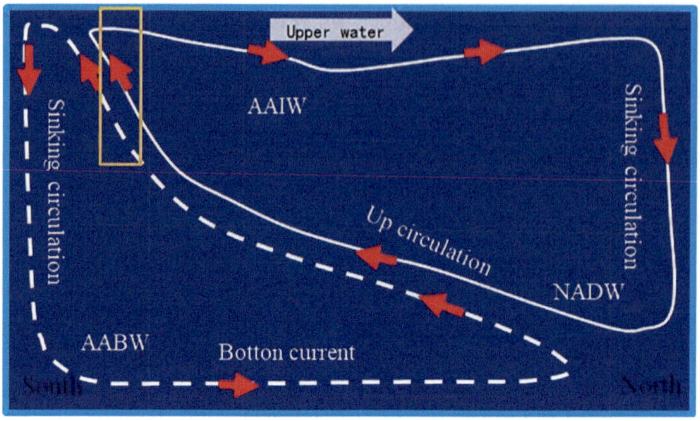

Fig. 8.4 Schematic diagram of the formation of ocean thermohaline circulation, illustrating the key water masses: Antarctic Intermediate Water (AAIW), Antarctic Bottom Water (AABW), and North Atlantic Deep Water (NADW)

behind thermohaline circulation is the cooling and increased density of seawater in high-latitude regions due to the input of meltwater, which initiates the sinking of water masses. To compensate for the water lost in this sinking circulation, surrounding seawater flows in, maintaining the global thermohaline circulation. This circulation is also characterized by the movement of warmer, less dense surface water from lower latitudes to higher latitudes, which forms the theoretical basis for the conceptual model of the oceanic conveyor belt.

An important component of ocean thermohaline circulation is the Meridional Overturning Circulation (MOC), which involves the movement of warm surface water from low to high latitudes, where it cools and sinks to form the sinking circulation. This deep-water circulation then reverses direction at high latitudes and moves toward lower latitudes. The entire process is governed by the differences in water density caused by temperature and salinity variations. The Atlantic Meridional Overturning Circulation (AMOC) is a particularly well-known example of MOC and has garnered significant attention from marine climatologists. The AMOC involves the northward transport of warmer, saltier surface water in the Atlantic at low latitudes and the southward transport of cooler, denser deep water from the North Atlantic at higher latitudes. This circulation system plays a crucial role in transferring heat from low to high latitudes, making winter temperatures in northern Europe approximately 10 °C warmer than in other regions at the same latitude.

The changes in the cryosphere impact the meridional overturning circulation, which plays a crucial role in the global water cycle. As the cryosphere continues to warm, extensive meltwater is generated. The cooling effect of this cold, fresh water influences the density of ocean water at high latitudes, while its freshening effect reduces this density. In a warming climate scenario, if the freshening effect outweighs the cooling effect, surface seawater density in high latitudes may become less than that of bottom seawater, potentially hindering or even reversing thermohaline circulation. Conversely, if the cooling effect predominates, surface seawater density increases relative to bottom water, thereby intensifying thermohaline circulation. Studies suggest that ongoing climate warming could lead to accelerated melting of fresh water from the Greenland ice sheet, decreasing seawater density and impeding its sinking in high latitudes. This weakening of the Atlantic Meridional Overturning Circulation (AMOC) could have cooling effects on northern Europe's climate.

The theory of meridional overturning circulation (MOC) provides insights into climate change driven by cryospheric shifts. Ice core records unveil millennia-scale climate fluctuations during the last ice age, revealing interglacial rapid warming events spanning hundreds to thousands of years in sedimentary records of the North Atlantic and Pacific Oceans. To understand the driving mechanisms behind these warming events, it is posited that MOC was unstable during the ice age. As ice sheets began to expand, MOC weakened or ceased altogether, reducing the transport of salty water from the Atlantic to other ocean basins. For instance, if there was a net evaporation in the North Atlantic, leading to moisture accumulation on land as snow, this would increase ice sheet replenishment and thus raise ocean salinity. Upon reaching a critical salinity level, deep ocean convection would initiate, activating the Atlantic meridional overturning circulation. This circulation would then transfer heat from lower latitudes, releasing it into the North Atlantic and initiating gradual ice sheet melting. As ice sheets melted, freshwater influx into the North Atlantic from melting ice or icebergs would gradually weaken or halt MOC, potentially leading it back to a weakened or closed state.

Water-hosing experiments have confirmed the critical role of cryospheric freshwater effects in climate change. These experiments involve artificially adding freshwater to the North Atlantic region between 50° and 70° N to simulate the response of global climate when the Atlantic meridional overturning circulation (AMOC) is weakened or interrupted. The results of these tests indicate that the addition of freshwater at high latitudes in the North Atlantic rapidly weakens or halts AMOC, leading to significant cooling in the North Atlantic region, slight warming in the South Atlantic, and a southward shift of the Intertropical Convergence Zone (ITCZ). This demonstrates that when AMOC weakens or stops, temperatures in the North Atlantic decrease notably.

Projections suggest that AMOC will weaken in the twenty-first century, with estimated reductions ranging optimally from 11% under the RCP2.6 scenario to 34% under the more severe RCP8.5 scenario, although no abrupt changes or collapses are anticipated (IPCC 2013).

8.3 Cryosphere and Sea Level Change

Cryospheric changes have been closely linked to sea level fluctuations throughout the historical evolution of global climate alternations between cold and warm periods. When climate warms, the melting of ice contributes to sea level rise, whereas during colder periods, the freezing of marine freshwater leads to a decline in sea level.

8.3.1 Factors of Sea Level Change

Sea level changes occur across various spatial and temporal scales, driven by a complex array of factors, including cryospheric changes, volcanic activity, tectonic movements, geoid variations, astronomical influences, geophysical processes, greenhouse gas emissions, and shifts in land water storage. These factors can be categorized into changes in seawater volume, ocean basin volume, and geoid alterations.

Seawater volume changes are influenced by factors such as glacier retreat and advance, density-volume effects, and the discharge and infiltration of water from the Earth's mantle. Among these, glacial changes have the most significant impact on sea level fluctuations. During glacial periods, sea levels fall as seawater is frozen and stored on land, while during interglacial stages, sea levels rise dramatically due to the influx of meltwater from the cryosphere into the oceans. Throughout the Quaternary glaciation-interglacial cycles, sea levels have varied by as much as 100–200 m.

The density-volume effect also plays a role in sea level changes, manifesting in thermal expansion and contraction as well as changes in ocean thermohaline circulation. When ocean temperatures rise, the expansion of seawater leads to a rise in sea level, while cooling causes the water to contract and sea levels to fall. It is estimated that a 1 °C increase in global ocean temperatures could result in a sea level rise of approximately 0.6 m. The discharge and infiltration of mantle water further influence sea level by altering seawater volume. Additionally, changes in ocean salinity can affect sea levels; for example, a decrease in salinity from 35 to 25 parts per thousand would lower sea levels by 7.6 m. Over the past 20 million years, the combined effects of cooler temperatures and lower salinity have contributed to a 5.3-m decrease in sea level.

The variation in ocean basin volume is influenced by deep marine deposits, volcanic eruptions, ocean floor spreading, subduction rates, crustal isostatic compensation, and tectonic movements, with tectonic activity and seabed expansion exerting the most significant impact on sea level changes.

Tectonic movements, such as the uplift of the Himalayas on the Eurasian plate, have caused sea levels to drop by approximately 10 m. When the rate of seabed expansion increases, the newly formed ocean floor near mid-ocean ridges does not have sufficient time to cool and sink, resulting in an elevated ocean basin floor. This elevation causes sea levels to rise, leading to transgression stages. Conversely, during periods of slower seabed expansion, the ocean floor has more time to cool and sink, lowering the basin floor, which in turn causes sea levels to fall and forms regression stages.

For instance, during the Cretaceous period of the Mesozoic era, rapid seabed expansion led to a sea level rise of approximately 300 m. In contrast, during the Paleogene and Neogene periods of the Cenozoic era, slower seabed expansion contributed to a sea level decrease of around 300 m.

The variation of the geoid is influenced by changes in Earth's orbital parameters and rotation rate. Earth's orbit and rotation are typically in relative equilibrium with the geoid, but any alterations in these parameters disrupt this balance, necessitating adjustments in the geoid's shape to accommodate the new orbital and rotational dynamics.

Among the factors contributing to sea level change, the time scales of their effects vary significantly, spanning tectonic, orbital, millennial, and centennial scales (Fig. 8.5). Of particular concern is sea level change on a centennial scale, driven by global warming due to anthropogenic greenhouse gas emissions since the Industrial Revolution. This has become a serious threat to the human living environment, attracting widespread attention. The primary characteristic of this change is a notable rise in sea levels over a relatively short period, which has triggered a cascade of climate, environmental, and disaster-related issues.

The main drivers of sea level rise on a centennial scale are the accelerated melting of the cryosphere, the thermal expansion of the oceans, and changes in land water storage. Notably, cryospheric changes have had the most significant impact on sea level, particularly the recent rapid melting of polar ice, which has offset the reduction in sea level rise caused by ocean thermal expansion, resulting in a continued rise at nearly the same rate (Table 8.1). For instance, observations from 1993 to 2010

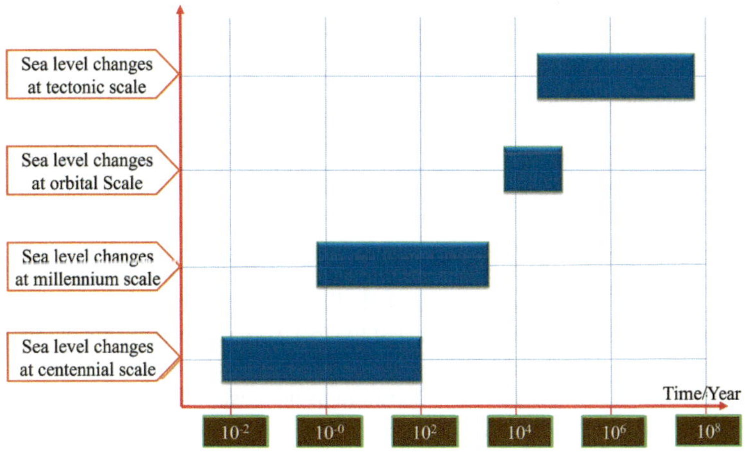

Fig. 8.5 Time scales of sea level changes under different mechanisms, illustrating the varying impacts of tectonic activity, orbital parameters, and anthropogenic factors on sea level over tectonic, orbital, millennial, and centennial scales

Table 8.1 Global sea level changes observed and simulated during different periods in the past (Data sourced from IPCC Special Report on the Ocean and Cryosphere in a Changing Climate [SROCC] 2019) (Unit: mm/a)

		Source	1901–1990	1970–2015	1993–2015	2006–2015
Contributions' source	Observations	Thermal expansion		0.89 (0.84–0.94)	1.36 (0.96–1.76)	1.40 (1.08–1.72)
		Glaciers	0.49 (0.34–0.64)	0.46 (0.21–0.72)	0.56 (0.34–0.78)	0.61 (0.53–0.69)
		Antarctica ice sheet			0.29 (0.11–0.47)	0.43 (0.34–0.52)
		Land water storage	−0.12	−0.07	0.09	−0.21 (−0.36–0.06)
		Observed GMSL	1.38 (0.81–1.95)	2.06 (1.77–2.34)	3.16 (2.79–3.53)	3.58 (3.10–4.06)
	Simulations	Thermal expansion	0.32 (0.04–0.60)	0.97 (0.45–1.48)	1.48 (0.86–2.11)	1.52 (0.96–2.09)
		Glaciers	0.53 (0.38–0.68)	0.73 (0.50–0.95)	0.99 (0.60–1.38)	1.10 (0.64–1.56)
		Greenland SMB	−0.02 (–0.05–0.02)	0.03 (–0.01–0.07)	0.08 (–0.01–0.16)	0.12 (–0.02–0.26)
		Total including land water storage and ice discharge	0.67 (0.02–1.32)	0.18 (–0.46–0.82)	0.03 (–0.81–0.87)	0.04 (–0.85–0.93)

indicate that the contribution of ice sheets and glaciers was 1.46 mm per year, which is comparable to the total contribution from other factors.

8.3.2 Contribution of Cryosphere to Sea Level Change Since 1900

The cryosphere and ocean thermal expansion are among the most critical factors contributing to sea level change. As global climate warming continues, the relative contributions of these factors are expected to shift. Ocean thermal expansion, which occurs as seawater warms and expands, has historically been a significant driver of sea level rise. However, as temperatures continue to increase, the role of ocean thermal expansion is projected to weaken. In contrast, the contribution from the cryosphere, particularly the melting of the Greenland and Antarctic ice sheets, is anticipated to increase substantially.

1. **Mountain Glaciers**

Mountain glaciers, which constitute approximately 1% of the global ice mass, are particularly sensitive to climate warming due to their smaller size and location in warmer regions compared to ice sheets. In the current climate, mountain glaciers are retreating rapidly, and the resulting influx of glacial meltwater into the oceans significantly contributes to sea level rise. The total area of mountain glaciers (excluding those in Antarctica and Greenland) is approximately 706,700 km^2 (RGI7.0), with a total ice mass ranging between 51,000 and 133,000 km^3. If all these glaciers were to melt, it would result in a sea level rise of approximately 0.15 to 0.37 m.

An important factor to consider when assessing the contribution of mountain glaciers to sea level rise is the role of annual precipitation, which replenishes glacier mass. While the total amount of glacial meltwater corresponds to the total glacier loss, the accumulation of glacier mass through precipitation plays a mitigating role in the ocean-land water cycle and, by extension, sea level rise. Therefore, the contribution of mountain glaciers to sea level change is determined by the net mass balance—the difference between glacier mass accumulation and ablation. It is also important to note that glacier changes reflect the cumulative effect of multi-year mass balance; a positive or negative mass balance for only one year or a few years does not immediately cause glaciers to advance or retreat, as there is often a lag period before these changes manifest.

Glacier mass balance calculations indicate that the majority of glaciers worldwide have experienced negative mass balances over the past 40 years, consistent with the global warming trend observed in recent decades. While some regions showed intermittent positive balances in the late 1980s and early 1990s, the cumulative loss remained substantial. Since the 1990s, however, glacier mass balances have predominantly trended negatively, exacerbating glacier retreat and meltwater runoff. The IPCC estimates that mountain glaciers contributed between 0.41 to 0.83 mm/year to sea level rise (Table 8.2), although current evaluation results exhibit notable uncertainties. For instance, the average sea level contribution from 1895 to 2005 (0.83 mm/year) exceeded that from 2003 to 2009 (0.59 mm/year), contrasting with recent global warming trends. Satellite data from 2000 to 2019 suggests glaciers lost an average of 267 $\pm$ 16 Gt/year, equivalent to 0.74 $\pm$ 0.04 mm/year sea-level rise (21 $\pm$ 3% of observed sea-level rise) (Hugonnet et al. 2021). Overall, mountain glaciers' contribution to sea level rise underscores their pivotal role amidst ongoing climate change.

Based on observed mass balance data, the contribution of glaciers to sea level change is estimated at 0.27 mm/year. This figure is relatively small as it excludes glaciers in Alaska, Patagonia, and Central Asia. Monitoring data indicate that ice cap melting in Patagonia from 1995 to 2000 contributed to a sea level rise of 0.10 $\pm$ 0.01 mm/year. Additionally, laser monitoring reveals that glaciers in Alaska experienced melting from the mid-1950s to the mid-1990s, resulting in a sea level rise of approximately 0.14 $\pm$ 0.04 mm/year. Combining these findings, the overall contribution of mountain glaciers to sea level rise is approximately 0.66 mm/year.

Mountain glaciers, while significant, represent only a small fraction of the total land ice and their influence on sea level change is secondary to that of thermal expansion of seawater. The general trend of retreat in mountain glaciers is expected

Table 8.2 Contributions of mountain glaciers to sea level change during different periods

Time	Contribution (mm/a)	Data source
1800–2005	0.41 ± 0.10	Leclercq et al. (2011)
1895–2005	0.83 ± 0.21	
1901–1990	0.54 ± 0.07	IPCC AR5 (2013), Ding et al. (2017), IPCC AR6 (2019), IPCC SROCC (2019)
1971–2009	0.62 ± 0.37	
1993–2009	0.76 ± 0.37	
2005–2009	0.83 ± 0.37	
2003–2009	0.59 ± 0.07	
2006–2018	0.61 ± 0.16	

to persist over the long term. Particularly for small-scale glaciers, as their areas continue to shrink, the volume of meltwater will eventually decrease after an initial increase, marking an "inflection point" in glacier runoff. Overall, while mountain glaciers contribute to sea level change, their impact is limited and is expected to diminish as glacier areas continue to decrease.

2. **Greenland Ice Sheet and Antarctic Ice Sheet**

The melting trends of the Greenland and Antarctic ice sheets have been progressively increasing, leading to a gradual rise in their contributions to sea level change. According to the IPCC's Fifth Assessment Report, the contributions to sea level rise from the Greenland Ice Sheet and the Antarctic Ice Sheet during the most recent decade (2002–2011) were 0.59 mm/year and 0.40 mm/year, respectively. These figures represent a significant increase compared to their contributions during the previous decade (1992–2001) (Table 8.3). Although the contributions from the Greenland and Antarctic ice sheets were similar between 1992 and 2001, the contribution from the Greenland Ice Sheet surged in comparison to the Antarctic Ice Sheet from 2002 to 2011. This indicates that the Greenland Ice Sheet has been melting at a faster rate and contributing more significantly to sea level rise than the Antarctic Ice Sheet.

At a regional scale, sea level change is influenced not only by volume effects but also by ocean dynamics, such as ocean circulation. The reciprocal of sea water density is known as the sea water specific volume, which represents the volume of sea water per unit mass and is a function of salinity and temperature. As global temperatures rise, the ocean's volume expands due to thermal expansion, contributing to sea level rise. Additionally, increased salinity can further enhance sea water volume. Climate warming affects ocean volume by altering water temperature and salinity, leading to changes in sea level.

From an ocean dynamics perspective, changes in meridional overturning circulation can lead to the accumulation of local seawater, resulting in regional sea level variations. On a global scale, sea level change due to ocean dynamics remains relatively constant. Dynamic sea level is primarily influenced by the redistribution of

Table 8.3 The contributions of Greenland Ice Sheet and Antarctic Ice Sheet to sea level change (IPCC 2013, 2021; IPCC SROCC 2019)

Contribution factor	Contribution/mm·a^{-1}				
	IPCC AR4		SROCC		
	1961–2003	1993–2003	1992–2001	2002–2011	2012–2016
Greenland ice Sheet	0.05 ± 0.12	0.21 ± 0.07	0.02 [0.21–0.25]	0.73 [0.67–0.79]	0.68 [0.64–0.72]
The Antarctic ice sheet	0.14 ± 0.41	0.21 ± 0.35	0.14 [0.12–0.16]	0.23 [0.16–0.30]	0.55 [0.48–0.62]

Note positive value represents sea level rise, negative value represents sea level decrease

ocean water, atmospheric circulation, temperature, and salinity. Under the current melting rates of the Greenland Ice Sheet, regional sea level changes are predominantly driven by dynamic sea level variations. At higher melting rates, the impact on dynamic sea level is most pronounced in the northwest Atlantic.

Large volumes of meltwater from the Greenland Ice Sheet entering the North Atlantic can significantly weaken the meridional overturning circulation. The influx of meltwater causes the surface waters of the North Atlantic to become colder and less saline. As the heat from these waters is transferred to the atmosphere, the water density increases. The denser water then sinks deeper into the ocean and flows southward, eventually rising in other parts of the ocean. This disruption in the meridional overturning circulation can have widespread effects on global ocean circulation patterns and climate.

The impact of Greenland Ice Sheet melting rates on sea level change varies significantly across different regions, particularly affecting thermal specific capacity. When the melting rate is low, its impact on sea level change is primarily confined to the North Atlantic Ocean. As the melting rate increases, the effects extend to the Arctic, South Atlantic, Pacific, and Indian Oceans. Increased meltwater volume leads to higher sea levels in the Arctic Ocean, southern Greenland, and the eastern subtropical Atlantic, with the most pronounced rises observed along the northern European coast and southern Greenland. This is due to the substantial influx of freshwater from the rapidly melting Greenland Ice Sheet.

Conversely, in the vicinity of 45° N latitude in the Atlantic Ocean and the tropical Atlantic, salinity levels and sea levels have decreased significantly. This decline is mainly attributed to the weakening of thermohaline circulation, which is a consequence of the reduced Atlantic meridional overturning circulation (Li et al. 2015).

The IPCC assessment is considered the most authoritative source on the impact of the cryosphere on sea level change. Since the Industrial Revolution, ocean thermal expansion and cryosphere contributions have accounted for approximately half of the observed sea level rise, excluding the effects of land water storage. Looking ahead, as the contribution of ocean thermal expansion to sea level rise diminishes and the impact of ice sheets increases, the cryosphere—particularly at high latitudes—is expected to play a more significant role in sea level change compared to thermal expansion.

Questions

1. What is the influence mechanism of cryospheric changes on ocean circulation?
2. How do mountain cryosphere changes contribute to global sea level rise?

References

Broecker WS (1987) The biggest chill. Nat Hist Mag 96:74–82
Broecker WS (1991) The great ocean conveyor. Oceanography 4(2):79–89
Ding YJ, Zhang SQ (2015) Hydrological effects of the cryospheric water cycle on a global scale. Chin Sci Bull 60:593–602
Ding YJ, Zhang SQ, Chen RS (2017) Introduction to cold region hydrology. Science Press, Beijing
Hugonnet R et al (2021) Accelerated global glacier mass loss in the early twenty-first century. Nature 592(7856):726–731
IPCC (2013) Climate change 2013: the physical science basis. Cambridge University Press
IPCC Special Report on the Ocean and Cryosphere in a Changing Climate (2019) In: Pörtner H-O, Roberts DC, Masson-Delmotte V, Zhai P, Tignor M, Poloczanska E, Mintenbeck K, Alegría A, Nicolai M, Okem A, Petzold J, Rama B, Weyer NM (eds) [M] Cambridge University Press, Cambridge
IPCC (2021) Climate Change 2021: the physical science basis. Contribution of working group I to the sixth assessment report of the intergovernmental panel on climate change [M]. Cambridge University Press, Cambridge
Leclercq PW, Oerlemans J, Cogley JG (2011) Estimating the glacier contribution to sea-level rise for the period 1800–2005. Surv Geophys 32:519–535
Lehner F, Raible CC, Hofer D, Stocker TF (2011) The freshwater balance of polar regions in transient simulations from 1500 to 2100 AD using a comprehensive coupled climate model. Clim Dyn 39(1–2):347–363. https://doi.org/10.1007/s00382-011-1199-6
Li J, Zuo JC, Tan W et al (2015) Impact of Greenland glacier melting rate on sea level change in the 21st century. Acta Oceanologica Sinica (in Chinese) 37(7):22–32

Extended Reading

Ding YJ, Xiao CD (2019) Comprehensive report on cryosphere changes and their impacts. Science Press, Beijing

GPSR Compliance

The European Union's (EU) General Product Safety Regulation (GPSR) is a set of rules that requires consumer products to be safe and our obligations to ensure this.

If you have any concerns about our products, you can contact us on ProductSafety@springernature.com

In case Publisher is established outside the EU, the EU authorized representative is:

Springer Nature Customer Service Center GmbH
Europaplatz 3
69115 Heidelberg, Germany

Batch number: 10704307

Printed by Printforce, the Netherlands